Analysis and Synthesis for Interval Type-2 Fuzzy-Model-Based Systems

Hongyi Li · Ligang Wu · Hak-Keung Lam
Yabin Gao

Analysis and Synthesis for Interval Type-2 Fuzzy-Model-Based Systems

Springer

Hongyi Li
College of Engineering
Bohai University
Jinzhou
China

Hak-Keung Lam
Division of Engineering
King's College London
London
UK

Ligang Wu
Space Control and Inertial Technology Research Center
Harbin Institute of Technology
Harbin
China

Yabin Gao
Space Control and Inertial Technology Research Center
Harbin Institute of Technology
Harbin
China

ISBN 978-981-10-0592-3 ISBN 978-981-10-0593-0 (eBook)
DOI 10.1007/978-981-10-0593-0

Library of Congress Control Number: 2016932340

Printed on acid-free paper

This Springer imprint is published by Springer Nature
The registered company is Springer Science+Business Media Singapore Pte Ltd.

To my family

Hongyi Li

To my family

Ligang Wu

To my family

Hak-Keung Lam

To my family

Yabin Gao

Preface

Problem formulations of physical systems and processes can often lead to complex nonlinear systems, which may cause analysis and synthesis difficulties. Study of nonlinear systems is often problematic due to their complexities. One effective way of representing a complex nonlinear dynamic system is the so-called Takagi-Sugeno (T-S) fuzzy model, which is governed by a family of fuzzy IF-THEN rules that represent local linear input–output relations of the system. It incorporates a family of local linear models that smoothly blend together through fuzzy membership functions. This in essence, is a multi-model approach in which simple sub-models (typically linear models) are fuzzily combined to describe the global behavior of a nonlinear system. Based on the fuzzy model, the control design is carried out by using the parallel distributed compensation (PDC) scheme. The strategy is that a linear state-feedback controller or filter is designed for each local linear model. The obtained overall controller or filter is nonlinear in general, and is also a fuzzy 'blending' of each individual linear controller or filter.

Analysis and synthesis including state-feedback control, output-feedback control, tracking control, optical control, filtering, fault detection, and model reduction for a class of Interval Type-2 (IT2) T-S fuzzy systems are all thoroughly studied. Fresh novel techniques, including the linear matrix inequality (LMI) techniques, the slack matrix method, and so on, are applied to such systems. This monograph is divided into two sections. First, we focus on IT2 fuzzy controller and filter design for continuous-time IT2 T-S fuzzy systems. The following problems are investigated in this book: (1) the problem of stability and stabilization for IT2 fuzzy-model-based systems subject to parameter uncertainties; (2) the problems of state-feedback control and the output-feedback based control for IT2 T-S fuzzy systems under a new extended dissipativity performance; (3) the sampled-data control problem for IT2 fuzzy systems with actuator fault; (4) the output tracking control problem for nonlinear systems with actuator fault; (5) the switched output-feedback control problem for IT2 fuzzy systems; (6) the problem of filter design for IT2 fuzzy systems with $\mathcal{D}$ stability constraints and new performance index; (7) the fault detection problem for IT2 fuzzy systems subject to sensor nonlinearities; (8) the model reduction problem for IT2 fuzzy systems. Secondly, the theories and techniques developed in the previous part are extended to the stability analysis and controller design problems of discrete-time IT2 T-S fuzzy systems. The below problems are studied: (1) the optimal control problem of discrete-time IT2 fuzzy time delay systems; (2) the fault-tolerant control problem for discrete-time IT2 fuzzy time delay systems with time-varying delay and actuator faults; (3) the static output-feedback control problem for discrete-time IT2 fuzzy systems; (4) the guaranteed cost output tracking control problem for IT2 fuzzy systems. Among the topics, simulation results including some typical real applications are presented to illustrate the effectiveness and the practicability of the fuzzy control design methods proposed in the previous parts.

Jinzhou, China Hongyi Li
Harbin, China Ligang Wu
London, UK Hak-Keung Lam
Harbin, China Yabin Gao
October 2015

Acknowledgments

There are numerous individuals without whose help this book would not have been completed. Special thanks go to Prof. Peng Shi from the University of Adelaide, Prof. Hamid Reza Karimi from University of Agder, Dr. Christian Deters from King's College London, Dr. Emanuele Lindo Secco from King's College London, Dr. Helge A Wurdemann from King's College London, Prof. Kaspar Althoefer from King's College London, Dr. Qi Zhou from Bohai University, Dr. Ramasamy Sakthivel from Sungkyunkwan University, and Dr. Mohammed Chadli from University of Picardie Jules Verne, for their valuable suggestions, constructive comments and support.

Our acknowledgments also go to our fellow colleagues who have offered invaluable support and encouragement throughout this research effort. Thanks go to our students, Yingnan Pan, Xingjian Sun, Yabin Gao, Di Liu, and Chengwei Wu for their commentary. The authors are especially grateful to their families for their encouragement and never-ending support when it was most required. Finally, we would like to thank the editors at Springer for their professional and efficient handling of this project.

The writing of this book was supported in part by the National Natural Science Foundation of China (61573070, 61525303, 61333012), the Program for New Century Excellent Talents in University (NCET-13-0696), the Top-Notch Young Talents Program of China (L. Wu), the Fok Ying Tung Education Foundation (141059), the Heilongjiang Outstanding Youth Science Fund (JC201406), the Program for Liaoning Excellent Talents in University (LJQ20141126), the Natural Science Foundation of Liaoning Province (2015020049), the Special Chinese National Postdoctoral Science Foundation (2015T80262), the Chinese National Postdoctoral Science Foundation (2014M551111) and the Self-Planned Task of State Key Laboratory of Robotics and System (HIT) (201505B).

Acknowledgements

[illegible]

Contents

Notations and Acronyms

$\triangleq$	Is defined as
$\in$	Belongs to
$\forall$	For all
$\sum$	Sum
$\mathbf{R}$	Field of real numbers
$\mathbf{R}^n$	space of n-dimensional real vectors
$\mathbf{R}^{n\times m}$	space of $n \times m$ real matrices
$\mathbf{E}\{\cdot\}$	Mathematical expectation operator
$\mathbf{He}(A)$	$A+A^T$
lim	Limit
max	Maximum
min	Minimum
sup	Supremum
inf	Infimum
$rank(\cdot)$	Rank of a matrix
$trace(\cdot)$	Trace of a matrix
$\lambda_{\min}(\cdot)$	Minimum eigenvalue of a real symmetric matrix
$\lambda_{\max}(\cdot)$	Maximum eigenvalue of a real symmetric matrix
I	Identity matrix
I_n	$n \times n$ identity matrix
0	Zero matrix
$0_{n\times m}$	Zero matrix of dimension $n \times m$
X^T	Transpose of matrix X
X^*	Conjugate transpose of matrix X
X^{-1}	Inverse of matrix X
$X > (<) 0$	X is real symmetric positive (negative) definite
$X \geq (\leq) 0$	X is real symmetric positive (negative) semi-definite
$\mathcal{L}_2\{[0,\infty),[0,\infty)\}$	Space of square summable sequences on $\{[0,\infty),[0,\infty)\}$ (continuous case)

$\ell_2\{[0,\infty),[0,\infty)\}$	Space of square summable sequences on $\{[0,\infty),[0,\infty)\}$ (discrete case)
$\lvert\cdot\rvert$	Euclidean vector norm
$\lVert\cdot\rVert$	Euclidean matrix norm (spectral norm)
$\lVert\cdot\rVert_2$	$\mathcal{L}_2$-norm: $\sqrt{\int_0^\infty \lvert\cdot\rvert^2 dt}$ (continuous case) ℓ_2-norm: $\sqrt{\sum_0^\infty \lvert\cdot\rvert^2}$ (discrete case)
$\lVert\cdot\rVert_E$	$\mathbf{E}\{\lVert\cdot\rVert_2\}$
$\lVert\mathbf{T}\rVert_\infty$	$\mathcal{H}_\infty$ norm of transfer function $\mathbf{T}$: $\sup_{\omega\in[0,\infty)}\lVert\mathbf{T}(j\omega)\rVert$ (continuous case) $\sup_{\omega\in[0,2\pi)}\lVert\mathbf{T}(e^{j\omega})\rVert$ (discrete case)
diag	Block diagonal matrix with blocks $\{X_1,\ldots,X_m\}$
*	Symmetric terms in a symmetric matrix
DOF	Dynamic output-feedback
LMI	Linear matrix inequality
LKF	Lyapunov-Krasovskii functional
LTI	Linear time-invariant
LMF	Lower membership function
PDC	Parallel distributed compensation
SOF	Static output-feedback
T-S	Takagi-Sugeno
UMF	Upper membership function

List of Figures

List of Tables

Chapter 1
Introduction

Modeling practical physical systems frequently results in complex nonlinear systems, which poses great difficulties regarding system analysis and synthesis. Local linearization is a typical method used for the analysis and synthesis of nonlinear systems. However, it has been well recognized that the local linearization model is valid only for a certain range of operating conditions, and can only guarantee the local stability of the original nonlinear system. Another approach, fuzzy control, emerged and developed following the first paper on fuzzy sets [227], has attracted great attention from both the academic and industrial communities. The reason lies much in its effectiveness in obtaining nonlinear control systems, especially when the knowledge of the plant or even the precise control action of the situation is unknown. Thus, fuzzy control has even been found to have many applications in industrial systems and processes, see for example, [6, 9, 12–15, 152]. Bonissone et al. considered industrial applications of fuzzy logic at general electric [13]. The authors in [6] considered the fuzzy logic control to suppress noises and coupling effects in a laser tracking system. The authors in [152] presented a survey on industrial applications of fuzzy control. In fact, fuzzy control has proved to be a successful control approach for complex nonlinear systems. Fuzzy control has even been suggested as an alternative approach to conventional control techniques. Furthermore, stability analysis is an important issue in the field of fuzzy control systems.

The past decades have seen fuzzy rule-based modeling become an active research field due to its unique merits in solving complex nonlinear system identification and control problems [10, 100, 175, 177, 178, 186, 187, 189]. The authors [177] presented trajectory stabilization of a model car via fuzzy control. In an attempt to obtain more flexibility and more effective means of handling and processing uncertainties in complicated and ill-defined systems, Zadeh proposed a linguistic approach as the model of human thinking, introducing the fuzziness into systems theory [227]. Different from conventional modeling, fuzzy rule-based modeling is essentially a multi-model approach in which individual rules (where each rule acts like a 'local model') are combined to describe the global behavior of the system.

© Springer Science+Business Media Singapore 2016
H. Li et al., *Analysis and Synthesis for Interval Type-2
Fuzzy-Model-Based Systems*, DOI 10.1007/978-981-10-0593-0_1

Among the array of model-based fuzzy systems, the Takagi–Sugeno (T–S) fuzzy system [174] is one of the most popular. In terms of fuzzy sets and fuzzy reasoning applied to a set of linear input-output subsystems, T–S fuzzy systems effectively represent complex nonlinear systems, such as [18, 30, 32, 40, 41, 57, 61, 63, 69, 76, 81, 89, 101, 102, 105, 119, 126–128, 135, 144, 149, 150, 156–159, 166, 167, 176, 183–185, 197, 198, 213, 217, 221, 222, 224, 235]. Feng in [49] provided a survey on analysis and design of model-based fuzzy control systems. Gao et al. in [56] provided a novel stability analysis and stabilization for discrete-time fuzzy systems with time-varying delay. Reference [166] presented a novel approach to filter design for T–S fuzzy discrete-time systems with time-varying delay. Choi and Park in [41] considered the guaranteed cost control problem for discrete-time switching fuzzy systems. Reference [176] designed guaranteed cost controller of polynomial fuzzy systems via a sum of squares approach. The authors in [127] consider the problems of fault estimation and tolerant control for fuzzy stochastic systems. Reference [101] studied the stability analysis and nonlinear observer design using T–S fuzzy models. The authors in [30] designed a robust observer for unknown inputs T–S fuzzy models. Using a T–S fuzzy plant model enables the description of a nonlinear system as a weighted sum of combined simple linear subsystems. This fuzzy model is made up of a family of fuzzy IF-THEN rules representing local linear input/output relations of the system. The overall fuzzy model of the system is achieved by smoothly blending these local linear models together through membership functions. Upon obtaining the fuzzy model, the control design is carried out via the parallel distributed compensation (PDC) approach [179, 192], which employs multiple linear controllers corresponding to the locally linear plant models with automatic scheduling performed via fuzzy rules.

Lyapunov stability theory is one of the most popular methods to investigate the stability of fuzzy control systems. Some fundamental stability results in terms of linear matrix inequalities (LMIs) [17] were achieved in [4, 5, 16, 43, 64, 107, 109, 151, 172, 187, 192, 200]. The authors in [192] provided a novel stability analysis approach to fuzzy control of nonlinear systems. Reference [4] designed an H_∞ fuzzy state-feedback controller for nonlinear systems with-stability constraints. Li et al. in [109] considered the dynamic parallel distributed compensation for T–S fuzzy systems based on LMI approach. Guerra and Vermeiren in [64] proposed LMI-based relaxed non-quadratic stabilization conditions for nonlinear systems in the T–S form; Tseng et al. designed fuzzy tracking controller for nonlinear dynamic systems via T–S fuzzy model in [187]. The fuzzy-model-based (FMB) control system is guaranteed to be asymptotically stable if there exists a common solution to a set of Lyapunov inequalities in terms of LMIs. With the proposed PDC design concept, some stability conditions were relaxed in [172]. More relaxed stability conditions under PDC can be found in [29, 47, 58, 60, 67, 88, 104, 160]. With the consideration of the information of the membership functions, stability conditions can be further relaxed [3, 145, 161]. Some relaxed stability conditions [47, 88, 132] have been proposed on the basis of [179]. T–S fuzzy model offers a fixed structure to some nonlinear systems and facilitates the related system analysis [42, 49]. The above results were obtained by using common quadratic Lyapunov function, which might lead to the conservativeness of corresponding results because the interactions of the fuzzy subsystems were not considered. Some non-quadratic Lyapunov functions have been proposed to reduce the conservativeness in stability analysis. Many relaxed stability results have been obtained by using piecewise-Lyapunov functions [38, 48, 77, 193] and fuzzy-Lyapunov functions [90, 98, 99, 110, 144, 171, 173]. The fuzzy control concept was extended to other stability/control problems such as output-feedback control [28, 45, 112–114, 133, 146, 154, 206, 219, 223], sampled-data control [53, 209], control systems with time delay [7, 23, 33, 34, 36, 56, 115, 122, 138, 199, 205, 208, 228, 229, 241], tracking control [129, 131, 207, 240], large scale fuzzy systems [123, 124, 130, 233, 239]. Moreover, many fuzzy filter design and fault detection (estimation) results based on Lyapunov stability theory have been developed [155, 168, 191, 225, 234, 236].

The aforementioned methods and results based on the T–S fuzzy model are obtained via the type-1 fuzzy model, which is based on type-1 fuzzy sets [226]. Type-1 fuzzy sets are able to effectively capture the system nonlinearities but not the uncertainties [73, 74, 194, 195]. Jafarzadeh et al. in [73, 74] considered stability analysis and control of discrete type-1 and type-2 TSK fuzzy systems; Wu provided the fundamental differences between type-1 interval type-2 fuzzy logic controllers in [194]. The system dynamics of the nonlinear systems can be represented as an average weighted sum of some local linear sub-systems, where the weightings are characterized by the type-1 membership functions. It has been shown in the literature that type-2 fuzzy sets [142], which extend the capability of type-1 fuzzy sets, are beneficial in representing and capturing uncertainties [70, 162], supported by a number of applications such as filtering [68, 117], analog module implementation and design [86, 87], active suspension systems [19], autonomous mobiles [66], electro hydraulic servo systems [83], extended Kalman filter [85], DC-DC power converters [125], nonlinear control [1, 97, 202], noise reduction [84], video streaming [75], inverted pendulum control [139] and so on. However, the type-2 fuzzy set theory was developed for a general type-2 fuzzy logic system but not mainly for FMB control scheme [24, 141]. In type-2 fuzzy logic systems [25, 82], the uncertainties in the nonlinear systems can be also described. Many remarkable results on type-2 fuzzy logic systems have revealed their dominant position in representing the uncertainties [2, 8, 11, 65, 70, 79, 140, 196]. Furthermore, the developed IT2 fuzzy logic systems make it simple in calculation, which have drawn much attention in many fields [20, 26, 31, 37, 50, 72, 78, 111, 118, 121, 136, 137, 143, 163–165, 182, 218, 232]. It can be seen that the parameter uncertainties were not considered in type-1 membership functions. The membership functions of T–S fuzzy systems may become uncertain if the original nonlinear systems have uncertain parameters. Hence, the stability results obtained by using the PDC design technique could no longer be valid. However, there are few research results about the IT2 fuzzy control systems in the literature. This motivates the investigation of the system stability and control design of IT2 fuzzy control systems.

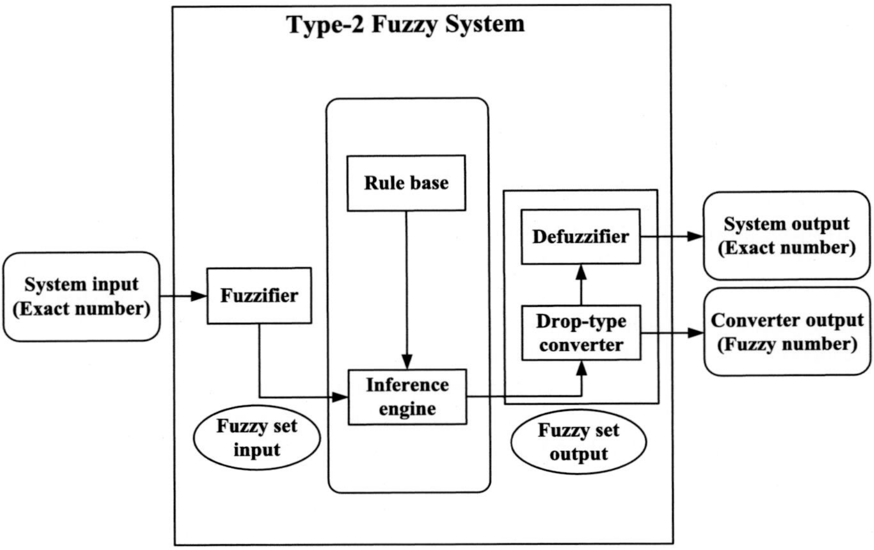

Fig. 1.1 Basic structure of type-2 fuzzy systems

The basic structure of a type-2 fuzzy system consists of five conceptual components: Rule base, fuzzifier, inference engine, drop-type converter and defuzzifier [49]. Different from the structure of the type-1 fuzzy systems, the type-2 fuzzy systems contain the drop-type converter component. Figure 1.1 shows the block diagram of a fuzzy system.

For demonstrating the differences between type-1 fuzzy model and IT2 fuzzy model, we introduce the two fuzzy models in the following context.

1.1 Type-1 T–S Fuzzy Systems

A simple continuous-time T–S fuzzy system is presented as follows:

♦ **Plant Form**:

Rule i: IF $\theta_1(t)$ is $\mathcal{M}_{i1}$ and $\theta_2(t)$ is $\mathcal{M}_{i2}$ and ... and $\theta_p(t)$ is $\mathcal{M}_{ip}$ THEN,

$$\dot{x}(t) = A_i x(t) + B_i u(t), \quad i = 1, 2, \ldots, r,$$

where $x\,(t) \in \mathbf{R}^n$ is the state vector; $u\,(t) \in \mathbf{R}^m$ is the input vector. $\mathcal{M}_{ij}$ is the fuzzy set and r is the number of IF-THEN rules; $\theta(t) = \begin{bmatrix} \theta_1(t) & \theta_2(t) & \cdots & \theta_p(t) \end{bmatrix}^T$ is the premise variable vector. $A_i \in \mathbf{R}^{n\times n}$ and $B_i \in \mathbf{R}^{n\times m}$ are system parameter matrices.

It is assumed that the premise variables are not dependent on the input variables $u(t)$. This assumption is needed to avoid a complicated defuzzification process of fuzzy controllers [180]. Given a pair of $(x(t), u(t))$, the final output of the T–S fuzzy systems is inferred as follows:

$$\dot{x}(t) = \sum_{i=1}^{r} h_i(\theta(t)) \left[A_i x(t) + B_i u(t)\right], \tag{1.1}$$

where $h_i(\theta(t))$, sometimes denoted by $h_i(\theta)$ for simplicity, is the normalized membership function with

$$h_i(\theta(t)) = \frac{\nu_i(\theta(t))}{\sum_{i=1}^{r} \nu_i(\theta(t))}, \quad \nu_i(\theta(t)) = \prod_{j=1}^{p} \mathcal{M}_{ij}(\theta_j(t)),$$

where $\mathcal{M}_{ij}(\theta_j(t))$ is the grade of membership of $\theta_j(t)$ in $\mathcal{M}_{ij}$. It is assumed that

$$\nu_i(\theta(t)) \geq 0, \quad i = 1, 2, \ldots, r,$$
$$\sum_{i=1}^{r} \nu_i(\theta(t)) > 0, \quad \forall t \geq 0.$$

Therefore,

$$h_i(\theta(t)) \geq 0, \quad i = 1, 2, \ldots, r; \quad \sum_{i=1}^{r} h_i(\theta(t)) = 1.$$

Similarly, a discrete-time T–S fuzzy system can be described by

♦ **Plant Form**:

Rule i: IF $\theta_1(k)$ is $\mathcal{M}_{i1}$ and $\theta_2(k)$ is $\mathcal{M}_{i2}$ and … and $\theta_p(k)$ is $\mathcal{M}_{ip}$, THEN,

$$x(k+1) = A_i x(k) + B_i u(k), \quad i = 1, 2, \ldots, r,$$

where $x(k) \in \mathbf{R}^n$ is the state vector; $u(k) \in \mathbf{R}^s$ is the input vector. $\mathcal{M}_{ij}$ is the fuzzy set and r is the number of IF-THEN rules; $\theta(k) = \begin{bmatrix} \theta_1(k) & \theta_2(k) & \cdots & \theta_p(k) \end{bmatrix}^T$ is the premise variables vector. A_i and B_i are known real constant matrices.

A more compact presentation of the discrete-time T–S fuzzy model can be given by

$$x(k+1) = \sum_{i=1}^{r} h_i(\theta(k)) \left[A_i x(k) + B_i u(k)\right]. \tag{1.2}$$

where the normalized membership function is

$$h_i(\theta(k)) = \frac{\nu_i(\theta(k))}{\sum_{i=1}^{r} \nu_i(\theta(k))}, \quad \nu_i(\theta(k)) = \prod_{j=1}^{p} \mathcal{M}_{ij}(\theta_j(k)),$$

where $\mathcal{M}_{ij}(\theta_j(k))$ is the grade of membership of $\theta_j(k)$ in $\mathcal{M}_{ij}$. It is assumed that

$$\nu_i(\theta(k)) \geq 0, \quad i = 1, 2, \ldots, r,$$

$$\sum_{i=1}^{r} \nu_i(\theta(k)) > 0, \quad \forall k \geq 0.$$

Therefore,

$$h_i(\theta(k)) \geq 0, \quad i = 1, 2, \ldots, r; \quad \sum_{i=1}^{r} h_i(\theta(k)) = 1.$$

1.2 Interval Type-2 T–S Fuzzy Model

A p-rule IT2 T–S fuzzy model [97, 117] is employed to descried the dynamics of the nonlinear plant. The rule is of the following format where the antecedent contains IT2 fuzzy sets and the consequent is a linear dynamical system.

♦ **Plant Form**:

Rule i: IF $f_1(x(t))$ is $\tilde{M}_1^i$ and ... and $f_\Psi(x(t))$ is $\tilde{M}_\Psi^i$, THEN,

$$\dot{x}(t) = A_i x(t) + B_i u(t), \tag{1.3}$$

where $\tilde{M}_\alpha^i$ is an IT2 fuzzy set of rule i corresponding to the function $f_\alpha(x(t))$, $\alpha = 1, 2, \ldots, \Psi$; $i = 1, 2, \ldots, p$; Ψ is a positive integer; $x(t) \in \mathbf{R}^n$ is the system state vector; $A_i \in \mathbf{R}^{n \times n}$ and $B_i \in \mathbf{R}^{n \times m}$ are the known system and input matrices, respectively; $u(t) \in \mathbf{R}^m$ is the input vector. The firing strength of the ith rule is of the following interval sets:

$$W_i(x(t)) = \left[\underline{w}_i(x(t)) \; \overline{w}_i(x(t)) \right], \quad i = 1, 2, \ldots, p,$$

where

$$\underline{w}_i\left(x(t)\right) = \prod_{\alpha=1}^{\Psi} \underline{\mu}_{\tilde{M}_\alpha^i}\left(f_\alpha\left(x(t)\right)\right) \geq 0,$$
$$\overline{w}_i\left(x(t)\right) = \prod_{\alpha=1}^{\Psi} \overline{\mu}_{\tilde{M}_\alpha^i}\left(f_\alpha\left(x(t)\right)\right) \geq 0,$$

and

$$\overline{\mu}_{\tilde{M}_\alpha^i}\left(f_\alpha\left(x(t)\right)\right) \geq \underline{\mu}_{\tilde{M}_\alpha^i}\left(f_\alpha\left(x(t)\right)\right) \geq 0,$$
$$\overline{w}_i\left(x(t)\right) \geq \underline{w}_i\left(x(t)\right) \geq 0, \quad \forall i,$$

in which $\underline{w}_i\left(x(t)\right)$, $\overline{w}_i\left(x(t)\right)$, $\underline{\mu}_{\tilde{M}_\alpha^i}\left(f_\alpha\left(x(t)\right)\right)$ and $\overline{\mu}_{\tilde{M}_\alpha^i}\left(f_\alpha\left(x(t)\right)\right)$ denote the lower grade of membership, upper grade of membership, LMF and UMF, respectively. The inferred IT2 T–S fuzzy model [97] is defined as follows:

$$\dot{x}\left(t\right) = \sum_{i=1}^{p} \tilde{w}_i\left(x\left(t\right)\right)\left(A_i x\left(t\right) + B_i u\left(t\right)\right), \tag{1.4}$$

where

$$\tilde{w}_i\left(x\left(t\right)\right) = \underline{\alpha}_i\left(x\left(t\right)\right)\underline{w}_i\left(x\left(t\right)\right) + \overline{\alpha}_i\left(x\left(t\right)\right)\overline{w}_i\left(x\left(t\right)\right) \geq 0, \quad \forall i,$$

with

$$\sum_{i=1}^{p} \tilde{w}_i\left(x\left(t\right)\right) = 1,$$
$$0 \leq \underline{\alpha}_i\left(x\left(t\right)\right) \leq 1, \quad \forall i,$$
$$0 \leq \overline{\alpha}_i\left(x\left(t\right)\right) \leq 1, \quad \forall i,$$
$$\underline{\alpha}_i\left(x\left(t\right)\right) + \overline{\alpha}_i\left(x\left(t\right)\right) = 1, \quad \forall i,$$

in which $\underline{\alpha}_i\left(x\left(t\right)\right)$ and $\overline{\alpha}_i\left(x\left(t\right)\right)$ are nonlinear functions which are not necessarily known but exist; $\tilde{w}_i(x(t))$ can be regarded as the grades of membership of the embedded membership functions.

Similarly, the following form represents a discrete-time IT2 T–S fuzzy system.

♦ **Plant Form**:

Rule i: IF $f_1(x(k))$ is $\tilde{M}_1^i$ and ... and $f_\Psi(x(k))$ is $\tilde{M}_\Psi^i$, THEN,

$$x(k+1) = A_i x(k) + B_i u(k), \tag{1.5}$$

where $\tilde{M}_{\alpha}^{i}$ is an IT2 fuzzy set of rule i corresponding to the function $f_{\alpha}(x(k))$, $\alpha = 1, 2, \ldots, \Psi$; $i = 1, 2, \ldots, p$; Ψ is a positive integer; $x(k) \in \mathbf{R}^{n}$ is the system state vector; $A_i \in \mathbf{R}^{n\times n}$ and $B_i \in \mathbf{R}^{n\times m}$ are the known system and input matrices, respectively; $u(k) \in \mathbf{R}^{m}$ is the input vector. The firing strength of the ith rule is of the following interval sets:

$$W_i(x(k)) = \left[\, \underline{w}_i(x(k))\ \overline{w}_i(x(k)) \,\right], \quad i = 1, 2, \ldots, p,$$

where

$$\underline{w}_i(x(k)) = \prod_{\alpha=1}^{\Psi} \underline{\mu}_{\tilde{M}_{\alpha}^{i}}(f_{\alpha}(x(k))) \geq 0,$$
$$\overline{w}_i(x(k)) = \prod_{\alpha=1}^{\Psi} \overline{\mu}_{\tilde{M}_{\alpha}^{i}}(f_{\alpha}(x(k))) \geq 0,$$

and

$$\overline{\mu}_{\tilde{M}_{\alpha}^{i}}(f_{\alpha}(x(k))) \geq \underline{\mu}_{\tilde{M}_{\alpha}^{i}}(f_{\alpha}(x(k))) \geq 0,$$
$$\overline{w}_i(x(k)) \geq \underline{w}_i(x(k)) \geq 0, \quad \forall i,$$

in which $\underline{w}_i(x(k))$, $\overline{w}_i(x(k))$, $\underline{\mu}_{\tilde{M}_{\alpha}^{i}}(f_{\alpha}(x(k)))$ and $\overline{\mu}_{\tilde{M}_{\alpha}^{i}}(f_{\alpha}(x(k)))$ denote the lower grade of membership, upper grade of membership, LMF and UMF, respectively. The inferred IT2 T–S fuzzy model [97] is defined as follows:

$$x(k+1) = \sum_{i=1}^{p} \tilde{w}_i(x(k))(A_i x(k) + B_i u(k)),$$

where

$$\tilde{w}_i(x(k)) = \underline{\alpha}_i(x(k))\underline{w}_i(x(k)) + \overline{\alpha}_i(x(k))\overline{w}_i(x(k)) \geq 0, \quad \forall i,$$

with

$$\sum_{i=1}^{p} \tilde{w}_i(x(k)) = 1,$$
$$0 \leq \underline{\alpha}_i(x(k)) \leq 1, \quad \forall i,$$
$$0 \leq \overline{\alpha}_i(x(k)) \leq 1, \quad \forall i,$$
$$\underline{\alpha}_i(x(k)) + \overline{\alpha}_i(x(k)) = 1, \quad \forall i,$$

in which $\underline{\alpha}_i(x(k))$ and $\overline{\alpha}_i(x(k))$ are nonlinear functions which are not necessarily known but exist; $\tilde{w}_i(x(k))$ can be regarded as the grades of membership of the embedded membership functions and (2.3) defines the type reduction.

Example 1.1 A simple example is given below to illustrate the IT2 T–S fuzzy model [97]. Consider the following simple scalar system subject to the uncertain parameter of $a(t)$:

$$\dot{x}(t) = \sin\left(a(x(t))x(t)\right)x(t), \tag{1.6}$$

where $x(t) \in [-2, 2]$ is the system state. It is assumed that $a(x(t)) = \left(x(t)^2 + 1\right)/10$ has the known bounds of $\underline{a} \leq a(x(t)) \leq \overline{a}$, where $\underline{a} = 0.1$ and $\overline{a} = 0.5$ are the constant lower and upper bounds of $a(x(t))$, respectively. From (1.6), it can be seen that the lower and upper bounds for $\sin\left(a(x(t))x(t)\right)$ are $\sin(1)$ and $\sin(-1)$ (i.e., 0.8415 and −0.8415), respectively. By following the way proposed for deriving the type-1 T–S fuzzy model and denoting the membership functions as $\mu_{M_1^1} = (x(t), a(x(t)))$ and $\mu_{M_1^2} = (x(t), a(x(t))) = 1 - \mu_{M_1^1}(x(t), a(x(t)))$, we have $\mu_{M_1^1}(x(t), a(x(t)))\sin(1) + \mu_{M_1^2}(x(t), a(x(t)))\sin(-1) = \sin\left(a(x(t))x(t)\right)$. Reshuffling the terms, we have $\mu_{M_1^1}(x(t), a(x(t))) = \left(1 - \sin(a(x(t))x(t))\right)/2$ and $\mu_{M_1^2}(x(t), a(x(t))) = \left(1 + \sin\left(a(x(t))x(t)\right)\right)/2$. Considering $a(x(t))$ to be a constant, we have the following type-1 fuzzy rule to describe the system of (1.6):

$$\textbf{Rule } i\text{: IF } x(t) \text{ is } M_1^i, \text{ THEN } \dot{x}(t) = A_i x(t), \; i = 1, 2, \tag{1.7}$$

where M_1^1 and M_1^2 are type-1 fuzzy sets; $A_1 = -1$ and $A_2 = 1$. The type-1 T–S fuzzy model is defined as

$$\dot{x}(t) = \sum_{i=1}^{2} w_i A_i x(t),$$

where the normalized grades of membership are defined as $w_i(x(t)) = \mu_{M_1^i}(x(t))/(\mu_{M_1^1}(x(t)) + \mu_{M_1^2}(x(t)))$, $i = 1, 2$ (as $a(x(t))$ is assumed to be a constant, it is not a parameter of the type-1 membership functions). Figure 1.2 shows the type-1 membership functions of $\mu_{M_1^1}(x(t), a(x(t))) = \left(1 - \sin(a(x(t))x(t))\right)/2$ subject to different values of $a(x(t))$. When $a(x(t))$ is considered as an uncertain value in the range of $\underline{a}$ and $\overline{a}$, it can be imagined that $\mu_{M_1^1}(x(t))$ is no longer a crisp membership function but characterized by the lower and upper memberships of $\underline{\mu}_{\tilde{M}_1^1}(x(t))$ and $\bar{\mu}_{\tilde{M}_1^1}(x(t))$, respectively. Under such a situation, an IT2 T–S fuzzy model is employed to represent the nonlinear plant of (1.6) subject to parameter uncertainty of $a(x(t))$. With the information of the type-1 membership functions, the LMFs and UMFs can be obtained as follows. We have $\underline{\mu}_{\tilde{M}_1^1}(x(t)) \leq \mu_{M_1^1}(x(t), a(x(t))) \leq \bar{\mu}_{\tilde{M}_1^1}(x(t))$, where

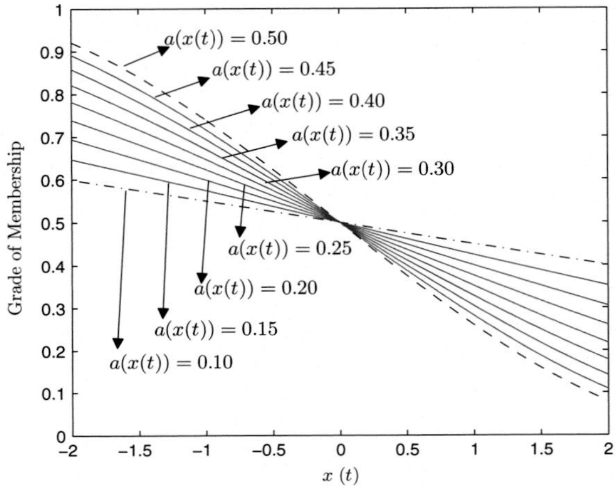

Fig. 1.2 Plot of $\mu_{M_1^1}(x(t), a(x(t))) = (1 - \sin(a(x(t))x(t)))/2$ with various values of $a(x(t))$, and illustration of FOU, LMFs and UMFs

$$\underline{\mu}_{\tilde{M}_1^1}(x(t)) = \begin{cases} \dfrac{1 - \sin(0.1x(t))}{2}, & \text{for } x(t) < 0, \\ \dfrac{1 - \sin(0.5x(t))}{2}, & \text{for } x(t) \geq 0, \end{cases} \tag{1.8}$$

$$\bar{\mu}_{\tilde{M}_1^1}(x(t)) = \begin{cases} \dfrac{1 - \sin(0.5x(t))}{2}, & \text{for } x(t) < 0, \\ \dfrac{1 - \sin(0.1x(t))}{2}, & \text{for } x(t) \geq 0. \end{cases} \tag{1.9}$$

Similarly, we have $\underline{\mu}_{\tilde{M}_1^2}(x(t)) \leq \mu_{M_1^2}(x(t), a(x(t))) \leq \bar{\mu}_{\tilde{M}_1^2}(x(t))$, where

$$\underline{\mu}_{\tilde{M}_1^2}(x(t)) = \begin{cases} \dfrac{1 + \sin(0.5x(t))}{2}, & \text{for } x(t) < 0, \\ \dfrac{1 + \sin(0.1x(t))}{2}, & \text{for } x(t) \geq 0, \end{cases} \tag{1.10}$$

$$\bar{\mu}_{\tilde{M}_1^2}(x(t)) = \begin{cases} \dfrac{1 + \sin(0.1x(t))}{2}, & \text{for } x(t) < 0, \\ \dfrac{1 + \sin(0.5x(t))}{2}, & \text{for } x(t) \geq 0. \end{cases} \tag{1.11}$$

The following rules for the IT2 T–S fuzzy model can be achieved:

Rule i: IF $x(t)$ is $\tilde{M}_1^i$, THEN $\dot{x}(t) = A_i x(t)$, $i = 1, 2$. (1.12)

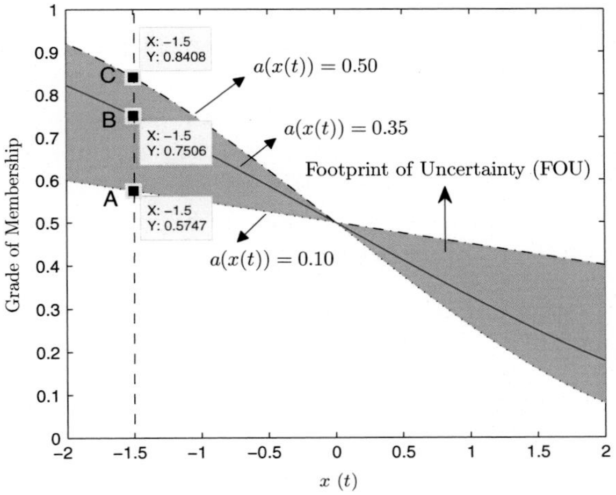

Fig. 1.3 LMF $\underline{\mu}_{\tilde{M}_1^1}(x(t))$ (*dotted line*), UMF $\bar{\mu}_{\tilde{M}_1^1}(x(t))$ (*dash-dot line*), and membership function $\mu_{M_1^1}(a(x(t)), x(t))$ (*solid line*)

From (1.4), the IT2 T–S fuzzy model for the nonlinear plant of (1.6) is defined as

$$\dot{x}(t) = \sum_{i=1}^{2} \left[\underline{\alpha}_i(x(t))\, \underline{w}_i(x(t)) + \overline{\alpha}_i(x(t))\, \overline{w}_i(x(t)) \right] A_i x(t), \tag{1.13}$$

where $\underline{w}_i(x(t)) = \underline{\mu}_{\tilde{M}_1^i}(x(t))$ and $\overline{w}_i(x(t)) = \bar{\mu}_{\tilde{M}_1^i}(x(t))$ for all i. Figure 1.3 shows the type-1 membership function $\underline{\mu}_{\tilde{M}_1^1}(x(t))$ with $a(t) = 0.35$, the LMFs and UMFs. The area in between the LMFs and UMFs is the FOU. Based on the LMFs and UMFs, the type-1 membership function in between can be reconstructed with the introduction of $\underline{\alpha}_i(x(t))$ and $\overline{\alpha}_i(x(t))$. Referring to Fig. 1.3, for $x(t) = -1.5$, it can be seen that the grades of membership at points **A**, **B**, and **C** are 0.8408, 0.7506 and 0.5747, respectively. The grade of membership at point **B** can be represented as $0.7506 = 0.5747 \times 0.3390 + 0.8408 \times (1 - 0.3390)$. By the same line of logic, every point of the membership function with $a(t) = 0.35$ can be determined based on the LMFs and UMFs. In general, the in-between membership functions can be reconstructed and represented in the form of a linear combination of the LMFs and UMFs, i.e., $\underline{\alpha}_i(x(t))\underline{w}_i(x(t)) + \overline{\alpha}_i(x(t))\overline{w}_i(x(t))$, where $\underline{\alpha}_i(x(t)) + \overline{\alpha}_i(x(t)) = 1$ for any $0.1 \le a(x(t)) \le 0.5$. In the above example, it can be seen that $\underline{\alpha}_i(x(t)) = 0.3390$ and $\overline{\alpha}_i(x(t)) = 0.6610$. In the stability analysis, we do know the exact values of $\underline{\alpha}_i(x(t))$ and $\overline{\alpha}_i(x(t))$, as they depend on the parameter uncertainty of $a(t)$. However, it can be seen that $\underline{\alpha}_i(x(t))$ and $\overline{\alpha}_i(x(t))$ exist.

With the IT2 fuzzy sets, the parameter uncertainties can be captured by the LMFs and UMFs. Consequently, an IT2 T–S fuzzy model of (1.13) can be used to describe the system of (1.6) with the uncertain parameter of $a(x(t))$. The IT2 T–S fuzzy model of (1.13) can be regarded as a collection of type-1 T–S fuzzy models with $a(x(t))$ taking various constant values.

Remark 1.2 The LMFs and UMFs for an IT2 T–S fuzzy model are not unique. Referring to Fig. 1.3, the FOU can be obtained by the area bounded by the LMFs and UMFs of $\underline{\mu}_{\tilde{M}_1^1}(x(t))$ and $\bar{\mu}_{\tilde{M}_1^1}(x(t))$, respectively. With the nonlinear functions $\underline{\alpha}_i(x(t))$ and $\overline{\alpha}_i(x(t))$, the in-between membership functions can be reproduced. Based on this concept, by considering any arbitrary LMFs and UMFs denoted by $\underline{\mu}_{\tilde{M}_1^1}(x(t))$ and $\bar{\mu}_{\tilde{M}_1^1}(x(t))$ satisfying the conditions of $\underline{\mu}_{\tilde{M}_1^1}(x(t)) \leq \bar{\mu}_{\tilde{M}_1^1}(x(t))$, the actual membership function can be reconstructed with the nonlinear functions $\underline{\alpha}_i(x(t))$ and $\overline{\alpha}_i(x(t))$ in other forms.

The following lemmas are introduced for the main results in the following context.

Lemma 1.3 ([28]) *Given appropriately dimensioned matrices Ξ, Q and P, with $\Xi^T = \Xi$, then*

$$\Xi + PF(t)Q^T + QF(t)P^T < 0, \tag{1.14}$$

holds for any F which $F(t)^T F(t) < I$ if and only if for some scalar $\varepsilon > 0$,

$$\Xi + \varepsilon^{-1}PP^T + \varepsilon QQ^T < 0. \tag{1.15}$$

Lemma 1.4 ([46]) *If $X = X^T$, $Y = Y^T$ and U and V are nonsingular, and satisfy*

$$UV^T + XY = I,$$

then $\begin{bmatrix} X & U \\ U^T & -V^{-1}YU \end{bmatrix}$ *and* $\begin{bmatrix} Y & V \\ V^T & -U^{-1}XV \end{bmatrix}$ *are symmetrical and satisfy*

$$\begin{bmatrix} X & U \\ U^T & -V^{-1}YU \end{bmatrix}\begin{bmatrix} Y & V \\ V^T & -U^{-1}XV \end{bmatrix} = \begin{bmatrix} I & 0 \\ 0 & I \end{bmatrix}.$$

Lemma 1.5 ([108]) *(Disc region): Let $\mho(q, r)$ denote any disc region center in q with radius r in the complex plane ($q, r \in \mathbf{R}$ and $r > 0$). Then, all the eigenvalues of A_i in (1.4) lie in the region $\mho(q, r)$ and only if there exists a matrix $P > 0$ satisfying*

$$\begin{bmatrix} -P & P(A_i - qI) \\ * & -r^2P \end{bmatrix} < 0. \tag{1.16}$$

Lemma 1.6 ([108]) *(Vertical strip region): Let $\Psi(v, u)$ denote a vertical strip region lying within the bounds v and u ($v < u$, $v, u \in \mathbf{R}$). Then, all the eigenvalues of A_i in (1.4) lie in the region $\Psi(v, u)$ if and only if there exists a matrix $P > 0$ satisfying*

$$(A_i - uI)^T P + P(A_i - uI) < 0, \tag{1.17}$$

$$-(A_i - vI)^T P - P(A_i - vI) < 0. \tag{1.18}$$

Lemma 1.7 *For any appropriate dimensioned real matrices M_{ij}, N_{ij} ($1 \le i \le p$ and $1 \le j \le c$), and matrix $Q > 0$, the following LMI holds:*

$$\left[\sum_{i=1}^{p}\sum_{j=1}^{c} h_{ij}(\xi(k)) M_{ij}\right]^T Q \left[\sum_{\kappa=1}^{p}\sum_{\iota=1}^{c} h_{\kappa\iota}(\xi(k)) N_{\kappa\iota}\right]$$
$$\le \frac{1}{2}\sum_{i=1}^{p}\sum_{j=1}^{c} h_{ij}(\xi(k))\left(M_{ij}^T Q M_{ij} + N_{ij}^T Q N_{ij}\right),$$

where $h_{ij}(\xi(k))$ for $i = 1, 2, \ldots, p$, $j = 1, 2, \ldots, c$, satisfy $h_{ij}(\xi(k)) \ge 0$, and $\sum\limits_{i=1}^{p}\sum\limits_{j=1}^{c} h_{ij}(\xi(k)) = 1$.

Proof Based on the fact that $2M^T QN \le \inf_{Q>0}\left\{M^T QM + N^T QN\right\}$, it is easily obtained that

$$2\left[\sum_{i=1}^{p}\sum_{j=1}^{c} h_{ij}(\xi(k)) M_{ij}\right]^T Q \left[\sum_{\kappa=1}^{p}\sum_{\iota=1}^{c} h_{\kappa\iota}(\xi(k)) N_{\kappa\iota}\right]$$
$$\le \sum_{i=1}^{p}\sum_{j=1}^{c}\sum_{\kappa=1}^{p}\sum_{\iota=1}^{c} h_{ij}(\xi(k)) h_{\kappa\iota}(\xi(k))\left(M_{ij}^T Q M_{ij} + N_{\kappa\iota}^T Q N_{\kappa\iota}\right)$$
$$= \sum_{i=1}^{p}\sum_{j=1}^{c} h_{ij}(\xi(k)) M_{ij}^T Q M_{ij} + \sum_{\kappa=1}^{p}\sum_{\iota=1}^{c} h_{\kappa\iota}(\xi(k)) N_{\kappa\iota}^T Q N_{\kappa\iota}$$
$$= \sum_{i=1}^{p}\sum_{j=1}^{c} h_{ij}(\xi(k))\left(M_{ij}^T Q M_{ij} + N_{ij}^T Q N_{ij}\right).$$

This completes the proof. □

Lemma 1.8 ([211]) *Given any matrices X, Y and $Z > 0$ with appropriate dimensions, then the inequality $X^T Y + Y^T X \le X^T Z X + Y^T Z^{-1} Y$ holds.*

1.3 Publication Contribution

This book represents the some attempts to reflect the state-of-the-art of the research area for handling stability/performance analysis and optimal synthesis problems for IT2 T–S fuzzy systems. The content of this book can be divided into two parts.

The first part will provide analysis and synthesis of continuous-time IT2 T–S fuzzy systems. Some sufficient conditions are derived for the stability, different controllers and filters design for the considered IT2 T–S fuzzy systems with different performances. The developed methodologies include the Lyapunov stability approach, LMI technique, etc. The main aim by using these advanced approaches is to effectively reduce the conservatism of the obtained results, and thus facilitate the design subsequently. Then, some optimal synthesis problems, including the stabilization, the state and output-feedback control with different system performances, the sampled-data control with actuator fault, the output tracking control with actuator fault, the switched output-feedback control, the filter design with $\mathcal{D}$ stability constraints, the fault detection with sensor nonlinearities, and the model reduction with $\mathcal{D}$ stability constraints, are investigated based on the analysis results. Focussing on the parallel theories and techniques developed in the previous part, the second section is extended to deal with discrete-time IT2 T–S fuzzy systems. Specifically, in this part, the main problems, including the optimal control with poles constraint, the fault-tolerant control, the reliable mixed H_2/H_∞ control, the output tracking control with predefined cost function, are investigated for the discrete-time IT2 T–S fuzzy systems. Based on the presented results, the corresponding simulation examples are provided to validate the effectiveness and the applicability of the design methods.

The features of this book can be highlighted as follows. (1) A unified framework is established for analysis and synthesis of T–S fuzzy systems with parameter uncertainties, where there are external perturbations and faults. (2) A series of problems are solved with new approaches for analysis and synthesis of continuous- and discrete-time fuzzy systems, including stability/performances analysis and stabilization, output-feedback control, tracking control, filtering, fault detection, and model approximation. (3) A set of newly developed techniques (e.g., the Lyapunov stability theory, the LMI technique, convex optimization) are exploited to handle the emerging mathematical/computational challenges.

This publication is a timely reflection of the developing new area of system analysis and synthesis theories for the so called IT2 T–S fuzzy systems. It is a collection of a series of latest research results and therefore serves as a useful textbook for senior and/or graduate students who are interested in knowing (1) the state-of-the-art of fuzzy systems and fuzzy control area; (2) recent advances in uncertain systems; (3) recent advances in stability/performance analysis, stabilization, output-feedback control, tracking control, fault-tolerant control, optical control, filtering, fault detection, and model approximation problems. Readers will also benefit from some new concepts, new models and new methodologies with theoretical significance in system analysis and control synthesis. It can also be used as a practical research reference for engineers dealing with stabilization, optimal control and filtering problems for IT2 T–S fuzzy systems. The aim of this book is to close the gap in literature by providing a unified yet neat framework for stability/performances analysis and synthesis of IT2 T–S fuzzy systems.

Generally, this is an advanced publication aimed at 3rd/4th-year undergraduates, postgraduates and academic researchers. Prerequisite knowledge includes fuzzy sets, linear algebra, matrix analysis, and linear control system theory.

Expected readers include (1) control engineers working on nonlinear control, fuzzy control and optimal control; (2) system engineers working on intelligent control and systems; (3) mathematicians and physician working on uncertain systems; (4) postgraduate students majoring in control engineering, system sciences and applied mathematics. This publication is also a useful reference for (1) mathematicians and physicians working on intelligent systems and nonlinear systems; (2) computer scientists working on algorithms and computational complexity; (3) 3rd/4th-year students who are interested in advanced control theory and its applications.

1.4 Publication Outline

The general layout of presentation of this monograph is divided into two parts. Part one focuses on the analysis and synthesis for continuous-time IT2 fuzzy systems, whilst second part studies the analysis and synthesis for discrete-time IT2 fuzzy systems. The organization structure of this monograph is shown in Fig. 1.4, and the main contents of this monograph are shown in Fig. 1.5.

Chapter 1 presents the research background, motivations and research problems, which involve state-feedback control design, output-feedback control design, tracking control design, filter design, and fault detection design for of IT2 T–S fuzzy systems, and then the outline of the monograph is listed.

Part I focuses on the analysis and synthesis for continuous-time IT2 T–S fuzzy systems. It begins with Chap. 2 and consists of eight chapters as follows.

Chapter 2 considers the stabilization problem of IT2 fuzzy systems. To facilitate the stabilization, an IT2 T–S fuzzy model is employed to represent the dynamics of nonlinear systems of which the parameter uncertainties are captured by IT2 membership functions characterized by the LMFs and UMFs. A novel IT2 fuzzy controller is proposed to perform the control process, where the membership functions and number of rules can be freely chosen and different from those of the IT2 T–S fuzzy model. To relax the stability analysis for this class of IT2 FMB control systems, the information of footprint of uncertainties (FOU), and the LMFs and UMFs are taken into account for the stability analysis. Based on the Lyapunov stability theory, some stability conditions in terms of LMIs are obtained to determine the system stability and achieve the control design.

Chapter 3 is concerned with the problems of state and output-feedback control for IT2 fuzzy systems with mismatched membership functions. The IT2 fuzzy model and the IT2 state and output-feedback controllers do not share the same membership functions. A novel performance index, which is expressed as an extended dissipativity performance, is introduced to be a generalization of H_∞, L_2-L_∞, passive and dissipativity performances indexes. Firstly, the IT2 T–S fuzzy model and the controllers are constructed by considering the mismatched membership functions. Secondly, on the basis of Lyapunov stability theory, the IT2 fuzzy state

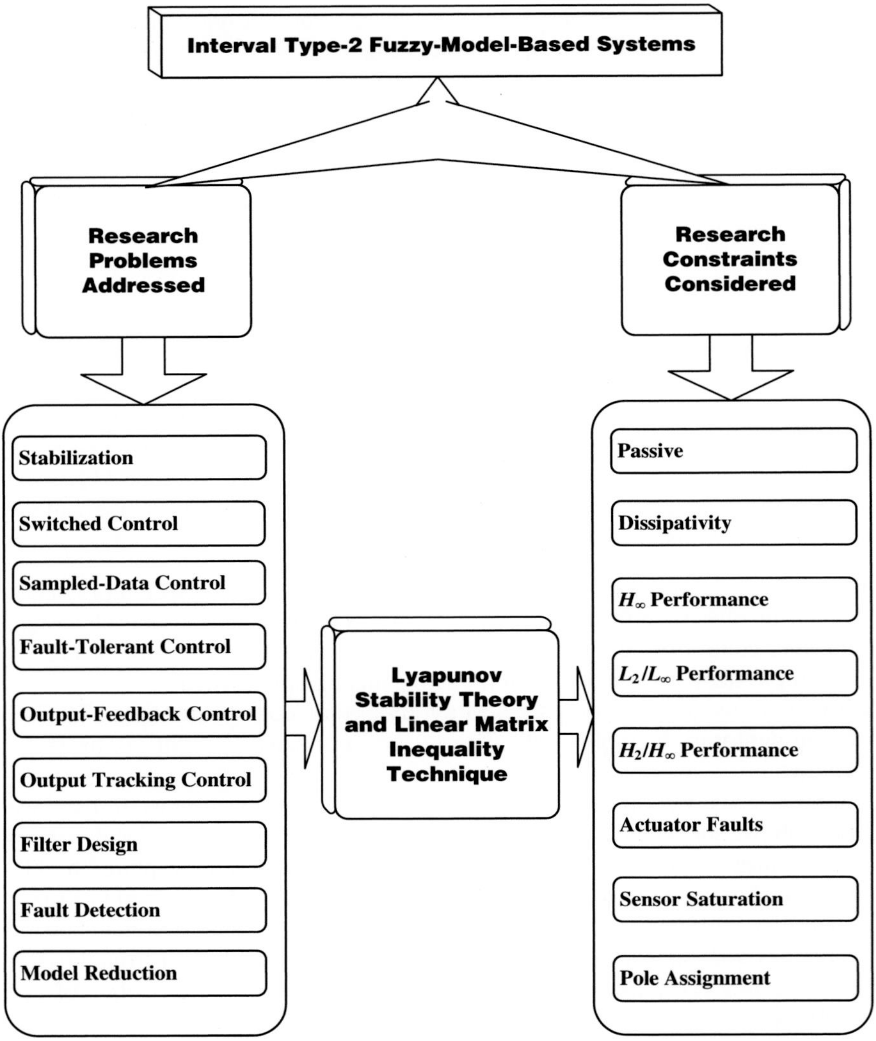

Fig. 1.4 Organizational structure of this publication

and output-feedback controllers are designed respectively to guarantee that the closed-loop system is asymptotically stable with extended dissipativity performance. The existence conditions of the two kinds of controllers are obtained in terms of convex optimization problems.

Chapter 4 focuses on designing sampled-data controller for IT2 fuzzy systems with actuator fault. The IT2 fuzzy system and the IT2 state-feedback controller share different membership functions. Firstly, considering the mismatched membership

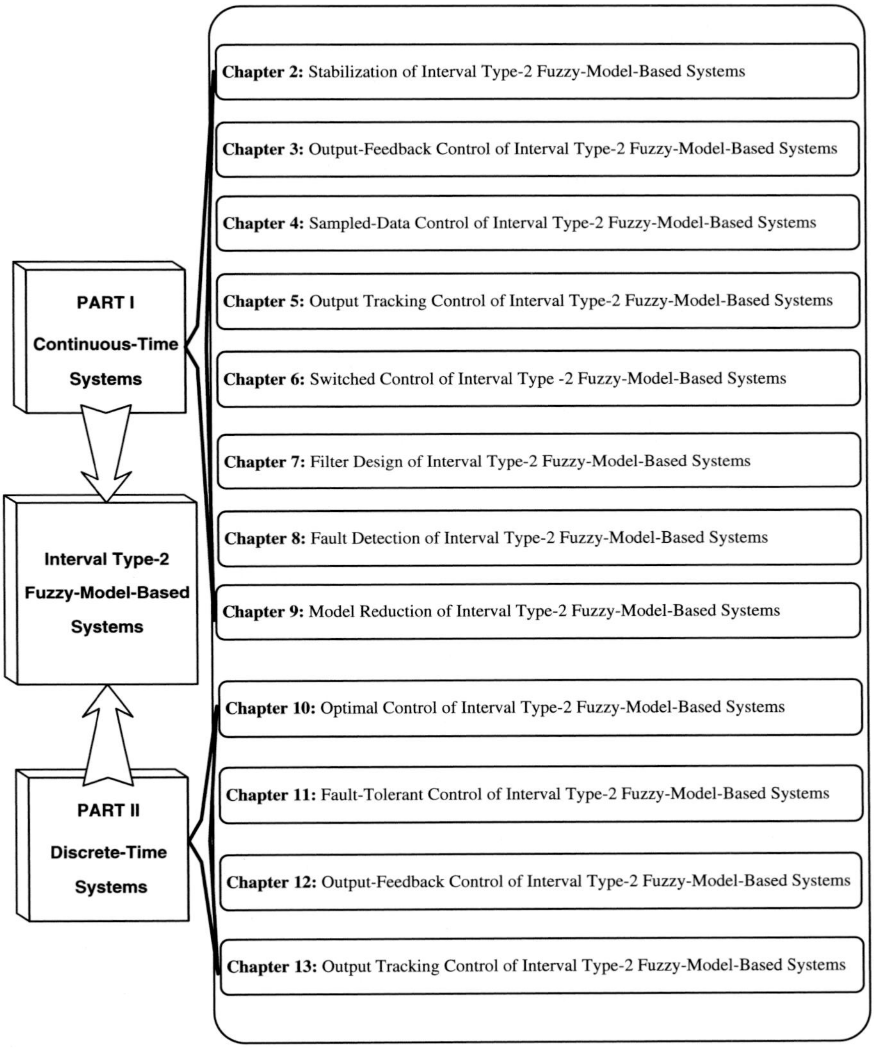

Fig. 1.5 Main contents of this publication

functions, the IT2 fuzzy model and the IT2 state-feedback sampled-data controller are constructed. Secondly, based on Lyapunov stability theory, an IT2 state-feedback sampled-data controller is designed such that the closed-loop system is asymptotically stable. The actuator failure is considered in the control systems. The resulting closed-loop system is reliable since the designed controller can guarantee the asymptotic stability and H_∞ performance when the actuator experiences failure.

Chapter 5 investigates the problem of output tracking control for IT2 fuzzy systems with actuator fault. An IT2 state-feedback fuzzy controller is designed to perform the tracking control problem, where the membership functions can be freely chosen since the number of fuzzy rules is different from that of the IT2 T–S fuzzy model. Based on Lyapunov stability theory, an existence condition of IT2 fuzzy H_∞ output tracking controller is obtained to guarantee that the output of the closed-loop system can track the output of a given reference model well in the H_∞ sense.

Chapter 6 studies the switched control problem for IT2 fuzzy systems. A switched output-feedback controller, which depends on the values of membership functions, is constructed. The membership functions of IT2 fuzzy systems contain parameter uncertainties. Based on the type-2 fuzzy set theory, the parameter uncertainties can be effectively obtained. A novel type of IT2 switched output-feedback controller is designed to ensure that the closed-loop system is asymptotically stable with an H_∞ performance.

Chapter 7 investigates the problem of filter design for IT2 fuzzy systems with $\mathcal{D}$ stability constraints based on a new performance index. Attention is focused on solving the H_∞, L_2-L_∞, passive and dissipativity fuzzy filter design problems for IT2 fuzzy systems with $\mathcal{D}$ stability constraints in a unified frame. Under the new performance index frame, using Lyapunov stability theory, a novel type of IT2 filter is designed such that the filtering error system guarantees the prescribed H_∞, L_2-L_∞, passive and dissipativity performance levels with $\mathcal{D}$ stability constraints. The existence condition of the IT2 filter is expressed as the convex optimization problem and the filter parameters in the condition can be solved by the standard software.

Chapter 8 deals with the fault detection problem for IT2 fuzzy systems subject to sensor nonlinearities. By using a general observer-based fault detection filter as a residual generator, the fault detection problem is described as a filter design problem. The fault detection filter is designed to guarantee the prescribed H_∞ performance level. A decomposition approach is employed to handle the characteristic of sensor saturation. Using Lyapunov stability theory, a novel type of IT2 fault detection filter is designed to guarantee that the fault detection system is asymptotically stable with an H_∞ performance. In the design procedure, the parameters of the IT2 filter can be solved by the standard software.

Chapter 9 considers the problem of model reduction for IT2 fuzzy systems subject to $\mathcal{D}$ stability constraints. The membership functions and the number of rules can be freely chosen and they are different between the original system and the reduced-order system. By introducing some slack matrices and utilizing Lyapunov stability theory, the existence condition of model reduction is obtained to guarantee that the reduced-order model can approximate the original system with an H_∞ performance. The parameters of the reduced-order system in the condition can be obtained by standard software.

Part II is concerned with the analysis and synthesis for discrete-time IT2 T–S fuzzy systems. It begins with Chap. 10 and consists of four chapters as follows.

Chapter 10 investigates optimal control problem for discrete-time IT2 fuzzy systems with poles constraint. An IT2 fuzzy controller is characterized by two predefined functions, and the membership functions and the premise rules of the IT2 fuzzy controller can be chosen freely. The pole assignment is considered, which is constrained in a presented disk region. Based on Lyapunov stability theory, sufficient conditions of asymptotic stability with an H_∞ performance are obtained for the discrete-time IT2 fuzzy system. Based on the criterion, the desired IT2 state-feedback controller is designed to guarantee that the closed-loop system is asymptotically stable with a prescribed H_∞ performance and all the poles rest in the disk region.

Chapter 11 is concerned with the problem of fault-tolerant control for discrete-time IT2 fuzzy time delay system with actuator faults under imperfect premise matching. The time-varying delay and actuator fault are first taken into account for the discrete-time IT2 fuzzy systems. The fault-tolerant controller is designed to compensate for the effect of faults by stabilizing the closed-loop system under the actuator failures. Furthermore, the standard IT2 state-feedback controller is designed such that the closed-loop system is asymptotically stable and has an H_∞ performance. The obtained conditions of the fault-tolerant controller and the standard IT2 controller can be expressed by the convex optimization problems.

Chapter 12 investigates the problem of reliable mixed H_2/H_∞ control for discrete-time IT2 fuzzy systems via static output-feedback control method. The number of fuzzy rules and the membership functions for the static output-feedback controller are different from those for the plant. A sufficient criterion of reliable stability with mixed H_2/H_∞ performance is derived for the closed-loop system with sensor failure. The static output-feedback controller is designed for two different cases (known sensor failure case and unknown sensor failure case) to guarantee the reliable stability with mixed H_2/H_∞ performance. Moreover, a novel criteria are presented to obtain the optical H_2 performance for the closed-loop system.

Chapter 13 investigates the problem of guaranteed cost output tracking control for discrete-time IT2 fuzzy systems subject to mismatched premise variables. Based on the IT2 T–S fuzzy model, the criterion to design the desired controller is obtained, which guarantees the closed-loop system to be asymptotically stable and satisfies the predefined cost function. Moreover, the controller to be designed does not need to share the same premise variables of the system, which enhances the flexibility of controller design and reduces the conservativeness.

Chapter 14 draws conclusions on the book, and points out some possible research directions related to the work done in this book.

Part I
Continuous-Time Systems

Chapter 2
Stabilization of Interval Type-2 Fuzzy-Model-Based Systems

2.1 Introduction

This chapter investigates the stability of IT2 FMB control systems under imperfect premise matching. Unlike the authors' work in [97] under PDC design concept, it was required that the IT2 fuzzy controller shares the same premise membership functions and the same number of rules as those of the IT2 T–S fuzzy model. These limitations constrain the design flexibility and increase the implementation complexity of the IT2 fuzzy controller. This result of this chapter eliminates these limitations by proposing an IT2 fuzzy controller that the membership functions and the number of rules can be freely chosen enhancing the applicability of the IT2 FMB control scheme. By choosing simple membership functions and a smaller number of rules, it can reduce the implementation complexity of the IT2 fuzzy controller resulting in a lower implementation cost. However, the IT2 FMB control systems can be imperfectly matched membership functions, potentially leading to more difficult stability analysis as the favorable property of PDC design concept vanishes.

To carry out the stability analysis for IT2 FMB control system subject to imperfect premise membership functions, the LMFs and UMFs characterized the footprint of uncertainty (FOU) are chosen to be a favorable representation. This favorable representation allows the LMFs and UMFs to be taken in the stability analysis. Consequently, the stability conditions in terms of LMIs are membership-function-dependent, which are applied to the nonlinear plant under consideration, but not a family considered in some existing work. Preliminary result of the authors in [96] provides technical support to the work in this chapter. To further relax the stability conditions, the FOU is divided into a number of sub-FOUs. The information of the sub-FOUs along with those of LMFs and UMFs are brought to the stability analysis. Based on the Lyapunov stability theory, LMI-based stability conditions are obtained to guarantee the stability of the IT2 FMB control systems and synthesize the IT2 fuzzy controller.

© Springer Science+Business Media Singapore 2016
H. Li et al., *Analysis and Synthesis for Interval Type-2
Fuzzy-Model-Based Systems*, DOI 10.1007/978-981-10-0593-0_2

2.2 Problem Formulation and Preliminaries

Considering a nonlinear plant subject to parameter uncertainties represented by an IT2 T–S fuzzy model [97, 117], an IT2 fuzzy controller is proposed to perform the control process. An IT2 FMB control system is formed by connecting the IT2 T–S fuzzy model and the IT2 fuzzy controller in a closed loop. It is not required that both the IT2 T–S fuzzy model and the IT2 fuzzy controller share the same premise membership functions and the same number of rules.

A p-rule IT2 T–S fuzzy model [97, 117] is employed to describe the dynamics of the nonlinear plant. The rule is of the following format where the antecedent contains IT2 fuzzy sets and the consequent is a linear dynamical system.

♦ **Plant Form**:

Rule i: IF $f_1(x(t))$ is $\tilde{M}_1^i$ and ... and $f_\Psi(x(t))$ is $\tilde{M}_\Psi^i$, THEN

$$\dot{x}(t) = A_i x(t) + B_i u(t), \tag{2.1}$$

where $\tilde{M}_\alpha^i$ is an IT2 fuzzy set of rule i corresponding to the function $f_\alpha(x(t))$, $\alpha = 1, 2, \ldots, \Psi$; $i = 1, 2, \ldots, p$; Ψ is a positive integer; $x(t) \in \mathbf{R}^n$ is the system state vector; $A_i \in \mathbf{R}^{n \times n}$ and $B_i \in \mathbf{R}^{n \times m}$ are the known system and input matrices, respectively; $u(t) \in \mathbf{R}^m$ is the input vector. The firing strength of the ith rule is of the following interval sets:

$$W_i(x(t)) = \left[\underline{w}_i(x(t)) \;\; \overline{w}_i(x(t))\right], \quad i = 1, 2, \ldots, p,$$

where

$$\begin{aligned} \underline{w}_i(x(t)) &= \prod_{\alpha=1}^{\Psi} \underline{\mu}_{\tilde{M}_\alpha^i}(f_\alpha(x(t))) \geq 0, \\ \overline{w}_i(x(t)) &= \prod_{\alpha=1}^{\Psi} \overline{\mu}_{\tilde{M}_\alpha^i}(f_\alpha(x(t))) \geq 0, \\ \overline{\mu}_{\tilde{M}_\alpha^i}(f_\alpha(x(t))) &\geq \underline{\mu}_{\tilde{M}_\alpha^i}(f_\alpha(x(t))) \geq 0, \\ \overline{w}_i(x(t)) &\geq \underline{w}_i(x(t)) \geq 0, \quad \forall i, \end{aligned}$$

in which $\underline{w}_i(x(t))$, $\overline{w}_i(x(t))$, $\underline{\mu}_{\tilde{M}_\alpha^i}(f_\alpha(x(t)))$ and $\overline{\mu}_{\tilde{M}_\alpha^i}(f_\alpha(x(t)))$ denote the lower grade of membership, upper grade of membership, LMF and UMF, respectively. The inferred IT2 T–S fuzzy model [97] is defined as follows:

$$\dot{x}(t) = \sum_{i=1}^{p} \tilde{w}_i(x(t)) (A_i x(t) + B_i u(t)), \tag{2.2}$$

where

$$\tilde{w}_i\left(x\left(t\right)\right) = \underline{\alpha}_i\left(x\left(t\right)\right)\underline{w}_i\left(x\left(t\right)\right) + \overline{\alpha}_i\left(x\left(t\right)\right)\overline{w}_i\left(x\left(t\right)\right) \geq 0, \quad \forall i, \tag{2.3}$$

with

$$\sum_{i=1}^{p} \tilde{w}_i(x(t)) = 1, \tag{2.4}$$

and

$$\begin{aligned} 0 \leq \underline{\alpha}_i(x(t)) &\leq 1, \quad \forall i, \\ 0 \leq \overline{\alpha}_i(x(t)) &\leq 1, \quad \forall i, \\ \underline{\alpha}_i(x(t)) + \overline{\alpha}_i(x(t)) &= 1, \quad \forall i, \end{aligned}$$

in which $\underline{\alpha}_i(x(t))$ and $\overline{\alpha}_i(x(t))$ are nonlinear functions which are not necessarily known but exist; $\tilde{w}_i(x(t))$ can be regarded as the grades of membership of the embedded membership functions and (2.3) defines the type reduction.

Remark 2.1 It can be seen from (2.4) that the actual grades of membership, $\tilde{w}_i(x(t))$, can be reconstructed and expressed as a linear combination of $\underline{w}_i(x(t))$ and $\overline{w}_i(x(t))$, characterized by the LMFs and UMFs $\underline{\mu}_{\tilde{M}^i_\alpha}(f_\alpha(x(t)))$ and $\overline{\mu}_{\tilde{M}^i_\alpha}(f_\alpha(x(t)))$, which are scaled by the nonlinear functions $\underline{\alpha}_i(x(t))$ and $\overline{\alpha}_i(x(t))$, respectively. In other words, any membership functions within the FOU [97] can be reconstructed by the LMFs and UMFs. As the nonlinear plant is subject to parameter uncertainties, $\tilde{w}_i(x(t))$ will depend on the parameter uncertainties and thus leads to the values of $\underline{\alpha}_i(x(t))$ and $\overline{\alpha}_i(x(t))$ uncertain. It should be noted that the IT2 T–S fuzzy model (2.2) serves as a mathematical tool to facilitate the stability analysis and control synthesis, and is not necessarily implemented.

An IT2 fuzzy controller with c rules of the following format is proposed to stabilize the nonlinear plant represented by the IT2 T–S fuzzy model (2.2).

♦ **Controller Form**:

Rule i: IF $g_1(x(t))$ is $\tilde{N}^j_1$ and ... and $g_\Omega(x(t))$ is $\tilde{N}^j_\Omega$, THEN

$$u(t) = G_j x(t), \tag{2.5}$$

where $\tilde{N}^j_\beta$ is an IT2 fuzzy set of rule j corresponding to the function $g_\beta(x(t))$, $\beta = 1, 2, \ldots, \Omega$; $j = 1, 2, \ldots, c$; Ω is a positive integer; $G_j \in \mathbf{R}^{m\times n}$, $j = 1, 2, \ldots, c$, are the constant feedback gains to be determined. The firing strength of the jth rule is the following interval sets:

$$M_j(x(t)) = \left[\underline{m}_j(x(t)) \;\; \overline{m}_j(x(t))\right], \quad j = 1, 2, \ldots, c,$$

where

$$\underline{m}_j(x(t)) = \prod_{\beta=1}^{\Omega} \underline{\mu}_{\tilde{N}_\beta^j}(g_\beta(x(t))) \geq 0,$$
$$\overline{m}_j(x(t)) = \prod_{\beta=1}^{\Omega} \overline{\mu}_{\tilde{N}_\beta^j}(g_\beta(x(t))) \geq 0,$$
$$\overline{\mu}_{\tilde{N}_\beta^j}(g_\beta(x(t))) \geq \underline{\mu}_{\tilde{N}_\beta^j}(g_\beta(x(t))) \geq 0, \quad \forall j,$$

in which $\underline{m}_j(x(t))$, $\overline{m}_j(x(t))$, $\underline{\mu}_{\tilde{N}_\beta^j}(g_\beta(x(t)))$ and $\overline{\mu}_{\tilde{N}_\beta^j}(g_\beta(x(t)))$ stand for the lower grade of membership, upper grade of membership, LMF and UMF, respectively. The inferred IT2 fuzzy controller is defined as follows:

$$u(t) = \sum_{j=1}^{c} \tilde{m}_j(x(t))G_j x(t),$$

where

$$\tilde{m}_j(x(t)) = \frac{\underline{\beta}_j(x(t))\underline{m}_j(x(t)) + \overline{\beta}_j(x(t))\overline{m}_j(x(t))}{\sum_{k=1}^{c}\left[\underline{\beta}_k(x(t))\underline{m}_k(x(t)) + \overline{\beta}_k(x(t))\overline{m}_k(x(t))\right]} \geq 0, \quad \forall j, \quad (2.6)$$

with

$$\sum_{j=1}^{c} \tilde{m}_i(x(t)) = 1,$$
$$0 \leq \underline{\beta}_j(x(t)) \leq 1, \quad \forall j,$$
$$0 \leq \overline{\beta}_j(x(t)) \leq 1, \quad \forall j,$$
$$\underline{\beta}_j(x(t)) + \overline{\beta}_j(x(t)) = 1, \quad \forall j,$$

in which $\underline{\beta}_j(x(t))$ and $\overline{\beta}_j(x(t))$ are predefined functions; $\tilde{m}_j(x(t))$ can be regarded as the grades of membership of the embedded membership functions and (2.6) is the type reduction.

From (2.2) and (2.6), with the property of $\sum_{i=1}^{p} \tilde{w}_i(x(t)) = \sum_{j=1}^{c} \tilde{m}_j(x(t)) = \sum_{i=1}^{p}\sum_{j=1}^{c} \tilde{w}_i(x(t))\tilde{m}_j(x(t)) = 1$, we have the following IT2 FMB control system:

$$\dot{x}(t) = \sum_{i=1}^{p} \tilde{w}_i(x(t)) \left(A_i x(t) + B_i \sum_{j=1}^{c} \tilde{m}_j(x(t)) G_j x(t) \right)$$

$$= \sum_{i=1}^{p} \sum_{j=1}^{c} \tilde{w}_i(x(t)) \tilde{m}_j(x(t)) \left(A_i + B_i G_j \right) x(t). \tag{2.7}$$

The control objective of this chapter is to guarantee the system stability by determining the feedback gain G_j, such that the IT2 fuzzy controller (2.6) is able to drive the system states to the origin, i.e., $x(t) \to 0$ as time $t \to \infty$.

Basic LMI-based stability conditions guaranteeing the stability of the FMB based control system in the form of (2.7) are given in the following theorem.

Theorem 2.2 ([192]) *The FMB control system in the form of* (2.7) *is guaranteed to be asymptotically stable if there exist matrices* $N_j \in \mathbf{R}^{m \times n}$, $j = 1, 2, \ldots, c$, $X = X^T \in \mathbf{R}^{n \times n}$ *such that the following LMIs hold:*

$$X > 0,$$
$$Q_{ij} = A_i X + X A_i^T + B_i N_j + N_j^T B_i^T < 0, \quad \forall i, j,$$

where the feedback gains are defined as $G_j = N_j X^{-1}$ *for all* j.

Remark 2.3 The stability conditions in Theorem 2.2 are very conservative as the membership functions of both fuzzy model and fuzzy controller are not considered. The stability conditions can be reduced to $Q_{ij} = A_i X + X A_i^T + B_i N + N^T B_i^T < 0$ for all i by choosing a common feedback gain, i.e., $N = N_j$ for all j resulting in a linear controller.

To facilitate the stability analysis of the IT2 FMB control system (2.7), the state space of interest denoted as Φ is divided into q connected sub-state spaces denoted as Φ_k, $k = 1, 2, \ldots, q$ such that $\Phi = \bigcup_{k=1}^{q} \Phi_k$. Furthermore, to consider more information of the IT2 membership functions, local LMFs and UMFs within the FOU are introduced. Considering the FOU being divided into $\tau + 1$ sub-FOUs, in the lth sub-FOU, $l = 1, 2, \ldots, \tau + 1$, the LMFs and UMFs are defined as follows:

$$\underline{h}_{ijl}(x(t)) = \sum_{k=1}^{q} \sum_{i_1=1}^{2} \cdots \sum_{i_n=1}^{2} \prod_{r=1}^{n} v_{ri_r kl}(x_r(t)) \underline{\delta}_{iji_1 i_2 \ldots i_n kl}, \quad \forall i, j, k, l, \tag{2.8}$$

$$\overline{h}_{ijl}(x(t)) = \sum_{k=1}^{q} \sum_{i_1=1}^{2} \cdots \sum_{i_n=1}^{2} \prod_{r=1}^{n} v_{ri_r kl}(x_r(t)) \overline{\delta}_{iji_1 i_2 \ldots i_n kl}, \quad \forall i, j, k, l, \tag{2.9}$$

and

$$0 \le \underline{h}_{ijl}(x(t)) \le \overline{h}_{ijl}(x(t)) \le 1,$$
$$0 \le \underline{\delta}_{iji_1 i_2 \ldots i_n kl} \le \overline{\delta}_{iji_1 i_2 \ldots i_n kl} \le 1,$$

where $\underline{\delta}_{iji_1i_2\ldots i_nkl}$ and $\overline{\delta}_{iji_1i_2\ldots i_nkl}$ are constant scalars to be determined; $0 \leq v_{ri_skl}(x_r(t)) \leq 1$ and $v_{r1kl}(x_r(t)) + v_{r2kl}(x_r(t)) = 1$ for $r, s = 1, 2, \ldots, n$; $l = 1, 2, \ldots, \tau + 1$; $i_r = 1, 2$; $x(t) \in \Phi_k$; otherwise, $v_{ri_sk}(x_r(t)) = 0$. As a result, we have $\sum_{k=1}^{q}\sum_{i_1=1}^{2}\sum_{i_2=1}^{2}\cdots\sum_{i_n=1}^{2}\prod_{r=1}^{n} v_{ri_rkl}(x_r(t)) = 1$ for all l, which is used in the stability analysis.

We then express the IT2 FMB control system (2.7) in the following favorable form:

$$\dot{x}(t) = \sum_{i=1}^{p}\sum_{j=1}^{c}\tilde{h}_{ij}(x(t))\left(A_i + B_iG_j\right)x(t), \tag{2.10}$$

where

$$\begin{aligned}\tilde{h}_{ij}(x(t)) &= \tilde{w}_i(x(t))\tilde{m}_j(x(t)) \\ &= \sum_{l=1}^{\tau+1}\xi_{ijl}(x(t))\left[\underline{\gamma}_{ijl}(x(t))\underline{h}_{ijl}(x(t)) + \overline{\gamma}_{ijl}\overline{h}_{ijl}(x(t))\right],\end{aligned} \tag{2.11}$$

with

$$\sum_{i=1}^{p}\sum_{j=1}^{c}\tilde{h}_{ij}(x(t)) = 1. \tag{2.12}$$

In addition, $0 \leq \underline{\gamma}_{ijl}(x(t)) \leq \overline{\gamma}_{ijl}(x(t)) \leq 1$ are two functions, which are not necessary to be known, exhibiting the property that $\underline{\gamma}_{ijl}(x(t)) + \overline{\gamma}_{ijl}(x(t)) = 1$ for all i, j, l; $\xi_{ijl}(x(t)) = 1$ if the membership function $h_{ijl}(x(t))$ is within the sub-FOU l, otherwise, $\xi_{ijl}(x(t)) = 0$.

Remark 2.4 It should be noted that only one $\xi_{ijl}(x(t)) = 1$ for the fixed ijth membership function $\tilde{h}_{ij}(x(t))$ among the $\tau + 1$ sub-FOUs at any time instant and the rest equals to zero. It can be seen from (2.11) that the more the sub-FOUs are considered, the more information about the FOU is contained in the local LMFs and UMFs.

Remark 2.5 The local LMFs and UMFs can reconstruct $\tilde{h}_{ij}(x(t)) \equiv \tilde{w}_i(x(t))\tilde{m}_j(x(t))$ by representing it as a linear combination of $\underline{h}_{ijl}(x(t))$ and $\overline{h}_{ijl}(x(t))$ in sub-FOU l as shown in (2.11).

Remark 2.6 The IT2 FMB control system in (2.7) is a subset of (2.10). Comparing both the IT2 FMB control systems in (2.7) and (2.10), the one in (2.10) demonstrates some favorable properties to facilitate the stability analysis:

(1) The partial information of $\underline{h}_{ijl}(x(t))$ and $\overline{h}_{ijl}(x(t))$ is extracted and represented by the constant scalars $\underline{\delta}_{iji_1i_2\ldots i_nkl}$ and $\overline{\delta}_{iji_1i_2\ldots i_nkl}$, which are brought to the stability conditions.

(2) Referring to (2.8) and (2.9), the cross terms, $\prod_{r=1}^{n} v_{ri_r kl}(x_r(t))$, are independent of i and j thus can be collected in the stability analysis.
(3) With the nonlinear functions, $\underline{\gamma}_{ijl}(x(t))$ and $\overline{\gamma}_{ijl}(x(t))$, $\tilde{h}_{ijl}(x(t))$ can be reconstructed as shown in (2.11) as a linear combination of $\underline{h}_{ijl}(x(t))$ and $\overline{h}_{ijl}(x(t))$. Furthermore, with the expressions (2.8) and (2.9), the values of $\underline{h}_{ijl}(x(t))$ and $\overline{h}_{ijl}(x(t))$ are determined by the constant scalars $\underline{\delta}_{iji_1 i_2 \ldots i_n kl}$ and $\overline{\delta}_{iji_1 i_2 \ldots i_n kl}$ through $\prod_{r=1}^{n} v_{ri_r kl}(x_r(t))$. As a result, the stability of the IT2 FMB control system can be determined by $\underline{h}_{ijl}(x(t))$ and $\overline{h}_{ijl}(x(t))$ (the local lower and upper bounds of $\tilde{h}_{ij}(x(t))$) characterized by the constant scalars $\underline{\delta}_{iji_1 i_2 \ldots i_n kl}$ and $\overline{\delta}_{iji_1 i_2 \ldots i_n kl}$. These properties can be seen in the stability analysis carried out in the next section.

2.3 Main Results

The stability of the IT2 FMB control system (2.7) is investigated based on the Lyapunov stability theory with the consideration of the information of the LMFs and UMFs, and sub-FOUs. For brevity, in the following analysis, the time t associated with the variables is dropped for the situation without ambiguity, e.g., $x(t)$ is denoted as x. The variables $\underline{w}_i(x(t))$, $\overline{w}_i(x(t))$, $\tilde{w}_i(x(t))$, $\underline{m}_j(x(t))$, $\overline{m}_j(x(t))$, $\tilde{m}_j(x(t))$, $\tilde{h}_{ijl}(x(t))$, $v_{1i_1 kl}(x_1(t))$, $v_{2i_2 kl}(x_2(t))$, …, $v_{ni_n kl}(x_n(t))$ and $\xi_{ijl}(x(t))$ are denoted by $\underline{w}_i, \overline{w}_i, \tilde{w}_i, \underline{m}_j, \overline{m}_j, \tilde{m}_j, \tilde{h}_{ijl}, v_{1i_1 kl}, v_{2i_2 kl}, \ldots, v_{ni_n kl}$ and ξ_{ijl}, respectively. Furthermore, the property of $\sum_{i=1}^{p} \tilde{w}_i = \sum_{j=1}^{c} \tilde{m}_j = \sum_{i=1}^{p} \sum_{j=1}^{c} \tilde{w}_i \tilde{m}_j = \sum_{i=1}^{p} \sum_{j=1}^{c} \tilde{h}_{ij} = 1$ is utilized.

The stability analysis result is summarized in the following theorem to guarantee the asymptotic stability of the IT2 FMB control system (2.7) and facilitate the control synthesis.

Theorem 2.7 *Considering the FOU being divided into* $\tau + 1$ *sub-FOUs, the IT2 FMB control system* (2.7) *under imperfect premise matching, formed by a nonlinear plant (represented by the IT2 T–S fuzzy model* (2.2)*) and an IT2 fuzzy controller* (2.6) *connected in a closed loop, is guaranteed to be asymptotically stable if there exist matrices* $M = M^T \in \mathbf{R}^{n \times n}$, $N_j \in \mathbf{R}^{m \times n}$, $X = X^T \in \mathbf{R}^{n \times n}$, $W_{ijl} = W_{ijl}^T \in \mathbf{R}^{n \times n}$, $(i = 1, 2, \ldots, p;\ j = 1, 2, \ldots, c;\ l = 1, 2, \ldots, \tau + 1)$, *such that the following LMIs hold:*

$$X > 0, \tag{2.13}$$

$$W_{ijl} \geq 0, \quad \forall i, j, l, \tag{2.14}$$

$$Q_{ij} + W_{ijl} + M > 0, \quad \forall i, j, l, \tag{2.15}$$

$$\sum_{i=1}^{p}\sum_{j=1}^{c}\left[\overline{\delta}_{iji_1i_2\ldots i_nkl}Q_{ij} - \left(\underline{\delta}_{iji_1i_2\ldots i_nkl} - \overline{\delta}_{iji_1i_2\ldots i_nkl}\right)W_{ijl} + \overline{\delta}_{iji_1i_2\ldots i_nkl}M\right] - M < 0,\ \forall i_1, i_2, \ldots, i_n, k, l, \tag{2.16}$$

where $\underline{\delta}_{iji_1i_2\ldots i_nkl}$ *and* $\overline{\delta}_{iji_1i_2\ldots i_nkl}$, $i = 1, 2, \ldots, p$; $j = 1, 2, \ldots, c$; $i_1, i_2, \ldots, i_n = 1, 2$; $k = 1, 2, \ldots, q$; $l = 1, 2, \ldots, \tau + 1$ *are predefined constant scalars satisfying* (2.8) *and* (2.9); $Q_{ij} = A_iX + XA_i^T + B_iN_j + N_j^TB_i^T$ *for all* i *and* j; *and the feedback gains are defined as* $G_j = N_jX^{-1}$ *for all* j.

Proof We consider the following quadratic Lyapunov function candidate to investigate the stability of the IT2 FMB control systems (2.7) expressed in the form of (2.10).

$$V(t) = x^T(t)Px(t), \tag{2.17}$$

where $0 < P = P^T \in \mathbf{R}^{n\times n}$.

The main objective is to develop a condition guaranteeing that $V(t) > 0$ and $\dot{V}(t) < 0$ for all $x(t) \neq 0$. According to the Lyapunov stability theorem, by satisfying $V(t) > 0$ and $\dot{V}(t) < 0$ for all $x(t) \neq 0$, the IT2 FMB control system is guaranteed to be asymptotically stable, implying that $x(t) \to 0$ as $t \to \infty$.

Denote $z(t) = X^{-1}x(t)$ and $X = P^{-1}$. Define the feedback gains $G_j = N_jX^{-1}$ where $N_j \in \mathbf{R}^{m\times n}$, $j = 1, 2, \ldots, c$, are matrices to be determined. From (2.10) and (2.17), we have,

$$\begin{aligned}
\dot{V}(t) &= \dot{x}^T(t)Px(t) + x^T(t)P\dot{x}(t)\\
&= \sum_{i=1}^{p}\sum_{j=1}^{c}\tilde{h}_{ij}x^T(t)\left[\left(A_i + B_iG_j\right)^T P + P\left(A_i + B_iG_j\right)\right]x(t)\\
&= \sum_{i=1}^{p}\sum_{j=1}^{c}\tilde{h}_{ij}x^T(t)PP^{-1}\left[\left(A_i + B_iG_j\right)^T P + P\left(A_i + B_iG_j\right)\right]P^{-1}Px(t)\\
&= \sum_{i=1}^{p}\sum_{j=1}^{c}\sum_{l=1}^{\tau+1}\xi_{ijl}\left[\underline{\gamma}_{ijl}\underline{h}_{ijl} + \overline{\gamma}_{ijl}\overline{h}_{ijl}\right]z^T(t)Q_{ij}z(t),
\end{aligned} \tag{2.18}$$

where $Q_{ij} = A_iX + XA_i^T + B_iN_j + N_j^TB_i^T$.

Recalling the property that $0 \leq \underline{h}_{ijl} \leq \overline{h}_{ijl} \leq 1$, $0 \leq \underline{\gamma}_{ijl} \leq 1$, $0 \leq \overline{\gamma}_{ijl} \leq 1$ and $\underline{\gamma}_{ijl} + \overline{\gamma}_{ijl} = 1$ for all i, j, l, the information of the sub-FOUs is brought to the stability analysis with the introduction of some slack matrices through the following inequalities using the S-procedure [17]:

$$\left[\sum_{i=1}^{p}\sum_{j=1}^{c}\sum_{l=1}^{\tau+1}\xi_{ijl}\left(\underline{\gamma}_{ijl}\underline{h}_{ijl}+\overline{\gamma}_{ijl}\overline{h}_{ijl}\right)-1\right]M=0, \tag{2.19}$$

$$-\sum_{i=1}^{p}\sum_{j=1}^{c}\left(1-\underline{\gamma}_{ijl}\right)\left(\underline{h}_{ijl}-\overline{h}_{ijl}\right)W_{ijl}\geq 0, \tag{2.20}$$

where $M = M^T \in \mathbf{R}^{n\times n}$ ia an arbitrary matrix and $0 \leq W_{ijl} = W_{ijl}^T \in \mathbf{R}^{n\times n}$.

From (2.11), and (2.18)–(2.20), we have

$$\begin{aligned}\dot{V}(t) &= \sum_{i=1}^{p}\sum_{j=1}^{c}\sum_{l=1}^{\tau+1}\xi_{ijl}\left(\underline{\gamma}_{ijl}\underline{h}_{ijl}+\overline{\gamma}_{ijl}\overline{h}_{ijl}\right)z^T(t)Q_{ij}z(t)\\ &\leq \sum_{i=1}^{p}\sum_{j=1}^{c}\sum_{l=1}^{\tau+1}\xi_{ijl}\left[\underline{\gamma}_{ijl}\underline{h}_{ijl}+(1-\underline{\gamma}_{ijl})\overline{h}_{ijl}\right]z^T(t)Q_{ij}z(t)\\ &\quad-\sum_{i=1}^{p}\sum_{j=1}^{c}\sum_{l=1}^{\tau+1}\xi_{ijl}(1-\underline{\gamma}_{ijl})\left(\underline{h}_{ijl}-\overline{h}_{ijl}\right)z^T(t)W_{ijl}z(t)\\ &\quad+\left[\sum_{i=1}^{p}\sum_{j=1}^{c}\sum_{l=1}^{\tau+1}\xi_{ijl}\left(\underline{\gamma}_{ijl}\underline{h}_{ijl}+(1-\underline{\gamma}_{ijl})\overline{h}_{ijl}\right)-1\right]z^T(t)Mz(t)\\ &= z^T(t)\left[\sum_{i=1}^{p}\sum_{j=1}^{c}\sum_{l=1}^{\tau+1}\xi_{ijl}\left(\overline{h}_{ijl}Q_{ij}-\left(\underline{h}_{ijl}-\overline{h}_{ijl}\right)W_{ijl}+\overline{h}_{ijl}M\right)-M\right]z(t)\\ &\quad+\sum_{i=1}^{p}\sum_{j=1}^{c}\sum_{l=1}^{\tau+1}\xi_{ijl}\underline{\gamma}_{ijl}\left(\underline{h}_{ijl}-\overline{h}_{ijl}\right)z^T(t)\left(Q_{ij}+W_{ijl}+M\right)z(t).\end{aligned} \tag{2.21}$$

Referring to (2.21), $\dot{V}(t) < 0$ for $x(t) \neq 0$ is satisfied from

$$\sum_{i=1}^{p}\sum_{j=1}^{c}\sum_{l=1}^{\tau+1}\xi_{ijl}(x)\left[\overline{h}_{ijl}Q_{ij}-\left(\underline{h}_{ijl}-\overline{h}_{ijl}\right)W_{ijl}+\overline{h}_{ijl}M\right]-M<0, \tag{2.22}$$

and $Q_{ij} + W_{ijl} + M > 0$ (due to $\underline{h}_{ijl} - \overline{h}_{ijl} \leq 0$) for all i, j, l. Recalling that only one $\xi_{ijl} = 1$ for each fixed value of ij at any time instant such that $\sum_{l=1}^{\tau+1}\xi_{ijl} = 1$, the first set of inequalities is satisfied by $\sum_{i=1}^{p}\sum_{j=1}^{c}\left[\overline{h}_{ijl}Q_{ij}-\left(\underline{h}_{ijl}-\overline{h}_{ijl}\right)W_{ijl}+\overline{h}_{ijl}M\right]$ $-M < 0$ for all i, j, l. Expressing $\underline{h}_{ijl}$ and $\overline{h}_{ijl}$ with (2.8) and (2.9), respectively, and recalling that $\sum_{k=1}^{q}\sum_{i_1=1}^{2}\sum_{i_2=1}^{2}\cdots\sum_{i_n=1}^{2}\prod_{r=1}^{n}v_{ri_r kl} = 1$ for all l and $v_{ri_r kl} \geq 0$ for all r, i_r, k and l, the first set of inequalities will be satisfied if the following inequalities hold $\forall i_1, i_2, \ldots, i_n, k, l$:

$$\sum_{k=1}^{q}\sum_{i_1=1}^{2}\sum_{i_2=1}^{2}\cdots\sum_{i_n=1}^{2}\prod_{r=1}^{n} v_{ri_r kl}\left\{\sum_{i=1}^{p}\sum_{j=1}^{c}\left[\overline{\delta}_{iji_1i_2\ldots i_nkl}Q_{ij}\right.\right.$$

$$\left.\left.-(\underline{\delta}_{iji_1i_2\ldots i_nkl}-\overline{\delta}_{iji_1i_2\ldots i_nkl})W_{ijl}+\overline{\delta}_{iji_1i_2\ldots i_nkl}M\right]-M\right\}<0. \quad (2.23)$$

Consequently,

$$\sum_{i=1}^{p}\sum_{j=1}^{c}\left[\overline{h}_{ijl}Q_{ij}-(\underline{h}_{ijl}-\overline{h}_{ijl})W_{ijl}+\overline{h}_{ijl}M\right]-M<0,$$

can be guaranteed by

$$\sum_{i=1}^{p}\sum_{j=1}^{c}\left[\overline{\delta}_{iji_1i_2\ldots i_nkl}Q_{ij}-(\underline{\delta}_{iji_1i_2\ldots i_nkl}-\overline{\delta}_{iji_1i_2\ldots i_nkl})W_{ijl}\right.$$

$$\left.+\overline{\delta}_{iji_1i_2\ldots i_nkl}M\right]-M<0.$$

The LMI-based stability conditions above are summarized in Theorem 2.7. By satisfying those LMIs, the IT2 FMB control system (2.7) is guaranteed to be asymptotically stable.

Referring to (2.23), the advantages of representing the IT2 FMB control system (2.7) in the form of (2.10) can be seen. The membership functions $\tilde{h}_{ij}$ are reconstructed by the linear combination of the local LMFs and UMFs $\underline{h}_{ijl}$ and $\overline{h}_{ijl}$. Consequently, as seen from (2.21), the stability of the IT2 FMB control system is determined by the local LMFs and UMFs $\underline{h}_{ijl}$ and $\overline{h}_{ijl}$. By expressing $\underline{h}_{ijl}$ and $\overline{h}_{ijl}$ in the form of (2.8) and (2.9), respectively, they are characterized by the constant scalars $\underline{\delta}_{iji_1i_2\ldots i_nkl}$ and $\overline{\delta}_{iji_1i_2\ldots i_nkl}$. Furthermore, as the cross terms $\prod_{r=1}^{n} v_{ri_rkl}$ are independent of i and j, they can be extracted as shown in (2.23) to facilitate the stability analysis. With these favorable properties as previously stated in Remark 2.6, we only need to check $\sum_{i=1}^{p}\sum_{j=1}^{c}(\overline{\delta}_{iji_1i_2\ldots i_nkl}Q_{ijl}-(\underline{\delta}_{iji_1i_2\ldots i_nkl}-\overline{\delta}_{iji_1i_2\ldots i_nkl})W_{ijl}+\overline{\delta}_{iji_1i_2\ldots i_nkl}M)-M<0$ at some discrete points ($\underline{\delta}_{iji_1i_2\ldots i_nkl}$ and $\overline{\delta}_{iji_1i_2\ldots i_nkl}$) instead of every single point of the local LMFs and UMFs ($\underline{h}_{ijl}$ and $\overline{h}_{ijl}$) to guarantee (2.23). □

Remark 2.8 The stability conditions in Theorem 2.2 is a particular case of Theorem 2.7. If there exists a solution to the stability conditions in Theorem 2.2, $x>0$ and $Q_{ij}<0$ for all i and j can be achieved. Choosing $M=\varepsilon_1 I>0$ and $W_{ijl}=-Q_{ij}+(-\varepsilon_1+\varepsilon_2)I>0$ for all i, j, l with sufficiently small non-zero positive value of ε_1 and ε_2 in Theorem 2.7, LMIs (2.14) and (2.15) can be satisfied. As a result, recalling that $\overline{\delta}_{iji_1i_2\ldots i_nkl}\geq\underline{\delta}_{iji_1i_2\ldots i_nkl}\geq 0$, the LMIs in (2.16) become $\sum_{i=1}^{p}\sum_{j=1}^{c}\left(\overline{\delta}_{iji_1i_2\ldots i_nkl}\varepsilon_2 I-\underline{\delta}_{iji_1i_2\ldots i_nkl}W_{ijl}\right)-\varepsilon_1 I<0$ for all $i_1, i_2, \ldots, i_n, k$ and l, which will be satisfied by a sufficiently small value of ε_2. Consequently, the solution of the stability conditions in Theorem 2.2 is that of Theorem 2.7 but not on the other way round.

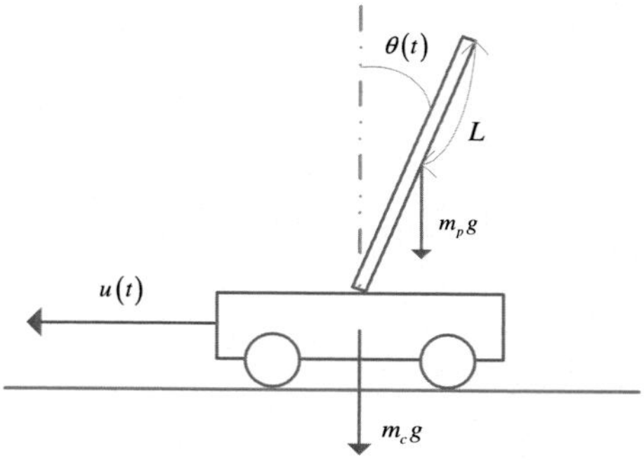

Fig. 2.1 An inverted pendulum system. © [2013] IEEE. Reprinted, with permission, from ref. [1]

2.4 Simulation Results

Example 2.9 In this example, we consider an inverted pendulum as shown in Fig. 2.1 subject to parameter uncertainties [97] as the nonlinear plant to be controlled. The dynamic equation for the inverted pendulum is given by

$$\ddot{\theta}(t) = \frac{g\sin(\theta(t)) - am_p L\dot{\theta}^2(t)\sin(2\theta(t))/2 - a\cos(\theta(t))u(t)}{4L/3 - am_p L\cos^2(\theta(t))}, \tag{2.24}$$

where $\theta(t)$ is the angular displacement of the pendulum, $g = 9.8\,\text{m/s}^2$ is the acceleration due to gravity, $m_p \in \begin{bmatrix} m_{p_{\min}} & m_{p_{\max}} \end{bmatrix} = [2 \quad 3]$ kg is the mass of the pendulum, $M_c \in [M_{\min} \quad M_{\max}] = [8 \quad 12]$ kg is the mass of the cart, $a = 1/(m_p + M_c)$, $2L = 1$ m is the length of the pendulum, and $u(t)$ is the force (N) applied to the cart. The inverted pendulum is considered working in the operating domain characterized by $x_1 = \theta(t) \in \begin{bmatrix} -\frac{5\pi}{12} & \frac{5\pi}{12} \end{bmatrix}$ and $x_2 = \dot{\theta}(t) \in \begin{bmatrix} -5 & 5 \end{bmatrix}$.

A 4-rule IT2 T–S fuzzy model in the form of (2.2) is employed to describe the inverted pendulum subject to parameter uncertainties with $x = \begin{bmatrix} x_1 \\ x_2 \end{bmatrix} = \begin{bmatrix} \theta(t) \\ \dot{\theta}(t) \end{bmatrix}$, $A_1 = A_2 = \begin{bmatrix} 0 & 1 \\ f_{1_{\min}} & 0 \end{bmatrix}$ and $A_3 = A_4 = \begin{bmatrix} 0 & 1 \\ f_{1_{\max}} & 0 \end{bmatrix}$; $B_1 = B_3 = \begin{bmatrix} 0 \\ f_{2_{\min}} \end{bmatrix}$, $B_2 = B_4 = \begin{bmatrix} 0 \\ f_{2_{\max}} \end{bmatrix}$; $f_{1_{\min}} = 10.0078$, $f_{1_{\max}} = 18.4800$, $f_{2_{\min}} = -0.1765$ and $f_{2_{\max}} = -0.0261$. The LMFs and UMFs are defined in Table 2.1.

A 2-rule IT2 fuzzy controller is employed to stabilize the inverted pendulum with the LMFs and UMFs chosen as

$$\underline{m}_1(x_1) = \underline{\mu}_{\tilde{N}_1^1}(x_1) = \overline{m}_1(x_1) = \overline{\mu}_{\tilde{N}_1^1}(x_1) = e^{\frac{-x_1^2}{0.35}},$$
$$\underline{m}_2(x_1) = \underline{\mu}_{\tilde{N}_1^2}(x_1) = \overline{m}_2(x_1) = \overline{\mu}_{\tilde{N}_1^2}(x_1) = 1 - \overline{\mu}_{\tilde{N}_1^1}(x_1),$$
$$\underline{\beta}_k = \overline{\beta}_k = \frac{1}{2}.$$

Table 2.1 LUMFs of the IT2 fuzzy systems.

LMFs	UMFs
$\underline{\mu}_{\tilde{M}_1^1}(x_1) = 1 - e^{-\frac{x_1^2}{1.2}}$	$\overline{\mu}_{\tilde{M}_1^1}(x_1) = 1 - 0.23e^{-\frac{x_1^2}{0.25}}$
$\underline{\mu}_{\tilde{M}_1^2}(x_1) = 1 - e^{-\frac{x_1^2}{1.2}}$	$\overline{\mu}_{\tilde{M}_1^2}(x_1) = 1 - 0.23e^{-\frac{x_1^2}{0.25}}$
$\underline{\mu}_{\tilde{M}_1^3}(x_1) = 0.23e^{-\frac{x_1^2}{0.25}}$	$\overline{\mu}_{\tilde{M}_1^3}(x_1) = e^{-\frac{x_1^2}{1.2}}$
$\underline{\mu}_{\tilde{M}_1^4}(x_1) = 0.23e^{-\frac{x_1^2}{0.25}}$	$\overline{\mu}_{\tilde{M}_1^4}(x_1) = e^{-\frac{x_1^2}{1.2}}$
$\underline{\mu}_{\tilde{M}_2^1}(x_1) = 0.5e^{-\frac{x_1^2}{0.25}}$	$\overline{\mu}_{\tilde{M}_2^1}(x_1) = e^{-\frac{x_1^2}{1.5}}$
$\underline{\mu}_{\tilde{M}_2^2}(x_1) = 1 - e^{-\frac{x_1^2}{1.5}}$	$\overline{\mu}_{\tilde{M}_2^2}(x_1) = 1 - 0.5e^{-\frac{x_1^2}{0.25}}$
$\underline{\mu}_{\tilde{M}_2^3}(x_1) = 0.5e^{-\frac{x_1^2}{0.25}}$	$\overline{\mu}_{\tilde{M}_2^3}(x_1) = e^{-\frac{x_1^2}{1.5}}$
$\underline{\mu}_{\tilde{M}_2^4}(x_1) = 1 - e^{-\frac{x_1^2}{1.5}}$	$\overline{\mu}_{\tilde{M}_2^4}(x_1) = 1 - 0.5e^{-\frac{x_1^2}{0.25}}$

In this example, we consider only one sub-FOU, i.e., $\tau = 0$. For simplicity, the subscript l is dropped for all variables. The number of equal-size regions for x_1 is arbitrarily chosen to be 500. The LMFs and UMFs $\underline{h}_{ij}(x_1)$ and $\overline{h}_{ij}(x_1)$ are defined by choosing

$$v_{11k}(x_1) = 1 - \frac{x_1 - \underline{x}_{1,k}}{\underline{x}_{1,k} - \overline{x}_{1,k}}, \quad v_{12k}(x_1) = 1 - v_{11k}(x_1),$$

where

$$\underline{x}_{1,k} = \frac{10\pi/12}{500}(k - 251), \quad \overline{x}_{1,k} = \frac{10\pi/12}{500}(k - 250), \quad k = 1, 2, \ldots, 500.$$

The constant scalars are chosen as

$$\begin{aligned}
\underline{\delta}_{ij1k} &= \underline{w}_i(\underline{x}_{1,k})\underline{m}_j(\underline{x}_{1,k}), \\
\underline{\delta}_{ij2k} &= \underline{w}_i(\overline{x}_{1,k})\underline{m}_j(\overline{x}_{1,k}), \\
\overline{\delta}_{ij1k} &= \overline{w}_i(\underline{x}_{1,k})\overline{m}_j(\underline{x}_{1,k}), \\
\overline{\delta}_{ij2k} &= \overline{w}_i(\overline{x}_{1,k})\overline{m}_j(\overline{x}_{1,k}).
\end{aligned}$$

Theorem 2.7 with $l = 1$ is employed to determine the system stability and synthesize the feedback gains. A feasible solution was found as

$$X = \begin{bmatrix} 0.0983 & -0.1870 \\ -0.1870 & 0.4989 \end{bmatrix},$$
$$G_1 = \begin{bmatrix} 1432.8239 & 653.0531 \end{bmatrix},$$
$$G_2 = \begin{bmatrix} 1845.9736 & 849.8562 \end{bmatrix}.$$

The IT2 fuzzy controller is employed to stabilize the inverted pendulum with $m_p =$ 3 kg and $M_c = 8$ kg. The state responses of the system with different initial sates are shown in Fig. 2.2, which shows that the inverted pendulum can be stabilized subject to different values of m_p and M_c, and different initial conditions.

For comparison purposes, considering the simulation result in [97], it can be seen that the IT2 fuzzy controller can also stabilize the inverted pendulum. However, the number of rule of the IT2 fuzzy controller is required to be 4 because of the PDC design concept. In this example, the IT2 T–S fuzzy model and fuzzy controller do not share the same premise membership functions and the same number of rules.

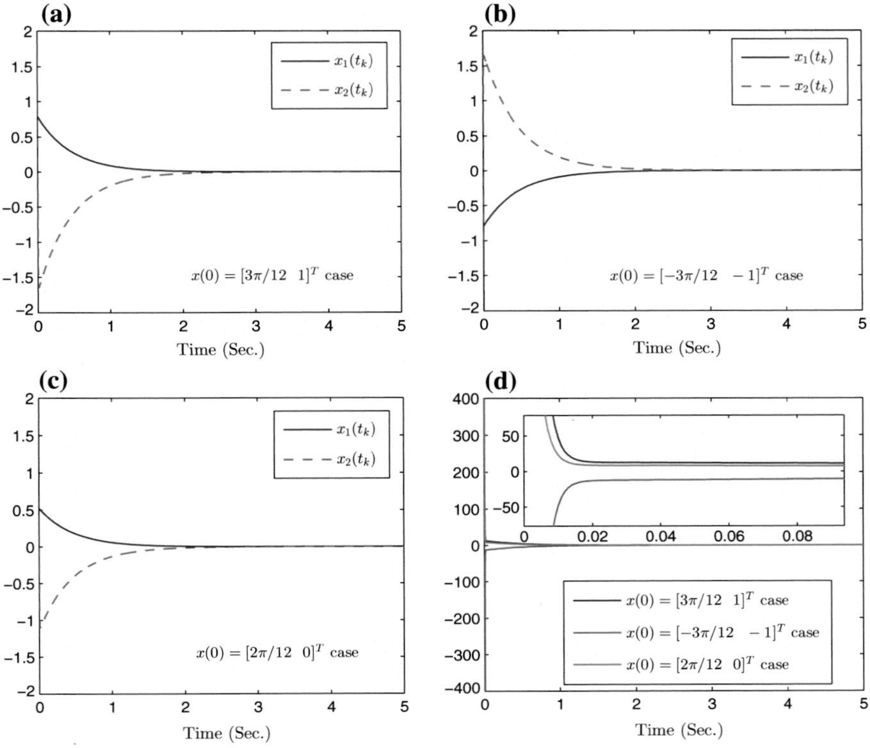

Fig. 2.2 States and control input of the closed-loop system

Consequently, the stability conditions proposed in [97] cannot be applied in this example. Furthermore, because the number of rules is 2 and simpler membership functions are used, the implementation complexity of the IT2 fuzzy controller are reduced.

2.5 Conclusion

The stability of IT2 FMB control systems subject to parameter uncertainties has been investigated. Under the imperfect premise matching, the IT2 fuzzy controller can choose freely the premise membership functions and the number of rules different from the IT2 T–S fuzzy model, enhancing the design flexibility and reducing the implementation complexity. To facilitate the stability analysis, a favorable form of LMFs and UMFs has been proposed and the information of sub-FOUs has been considered. The information of membership functions has been brought to the LMI-based stability conditions resulting in more relaxed stability analysis result. Simulation results have been given to illustrate the effectiveness of the proposed approach.

Chapter 3
Output-Feedback Control of Interval Type-2 Fuzzy-Model-Based Systems

3.1 Introduction

This chapter deals with the problems of state and output-feedback controllers design for IT2 fuzzy systems with mismatched membership functions based on a novel performance index. The IT2 fuzzy systems and the IT2 state and output-feedback controllers do not share the same membership functions. Firstly, the state-feedback and the output-feedback control systems are constructed. A new performance index, referred to extended dissipativity performance, is introduced. The extended dissipativity is a generalization of the H_∞ performance, the L_2-L_∞ performance, the passivity performance and dissipativity performance. Secondly, based on Lyapunov stability theory, the state and output-feedback controllers are designed respectively to guarantee that the closed-loop system is asymptotically stable with extended dissipativity performance. The existence conditions of the two kinds of controllers are obtained in terms of convex optimization problems, which can be solved by standard software.

3.2 Problem Formulation and Preliminaries

Consider the following IT2 fuzzy model with r rules that represents a continuous-time nonlinear system:

♦ **Plant Form**:

Rule i: IF $f_1(x(t))$ is W_{i1} and ... and $f_p(x(t))$ is W_{ip}, THEN

$$\begin{cases} \dot{x}(t) = A_i x(t) + B_i u(t) + D_{1i} w(t), \\ z(t) = C_i x(t) + D_{2i} w(t), \\ y(t) = C_{yi} x(t), \end{cases} \tag{3.1}$$

© Springer Science+Business Media Singapore 2016
H. Li et al., *Analysis and Synthesis for Interval Type-2
Fuzzy-Model-Based Systems*, DOI 10.1007/978-981-10-0593-0_3

where W_{is} stands for the ith IT2 fuzzy set of the function $f_s(x(t))$, $i = 1, 2, \ldots, r$, $s = 1, 2, \ldots p$; p is the number of premise variables; $x(t) \in \mathbf{R}^n$ is the system state vector, $u(t) \in \mathbf{R}^m$ is the input vector, $w(t) \in \mathbf{R}^h$ denotes the disturbance input which belongs to $\mathcal{L}_2[0, \infty)$, $z(t) \in \mathbf{R}^q$ is the control output and $y(t) \in \mathbf{R}^g$ is the measure output; A_i, B_i, C_i, D_{1i}, D_{2i} and C_{yi} are the known matrices with appropriate dimensions. The firing interval of the ith rule is as follows:

$$\begin{aligned} \tilde{\theta}_i(x(t)) &= \left[\prod_{s=1}^{p} \underline{\mu}_{W_{is}}(f_s(x(t))),\ \prod_{s=1}^{p} \overline{\mu}_{W_{is}}(f_s(x(t))) \right] \\ &= \left[\underline{\theta}_i(x(t)),\ \overline{\theta}_i(x(t)) \right], \end{aligned} \tag{3.2}$$

where $\underline{\theta}_i(x(t))$ denotes the lower grades of membership and $\overline{\theta}_i(x(t))$ denotes the upper grades of membership, $\underline{\mu}_{W_{is}}(f_s(x(t)))$ stands for the LMF and $\overline{\mu}_{W_{is}}(f_s(x(t)))$ stands for the UMF. Here, $\overline{\mu}_{W_{is}}(f_s(x(t))) \geq \underline{\mu}_{W_{is}}(f_s(x(t))) \geq 0$ and $\overline{\theta}_i(x(t)) \geq \underline{\theta}_i(x(t)) \geq 0$ for all i. Then the overall IT2 T–S fuzzy system is represented by

$$\begin{cases} \dot{x}(t) = \sum_{i=1}^{r} \theta_i(x(t)) \left[A_i x(t) + B_i u(t) + D_{1i} w(t) \right], \\ z(t) = \sum_{i=1}^{r} \theta_i(x(t)) \left[C_i x(t) + D_{2i} w(t) \right], \\ y(t) = \sum_{i=1}^{r} \theta_i(x(t)) C_{yi} x(t), \end{cases} \tag{3.3}$$

where

$$\theta_i(x(t)) = \underline{\lambda}_i(x(t))\underline{\theta}_i(x(t)) + \overline{\lambda}_i(x(t))\overline{\theta}_i(x(t)) \geq 0, \quad \forall i,$$

with

$$\begin{aligned} \sum_{i=1}^{r} \theta_i(x(t)) &= 1, \\ 0 \leq \underline{\lambda}_i(x(t)) &\leq 1, \quad \forall i, \\ 0 \leq \overline{\lambda}_i(x(t)) &\leq 1, \quad \forall i, \\ \underline{\lambda}_i(x(t)) + \overline{\lambda}_i(x(t)) &= 1, \quad \forall i, \end{aligned}$$

in which $\underline{\lambda}_i(x(t))$ and $\overline{\lambda}_i(x(t))$ are nonlinear functions, and $\theta_i(x(t))$ denotes the grades of membership of the embedded membership functions.

We first construct an IT2 fuzzy state-feedback controller [19] for the following control design. It is worth mentioning that the IT2 fuzzy system and the IT2 fuzzy state-feedback controller do not share the same membership functions. The jth rule of the fuzzy controller is of the following form:

♦ **Controller Form**:

Rule i: IF $g_1(x(t))$ is M_{j1} and ... and $g_p(x(t))$ is M_{jp}, THEN

$$u(t) = K_j x(t), \tag{3.4}$$

where M_{js} stands for the jth fuzzy set of the function $g_s(x(t))$, $j = 1, 2, \dots r$, $s = 1, 2, \dots p$; p is the number of premise variables; $K_j \in \mathbf{R}^{m\times n}$ is the state-feedback gain matrix of rule j. The firing interval of the jth rule is as follows:

$$\begin{aligned}\tilde{\eta}_j(x(t)) &= \left[\prod_{s=1}^{p} \underline{\mu}_{M_{js}}(g_s(x(t))),\ \prod_{s=1}^{p} \overline{\mu}_{M_{js}}(g_s(x(t)))\right] \\ &= \left[\underline{\eta}_j(x(t)),\ \overline{\eta}_j(x(t))\right],\end{aligned} \tag{3.5}$$

where $\underline{\eta}_j(x(t))$ denotes the lower grades of membership and $\overline{\eta}_j(x(t))$ denotes the upper grades of membership, $\underline{\mu}_{M_{js}}(g_s(x(t)))$ stands for the LMF and $\overline{\mu}_{M_{js}}(g_s(x(t)))$ stands for the UMF. $\overline{\mu}_{M_{js}}(g_s(x(t))) \geq \underline{\mu}_{M_{js}}(g_s(x(t))) \geq 0$ and $\overline{\eta}_j(x(t)) \geq \underline{\eta}_j(x(t)) \geq 0$ for all j. The overall IT2 fuzzy state-feedback control law is represented by

$$u(t) = \sum_{j=1}^{r} \eta_j(x(t)) K_j x(t), \tag{3.6}$$

where

$$\eta_j(x(t)) = \frac{\underline{\nu}_j(x(t))\underline{\eta}_j(x(t)) + \overline{\nu}_j(x(t))\overline{\eta}_j(x(t))}{\sum_{l=1}^{r}\left(\underline{\nu}_l(x(t))\underline{\eta}_l(x(t)) + \overline{\nu}_l(x(t))\overline{\eta}_l(x(t))\right)} \geq 0, \quad \forall j,$$

with

$$\begin{aligned}\sum_{j=1}^{r} \eta_j(x(t)) &= 1, \\ 0 \leq \underline{\nu}_j(x(t)) &\leq 1, \quad \forall j, \\ 0 \leq \overline{\nu}_j(x(t)) &\leq 1, \quad \forall j, \\ \underline{\nu}_j(x(t)) + \overline{\nu}_j(x(t)) &= 1, \quad \forall j,\end{aligned}$$

in which $\underline{\nu}_j(x(t))$ and $\overline{\nu}_j(x(t))$ are predefined functions, and $\eta_j(x(t))$ stands for the grades of membership of the embedded membership functions. For a simple description, we use the following notations: $\theta_i(x(t)) \triangleq \theta_i$ and $\eta_j(x(t)) \triangleq \eta_j$, where

$i, j = 1, 2, \ldots, r$. Applying the IT2 fuzzy controller (3.6) to system (3.3), the resulting IT2 fuzzy closed-loop system can be expressed as follows:

$$\begin{cases} \dot{x}(t) = \sum_{i=1}^{r}\sum_{j=1}^{r} \theta_i\eta_j \left[\left(A_i + B_iK_j\right)x(t) + D_{1i}w(t)\right], \\ z(t) = \sum_{i=1}^{r}\sum_{j=1}^{r} \theta_i\eta_j \left[C_ix(t) + D_{2i}w(t)\right], \end{cases} \tag{3.7}$$

where

$$\sum_{i=1}^{r}\theta_i = \sum_{j=1}^{r}\eta_j = \sum_{i=1}^{r}\sum_{j=1}^{r}\theta_i\eta_j = 1.$$

In this subsection, we will construct an IT2 fuzzy output-feedback controller in the following form:

Controller Rule k: IF $h_1(x(t))$ is N_{k1} and ... and $h_p(x(t))$ is N_{kp}, THEN

$$\begin{cases} \dot{\hat{x}}(t) = A_{ck}\hat{x}(t) + B_{ck}y(t), \\ u(t) = C_{ck}\hat{x}(t), \end{cases} \tag{3.8}$$

where $\hat{x}(t) \in \mathbf{R}^n$ is the state vector of the dynamic output-feedback controller; N_{ks} stands for the kth fuzzy set of the function $h_s(x(t))$, $k = 1, 2, \ldots, r$, $s = 1, 2, \ldots, p$; p is the number of premise variables; A_{ck}, B_{ck} and C_{ck} are control gain matrices with appropriate dimensions. The firing strength of the kth rule is the following interval set:

$$\begin{aligned} \tilde{\varpi}_k(x(t)) &= \left[\prod_{s=1}^{p}\underline{\mu}_{N_{ks}}(h_s(x(t))),\ \prod_{s=1}^{p}\overline{\mu}_{N_{ks}}(h_s(x(t)))\right] \\ &= \left[\underline{\varpi}_k(x(t)),\ \overline{\varpi}_k(x(t))\right], \end{aligned}$$

where $\underline{\varpi}_k(x(t))$ denotes the lower grades of membership and $\overline{\varpi}_k(x(t))$ denotes the upper grades of membership, $\underline{\mu}_{N_{ks}}(h_s(x(t)))$ stands for the LMF and $\overline{\mu}_{N_{ks}}(h_s(x(t)))$ stands for the UMF. Here, $\overline{\mu}_{N_{ks}}(h_s(x(t))) \geq \underline{\mu}_{N_{ks}}(h_s(x(t))) \geq 0$, and $\overline{\varpi}_k(x(t)) \geq \underline{\varpi}_k(x(t)) \geq 0$ for all k. The overall IT2 fuzzy output-feedback control law is represented by

$$\begin{cases} \dot{\hat{x}}(t) = \sum_{k=1}^{r} \varpi_k(x(t))\left[A_{ck}\hat{x}(t) + B_{ck}y(t)\right], \\ u(t) = \sum_{k=1}^{r} \varpi_k(x(t))C_{ck}\hat{x}(t), \end{cases} \tag{3.9}$$

where

$$\varpi_k(x(t)) = \frac{\underline{\kappa}_k(x(t))\underline{\varpi}_k(x(t)) + \overline{\kappa}_k(x(t))\overline{\varpi}_k(x(t))}{\sum_{p=1}^{r}\left(\underline{\kappa}_p(x(t))\underline{\varpi}_p(x(t)) + \overline{\kappa}_p(x(t))\overline{\varpi}_p(x(t))\right)} \geq 0, \quad \forall k, \quad (3.10)$$

with

$$\sum_{k=1}^{r} \varpi_k(x(t)) = 1,$$
$$0 \leq \underline{\kappa}_k(x(t)) \leq 1, \quad \forall k,$$
$$0 \leq \overline{\kappa}_k(x(t)) \leq 1, \quad \forall k,$$
$$\underline{\kappa}_k(x(t)) + \overline{\kappa}_k(x(t)) = 1, \quad \forall k,$$

in which $\underline{\kappa}_k(x(t))$ and $\overline{\kappa}_k(x(t))$ are predefined functions, $\varpi_k(x(t))$ denotes the grades of membership of the embedded membership functions. For a simple description, we define $\varpi_k(x(t)) \triangleq \varpi_k$, where $k = 1, 2, \ldots, r$. Under the property of $\sum_{i=1}^{r} \theta_i = \sum_{k=1}^{r} \varpi_k = \sum_{i=1}^{r} \sum_{k=1}^{r} \theta_i \varpi_k = 1$, it can be seen from (3.3) and (3.9) that the following closed-loop system is obtained:

$$\begin{cases} \dot{\bar{x}}(t) = \sum_{i=1}^{r}\sum_{k=1}^{r} \theta_i \varpi_k \left[\bar{A}_{ik}\bar{x}(t) + \bar{D}_{1i} w(t)\right], \\ z(t) = \sum_{i=1}^{r}\sum_{k=1}^{r} \theta_i \varpi_k \left[\bar{C}_i \bar{x}(t) + \bar{D}_{2i} w(t)\right], \end{cases} \quad (3.11)$$

where $\bar{x}(t) = \left[x^T(t)\ \hat{x}^T(t)\right]^T$ and

$$\bar{A}_{ik} = \begin{bmatrix} A_i & B_i C_{ck} \\ B_{ck} C_{yi} & A_{ck} \end{bmatrix}, \quad \bar{D}_{1i} = \begin{bmatrix} D_{1i} \\ 0 \end{bmatrix},$$
$$\bar{C}_i = \begin{bmatrix} C_i & 0 \end{bmatrix}, \quad \bar{D}_{2i} = D_{2i}.$$

The main purpose of this chapter is to design the IT2 fuzzy state-feedback controller (3.6) and output-feedback controller (3.9) such that the closed-loop system is asymptotically stable with the H_∞, L_2-L_∞, passive and dissipativity performance indexes. In [230], the authors introduced a new performance index, referred to extended dissipativity performance index, which is a generalization of H_∞, L_2-L_∞, passive and dissipativity performances indexes. In addition, the authors presented some new conditions for filter design of Markovian jump delay systems based on the new performance index. In the following part, we introduce the new performance index from the Ref. [230]. Firstly, the following assumption is given for developing the new performance index.

Assumption 3.1 *([230])* Let Φ, Ψ_1, Ψ_2 and Ψ_3 be matrices such that the following conditions hold:

(1) $\Phi = \Phi^T$, $\Psi_1 = \Psi_1^T$ and $\Psi_3 = \Psi_3^T$;

(2) $\Phi \geq 0$ and $\Psi_1 \leq 0$;

(3) $\|D_{2_i}\| \cdot \|\Phi\| = 0$;

(4) $(\|\Psi_1\| + \|\Psi_2\|) \cdot \|\Phi\| = 0$;

(5) $D_{2i}^T \Psi_1 D_{2i} + D_{2i}^T \Psi_2 + \Psi_2^T D_{2i} + \Psi_3 > 0$.

Definition 3.1 ([230]) For given matrices Φ, Ψ_1, Ψ_2 and Ψ_3 satisfying Assumption 3.1, system (3.7) (or system (3.11)) is said to be extended dissipative if there exists a scalar ρ such that the following inequality holds for any $t > 0$ and all $w(t) \in \mathcal{L}_2[0, \infty)$:

$$\int_0^t J(s)ds - z^T(t)\Phi z(t) \geq \rho, \tag{3.12}$$

where $J(t) = z^T(t)\Psi_1 z(t) + 2z^T(t)\Psi_2 w(t) + w^T(t)\Psi_3 w(t)$.

It can be seen from Definition 3.1 that the following performance indexes hold.

(1) Choosing $\Phi = 0$, $\Psi_1 = -I$, $\Psi_2 = 0$, $\Psi_3 = \gamma^2 I$ and $\rho = 0$, the inequality (3.12) reduces to the H_∞ performance [35].
(2) Let $\Phi = I$, $\Psi_1 = 0$, $\Psi_2 = 0$, $\Psi_3 = \gamma^2 I$ and $\rho = 0$, inequality (3.12) becomes the L_2-L_∞ (energy-to-peak) performance [39].
(3) If the dimension of output $z(t)$ is the same as that of disturbance $w(t)$, then the inequality in (3.12) with $\Phi = 0$, $\Psi_1 = 0$, $\Psi_2 = I$, $\Psi_3 = \gamma I$ and $\rho = 0$ becomes the passivity performance [216].
(4) Let $\Phi = 0$, $\Psi_1 = Q$, $\Psi_2 = S$, $\Psi_3 = R - \alpha I$ and $\rho = 0$, inequality (3.12) reduces to the strict (Q, S, R)-dissipativity [116].
(5) When $\Phi = 0$, $\Psi_1 = -\epsilon I$, $\Psi_2 = I$, $\Psi_3 = -\sigma I$ with $\epsilon > 0$ and $\sigma > 0$, inequality (3.12) becomes the very-strict passivity performance. In the definition of the very-strict passivity performance, the scalar ρ is not required to be zero. It was shown in [134] that ρ should be a non-positive scalar. This fact can also be seen from Assumption 3.1 and Definition 3.1. Indeed, when $w(t) = 0$, from (3.12), it follows that

$$\rho \leq \int_0^t e^T(s)\Psi_1 e(s)ds - e^T(t)\Phi e(t). \tag{3.13}$$

Noting from Assumption 3.1 that $\Phi \geq 0$ and $\Psi_1 \leq 0$. Thus, the above inequality implies that $\rho \leq 0$, and there always exist matrices $\tilde{\Phi}$ and $\tilde{\Psi}_1$ such that

$$\Phi = \tilde{\Phi}^T \tilde{\Phi}, \quad \Psi_1 = -\tilde{\Psi}_1^T \tilde{\Psi}_1. \tag{3.14}$$

Remark 3.2 The first item of Assumption 3.1 guarantees that the inequality (3.12) is well defined. The second item enables one to derive LMI based condition for the investigation of the dissipativity analysis problem. The conditions of Assumption 3.1 similar to (1), (2) and (5) were used in [52, 116]. On the other hand, when considering the L_2-L_∞ performance, it is well known that the output of the considered system should not include disturbance inputs [62]. Therefore, it should be assumed that $D_{2i} = 0$ when $\Phi \neq 0$, which justifies the need of the third item of Assumption 3.1. Finally, the fourth item of Assumption 3.1 is technically necessary for the development of our analysis and design methods.

In this chapter, our objective is to design the state-feedback controller in (3.6) and output-feedback controller in (3.9) for system (3.3) such that

(i) the closed-loop system (3.7) (or (3.11)) is asymptotically stable with $w(t) = 0$;
(ii) the closed-loop system (3.7) (or (3.11)) guarantees the new performance index (3.12).

3.3 Main Results

This section is concerned with the controllers design problem for IT2 T–S fuzzy system. The existence conditions of the controllers are given in the following theorems. We first present IT2 fuzzy state-feedback controller design results.

3.3.1 State-Feedback Control

Theorem 3.3 *For given matrices $\tilde{\Phi}$, $\tilde{\Psi}_1$, Ψ_2 and Ψ_3 satisfying (3.14) and Assumption 3.1, the system in (3.7) is asymptotically stable and satisfies the performance index in Definition 3.1, if there exist matrices $G = G^T > 0$, $Q = Q^T > 0$, $\Lambda_i^T = \Lambda_i$, M_j $(i, j = 1, 2, \ldots, r)$ with appropriate dimensions, and under the condition $\eta_j - \sigma_j \theta_j \geq 0$ $(0 < \sigma_j < 1)$ for all $j = 1, 2, \ldots, r$, such that the following LMIs hold:*

$$\Theta_{1ij} < 0, \tag{3.15}$$

$$\Theta_{2ij} < 0, \tag{3.16}$$

$$\Omega_{ij} - \Lambda_i < 0, \tag{3.17}$$

$$\sigma_i \Omega_{ii} - \sigma_i \Lambda_i + \Lambda_i < 0, \tag{3.18}$$

$$\sigma_j \Omega_{ij} + \sigma_i \Omega_{ji} - \sigma_j \Lambda_i - \sigma_i \Lambda_j + \Lambda_i + \Lambda_j \leq 0, \quad i < j, \tag{3.19}$$

where

$$\Omega_{ij} = \begin{bmatrix} \bar{\Omega}_{11ij} & \bar{\Omega}_{12ij} & \bar{\Omega}_{13ij} \\ * & \bar{\Omega}_{22ij} & \bar{\Omega}_{23ij} \\ * & * & -I \end{bmatrix},$$

$$\Theta_{1ij} = \begin{bmatrix} -Q & Q \\ * & G-2I \end{bmatrix}, \quad \Theta_{2ij} = \begin{bmatrix} -G & \tilde{C}_i^T \tilde{\Phi}^T \\ * & -I \end{bmatrix},$$

$$\tilde{C}_i = C_i Q, \quad \bar{\Omega}_{13ij} = \tilde{C}_i^T \tilde{\Psi}_1^T, \quad \bar{\Omega}_{11ij} = \mathbf{He}(A_i Q + B_i M_j),$$

$$\bar{\Omega}_{12ij} = D_{1i} - \tilde{C}_i^T \Psi_2, \quad \bar{\Omega}_{22ij} = -\mathbf{He}(D_{2i}^T \Psi_2) - \Psi_3, \quad \bar{\Omega}_{23ij} = D_{2i}^T \tilde{\Psi}_1^T.$$

Then the IT2 fuzzy state-feedback controller gain matrices are given as

$$K_j = M_j Q^{-1}.$$

In this case, the scalar ρ involved in Definition 3.1 can be chosen as

$$\rho = -V(x(0)). \tag{3.20}$$

Proof Choose a quadratic Lyapunov function for the stability analysis of system (3.7) as follows:

$$V(x(t)) = x^T(t) P x(t), \tag{3.21}$$

where $P = P^T > 0$. Then the time derivative of $V(t)$ is given by:

$$\begin{aligned} \dot{V}(x(t)) &= 2x^T(t) P \dot{x}(t) \\ &= \sum_{i=1}^{r} \sum_{j=1}^{r} \theta_i \eta_j \left[x^T(t) \mathbf{He}\left(P\left(A_i + B_i K_j \right) \right) x(t) + 2x^T(t) P D_{1i} w(t) \right]. \end{aligned}$$

Let $g(t) = Q^{-1} x(t)$, $\tilde{C}_i = C_i Q$ and $Q = P^{-1}$, then it can be obtained that

$$\dot{V}(x(t)) = \sum_{i=1}^{r} \sum_{j=1}^{r} \theta_i \eta_j \left[g^T(t) \mathbf{He}(A_i Q + B_i M_j) g(t) + 2g^T(t) D_{1i} w(t) \right].$$

From $\Psi_1 \leq 0$, it can be seen that

$$\begin{aligned} z^T(t) \Psi_1 z(t) &= \left[\sum_{i=1}^{r} \sum_{j=1}^{r} \theta_i \eta_j \left(\tilde{C}_i g(t) + D_{2i} w(t) \right) \right]^T \Psi_1 \\ &\quad \times \left[\sum_{l=1}^{r} \sum_{m=1}^{r} \theta_l \eta_m \left(\tilde{C}_l g(t) + D_{2l} w(t) \right) \right] \\ &\geq \sum_{i=1}^{r} \sum_{j=1}^{r} \theta_i \eta_j \left(\tilde{C}_i g(t) + D_{2i} w(t) \right)^T \Psi_1 \left(\tilde{C}_i g(t) + D_{2i} w(t) \right). \end{aligned}$$

Then

$$\dot{V}(x(t)) - J(t) \leq \xi^T(t) \sum_{i=1}^{r} \sum_{j=1}^{r} \theta_i \eta_j \tilde{\Omega}_{ij} \xi(t),$$

where

$$\begin{aligned}
\xi(t) &= \begin{bmatrix} g(t) \\ w(t) \end{bmatrix}, \quad \tilde{\Omega}_{ij} = \begin{bmatrix} \tilde{\Omega}_{1ij} & \tilde{\Omega}_{2ij} \\ * & \tilde{\Omega}_{3ij} \end{bmatrix}, \\
J(t) &= z^T(t)\Psi_1 z(t) + 2z^T(t)\Psi_2 w(t) + w^T(t)\Psi_3 w(t), \\
\tilde{\Omega}_{1ij} &= \mathbf{He}(A_i Q + B_i M_j) - \tilde{C}_i^T \Psi_1 \tilde{C}_i, \\
\tilde{\Omega}_{2ij} &= D_{1i} - \tilde{C}_i^T \Psi_1 D_{2i} - \tilde{C}_i^T \Psi_2, \\
\tilde{\Omega}_{3ij} &= -D_{2i}^T \Psi_1 D_{2i} - \mathbf{He}(D_{2i}^T \Psi_2) - \Psi_3.
\end{aligned}$$

Consider $\sum_{i=1}^{r} \sum_{j=1}^{r} \theta_i \left(\theta_j - \eta_j\right) \Lambda_i = 0$, where $\Lambda_i = \Lambda_i^T$ is an arbitrary matrix with appropriate dimensions. Then

$$\begin{aligned}
\sum_{i=1}^{r} \sum_{j=1}^{r} \theta_i \eta_j \Omega_{ij} &= \sum_{i=1}^{r} \sum_{j=1}^{r} \theta_i \eta_j \Omega_{ij} + \sum_{i=1}^{r} \sum_{j=1}^{r} \theta_i \left(\theta_j - \eta_j\right) \Lambda_i \\
&= \sum_{i=1}^{r} \sum_{j=1}^{r} \theta_i \left(\theta_j - \eta_j + \sigma_j \theta_j - \sigma_j \theta_j\right) \Lambda_i \\
&\quad + \sum_{i=1}^{r} \sum_{j=1}^{r} \theta_i \left(\eta_j + \sigma_j \theta_j - \sigma_j \theta_j\right) \Omega_{ij} \\
&= \sum_{i=1}^{r} \sum_{j=1}^{r} \theta_i \theta_j \left(\sigma_j \Omega_{ij} - \sigma_j \Lambda_i + \Lambda_i\right) \\
&\quad + \sum_{i=1}^{r} \sum_{j=1}^{r} \theta_i \left(\eta_j - \sigma_j \theta_j\right) \left(\Omega_{ij} - \Lambda_i\right) \\
&= \sum_{i=1}^{r} \sum_{j=1}^{r} \theta_i^2 \left(\sigma_i \Omega_{ii} - \sigma_i \Lambda_i + \Lambda_i\right) \\
&\quad + \sum_{i=1}^{r-1} \sum_{j=i+1}^{r} \theta_i \theta_j \left(\sigma_j \Omega_{ij} - \sigma_j \Lambda_i + \Lambda_i + \sigma_i \Omega_{ji} - \sigma_i \Lambda_j + \Lambda_j\right) \\
&\quad + \sum_{i=1}^{r} \sum_{j=1}^{r} \theta_i \left(\eta_j - \sigma_j \theta_j\right) \left(\Omega_{ij} - \Lambda_i\right). \qquad (3.22)
\end{aligned}$$

It can be seen from (3.17)–(3.19) that

$$\sum_{i=1}^{r}\sum_{j=1}^{r}\theta_i\eta_j\Omega_{ij} < 0.$$

By Schur complement, one can have

$$\sum_{i=1}^{r}\sum_{j=1}^{r}\theta_i\eta_j\tilde{\Omega}_{ij} < 0,$$

that is,

$$\dot{V}(t) - J(t) < \xi^T(t)\left(\sum_{i=1}^{r}\sum_{j=1}^{r}\theta_i\eta_j\tilde{\Omega}_{ij}\right)\xi(t) < 0.$$

Therefore, there is always a sufficiently small scalar $c > 0$ such that $\tilde{\Omega}_{ij} \leq -cI$. This means that

$$\dot{V}(x(t)) - J(t) \leq -c\,|\xi(t)|^2. \tag{3.23}$$

Thus $J(t) \geq \dot{V}(t)$ holds for any $t \geq 0$, which means

$$\int_0^t J(s)ds \geq V(x(t)) - V(x(0)). \tag{3.24}$$

It is shown from $(G - I)\,G^{-1}\,(G - I) \geq 0$ with $G > 0$ that

$$-G^{-1} \leq G - 2I. \tag{3.25}$$

From (3.15) and (3.25), we know that $P > G$, which means

$$V(x(t)) = x^T(t)Px(t) \geq x^T(t)Gx(t) \geq 0.$$

For the inequality (3.24), it is derived from (3.20) that

$$\int_0^t J(s)ds \geq x^T(t)Gx(t) + \rho, \quad \forall t \geq 0. \tag{3.26}$$

According to Definition 3.1, we need to prove that the following inequality holds for any matrices Φ, Ψ_1, Ψ_2 and Ψ_3 satisfying Assumption 3.1:

$$\int_0^t J(t)dt - z^T(t)\Phi z(t) \geq \rho. \tag{3.27}$$

To this end, we consider the two cases of $\|\Phi\| = 0$ and $\|\Phi\| \neq 0$, respectively.

Firstly, we consider the case when $\|\Phi\| = 0$. It follows from (3.26), for any $t \geq 0$,

$$\int_0^t J(s)ds \geq x^T(t)Gx(t) + \rho \geq \rho. \tag{3.28}$$

This implies (3.27) holds by noting that $z^T(t)\Phi z(t) \equiv 0$.

Secondly, we consider the case of $\|\Phi\| \neq 0$. In this case, it is required under Assumption 3.1 that $\|\Psi_1\| + \|\Psi_2\| = 0$ and $\|D_{2_i}\| = 0$, which implies that $\Psi_1 = 0$, $\Psi_2 = 0$ and $\Psi_3 > 0$. Thus, $J(s) = w^T(s)\Psi_3 w^T(s) \geq 0$. Then, using Schur complement to (3.16), it can be obtained that $\tilde{C}_i^T \Phi \tilde{C}_i \leq G$. For any $t \geq 0$, the following inequalities hold:

$$\begin{aligned}
\int_0^t J(s)ds - z^T(t)\Phi z(t) &\geq \int_0^t J(s)ds - \sum_{i=1}^r \sum_{j=1}^r \theta_i \eta_j \\
&\quad \times \left[(C_i x(t) + D_{2i} w(t))^T \Phi \, (C_i x(t) + D_{2i} w(t)) \right] \\
&= \int_0^t J(s)ds - \sum_{i=1}^r \sum_{j=1}^r \theta_i \eta_j \left(g^T(t) \tilde{C}_i^T \Phi \tilde{C}_i g(t) \right) \\
&\geq \int_0^t J(s)ds - \sum_{i=1}^r \sum_{j=1}^r \theta_i \eta_j x^T(t)Gx(t) \geq \rho.
\end{aligned}$$

Based on the two cases of $\|\Phi\| = 0$ and $\|\Phi\| \neq 0$, we know that the closed-loop system (3.7) is extended dissipative in the sense of Definition 3.1.

When $w(t) \equiv 0$, it follows from (3.23) that

$$\dot{V}(t) \leq z^T(t)\Psi_1 z(t) - c\,|\xi(t)|^2. \tag{3.29}$$

Noticing that $\Psi_1 < 0$ under Assumption 3.1, we have $\dot{V}(t) \leq -c\,|\xi(t)|^2$. Thus the closed-loop system (3.7) with $w(t) = 0$ is asymptotically stable. This completes the proof. □

3.3.2 Output-Feedback Control

In the following part, we will solve the problem of IT2 fuzzy output-feedback controller synthesis for the IT2 fuzzy system (3.3). By following the same line as the proof of Theorem 3.3, the following theorem is obtained directly.

Theorem 3.4 *For given matrices $\tilde{\Phi}$, $\tilde{\Psi}_1$, Ψ_2 and Ψ_3 satisfying (3.14) and Assumption 3.1, the closed-loop system in (3.11) is asymptotically stable and satisfies the performance index in Definition 3.1, if there exist matrices $P = P^T > 0$, $G > 0$ and*

$\tilde{\Lambda}_i^T = \tilde{\Lambda}_i$ $(i = 1, 2, \ldots, r)$ with appropriate dimensions, and under the condition $\varpi_k - \bar{\sigma}_k\theta_k \geq 0$ $(0 < \bar{\sigma}_k < 1)$ for all k, such that the following LMIs hold:

$$G - P < 0, \tag{3.30}$$

$$\tilde{\Theta}_2 < 0, \tag{3.31}$$

$$\Pi_{ik} - \tilde{\Lambda}_i < 0, \tag{3.32}$$

$$\bar{\sigma}_i\Pi_{ii} - \bar{\sigma}_i\tilde{\Lambda}_i + \tilde{\Lambda}_i < 0, \tag{3.33}$$

$$\bar{\sigma}_k\Pi_{ik} + \bar{\sigma}_i\Pi_{ki} - \bar{\sigma}_k\tilde{\Lambda}_i - \bar{\sigma}_i\tilde{\Lambda}_k + \tilde{\Lambda}_i + \tilde{\Lambda}_k \leq 0, \quad i < k, \tag{3.34}$$

where

$$\Pi_{ik} = \begin{bmatrix} \mathbf{He}(P\bar{A}_{ik}) & P\bar{D}_{1i} - \bar{C}_i^T\Psi_2 & \bar{C}_i^T\tilde{\Psi}_1^T \\ * & -\mathbf{He}(\bar{D}_{2i}^T\Psi_2) - \Psi_3 & \bar{D}_{2i}^T\tilde{\Psi}_1^T \\ * & * & -I \end{bmatrix},$$

$$\tilde{\Theta}_2 = \begin{bmatrix} -G & \bar{C}_i^T\tilde{\Phi}^T \\ * & -I \end{bmatrix}.$$

In the following theorem, the control gain matrices A_{ck}, B_{ck} and C_{ck} in (3.9) will be solved.

Theorem 3.5 *Considering the IT2 fuzzy system (3.3), for given matrices $\tilde{\Phi}$, $\tilde{\Psi}_1$, Ψ_2 and Ψ_3 satisfying (3.14) and Assumption 3.1, system (3.11) is asymptotically stable and satisfies the performance index in Definition 3.1, if there exists matrices $\bar{\Lambda}_i^T = \bar{\Lambda}_i$, $i = 1, 2, \ldots, r$, $\bar{G} = \begin{bmatrix} G_1 & G_2 \\ * & G_3 \end{bmatrix} > 0$, $\mathcal{R} > 0$, $\mathcal{S} > 0$, $\mathcal{A}_i$, $\mathcal{B}_i$ and $\mathcal{C}_i$ with appropriate dimensions, and under the condition $\varpi_k - \bar{\sigma}_k\theta_k \geq 0$ $(0 < \bar{\sigma}_k < 1)$ for all $k = 1, 2, \ldots, r$, such that the following LMIs hold:*

$$\begin{bmatrix} \mathcal{R} & I \\ I & \mathcal{S} \end{bmatrix} > 0, \tag{3.35}$$

$$\bar{G} - \begin{bmatrix} \mathcal{R} & I \\ I & \mathcal{S} \end{bmatrix} < 0, \tag{3.36}$$

$$\begin{bmatrix} -\bar{G} & \bar{\Theta}_2 \\ * & -I \end{bmatrix} < 0, \tag{3.37}$$

$$\check{\Pi}_{ik} - \hat{\Lambda}_i < 0, \tag{3.38}$$

$$\bar{\sigma}_i\tilde{\Pi}_{ii} - \bar{\sigma}_i\bar{\Lambda}_i + \bar{\Lambda}_i < 0, \tag{3.39}$$

$$\hat{\Pi}_{ik} - \bar{\sigma}_k\hat{\Lambda}_i - \bar{\sigma}_i\hat{\Lambda}_k + \hat{\Lambda}_i + \hat{\Lambda}_k \leq 0, \quad i < k, \tag{3.40}$$

where

$$\check{\Pi}_{ik} = \begin{bmatrix} \check{\Xi}_{1ik} & \check{\Xi}_{2ik} & \check{\Xi}_{3ik} & \check{\chi}_{1ik} & \check{\chi}_{2ik} & \check{\chi}_{3ik} \\ * & \check{\Xi}_{4ik} & \check{\Xi}_{5ik} & 0 & 0 & 0 \\ * & * & \check{\Xi}_{6ik} & 0 & 0 & 0 \\ * & * & * & \Xi_{7ik} & 0 & 0 \\ * & * & * & * & \Xi_{8ik} & 0 \\ * & * & * & * & * & \Xi_{8ik} \end{bmatrix},$$

$$\tilde{\Pi}_{ii} = \begin{bmatrix} \check{\Xi}_{1ii} & \check{\Xi}_{2ii} & \check{\Xi}_{3ii} \\ * & \check{\Xi}_{4ii} & \check{\Xi}_{5ii} \\ * & * & \check{\Xi}_{6ii} \end{bmatrix}, \quad \hat{\Lambda}_i = \begin{bmatrix} \bar{\Lambda}_i & 0 \\ 0 & 0_{2(n+2m)} \end{bmatrix},$$

$$\hat{\Pi}_{ik} = \begin{bmatrix} \hat{\Xi}_{1ik} & \hat{\Xi}_{2ik} & \hat{\Xi}_{3ik} & \hat{\chi}_{1ik} & \hat{\chi}_{2ik} & \hat{\chi}_{3ik} \\ * & \hat{\Xi}_{4ik} & \hat{\Xi}_{5ik} & 0 & 0 & 0 \\ * & * & \hat{\Xi}_{6ik} & 0 & 0 & 0 \\ * & * & * & \Xi_{7ik} & 0 & 0 \\ * & * & * & * & \Xi_{8ik} & 0 \\ * & * & * & * & * & \Xi_{8ik} \end{bmatrix},$$

$$\bar{\Theta}_2 = \begin{bmatrix} \mathcal{R}C_i^T\tilde{\Phi}^T \\ C_i^T\tilde{\Phi}^T \end{bmatrix}, \quad \check{\Xi}_{2ik} = \begin{bmatrix} D_{1i} - \mathcal{R}C_i^T\Psi_2 \\ \mathcal{S}D_{1i} - C_i^T\Psi_2 \end{bmatrix},$$

$$\check{\Xi}_{1ik} = \begin{bmatrix} \mathbf{He}(A_i\mathcal{R}+B_i\mathcal{C}_k) & A_i + \mathcal{A}_k^T \\ * & \mathbf{He}\left(\mathcal{S}A_i + \mathcal{B}_kC_{yi}\right) \end{bmatrix},$$

$$\check{\Xi}_{3ik} = \begin{bmatrix} \mathcal{R}C_i^T\tilde{\Psi}_1^T \\ C_i^T\tilde{\Psi}_1^T \end{bmatrix}, \quad \check{\Xi}_{4ik} = -\mathbf{He}\left(D_{2i}^T\Psi_2\right) - \Psi_3,$$

$$\check{\Xi}_{5ik} = D_{2i}^T\tilde{\Psi}_1^T, \quad \check{\Xi}_{6ik} = -I_m, \quad \hat{\Xi}_{1ik} = \bar{\sigma}_k\check{\Xi}_{1ik} + \bar{\sigma}_i\check{\Xi}_{1ki},$$

$$\hat{\Xi}_{2ik} = \bar{\sigma}_k\check{\Xi}_{2ik} + \bar{\sigma}_i\check{\Xi}_{1ki}, \quad \hat{\Xi}_{3ik} = \bar{\sigma}_k\check{\Xi}_{3ik} + \bar{\sigma}_i\check{\Xi}_{3ki},$$

$$\hat{\Xi}_{4ik} = \bar{\sigma}_k\check{\Xi}_{4ik} + \bar{\sigma}_i\check{\Xi}_{4ki}, \quad \hat{\Xi}_{5ik} = \bar{\sigma}_k\check{\Xi}_{5ik} + \bar{\sigma}_i\check{\Xi}_{5ki},$$

$$\hat{\Xi}_{6ik} = (\bar{\sigma}_k + \bar{\sigma}_i)\,\check{\Xi}_{6ii}, \quad \Xi_{7ik} = \begin{bmatrix} -I_n & 0 \\ 0 & -I_n \end{bmatrix},$$

$$\Xi_{8ik} = \begin{bmatrix} -I_m & 0 \\ 0 & -I_m \end{bmatrix}, \quad \check{\chi}_{1ik} = \begin{bmatrix} 0 & \mathcal{R}\,(A_i - A_k)^T \\ \mathcal{S} & 0 \end{bmatrix},$$

$$\hat{\chi}_{1ik} = \begin{bmatrix} 0 & \mathcal{R}\,(A_i - A_k)^T \\ (\bar{\sigma}_k - \bar{\sigma}_i)\,\mathcal{S} & 0 \end{bmatrix},$$

$$\check{\chi}_{2ik} = \begin{bmatrix} 0 & \mathcal{R}\left(C_{yi} - C_{yk}\right)^T \\ \mathcal{B}_k & 0 \end{bmatrix},$$

$$\check{\chi}_{3ik} = \begin{bmatrix} 0 & \mathcal{C}_k^T \\ \mathcal{S}\,(B_i - B_k) & 0 \end{bmatrix},$$

$$\hat{\chi}_{2ik} = \begin{bmatrix} 0 & \mathcal{R}\left(C_{yi} - C_{yk}\right)^T \\ \bar{\sigma}_k \mathcal{B}_k - \bar{\sigma}_i \mathcal{B}_i & 0 \end{bmatrix},$$
$$\hat{\chi}_{3ik} = \begin{bmatrix} 0 & \bar{\sigma}_k \mathcal{C}_k^T - \bar{\sigma}_i \mathcal{C}_i^T \\ \mathcal{S}\left(B_i - B_k\right) & 0 \end{bmatrix}.$$

Then, the IT2 fuzzy output-feedback controller gain matrices are given as follows:

$$\begin{aligned} C_{ci} &= \mathcal{C}_i \mathcal{M}^{-T}, \\ B_{ci} &= \mathcal{N}^{-1} \mathcal{B}_i, \\ A_{ci} &= \mathcal{N}^{-1}\left(\mathcal{A}_i - \mathcal{S} A_i \mathcal{R} - \mathcal{B}_i C_{yi} \mathcal{R} - \mathcal{S} B_i \mathcal{C}_i\right) \mathcal{M}^{-T}, \end{aligned}$$

where $\mathcal{M}$ and $\mathcal{N}$ are nonsingular matrices satisfying:

$$\mathcal{M}\mathcal{N}^T = I - \mathcal{R}\mathcal{S}. \tag{3.41}$$

Proof Using Schur complement, it can be seen from (3.40) that

$$\begin{bmatrix} \hat{\Xi}_{1ik} & \hat{\Xi}_{2ik} & \hat{\Xi}_{3ik} \\ * & \hat{\Xi}_{4ik} & \hat{\Xi}_{5ik} \\ * & * & \hat{\Xi}_{6ik} \end{bmatrix} - \sigma_k \bar{\Lambda}_i - \sigma_i \bar{\Lambda}_k + \bar{\Lambda}_i + \bar{\Lambda}_k$$
$$+ \sum_{j=1}^{3} \left(\hat{\Upsilon}_{jik} \hat{\Upsilon}_{jik}^T + \hat{\Gamma}_{jik} \hat{\Gamma}_{jik}^T \right) < 0, \quad \forall i, k,$$

where

$$\hat{\Upsilon}_{1ik} = \begin{bmatrix} 0 \\ \left(\bar{\sigma}_k - \bar{\sigma}_i\right) \mathcal{S} \\ 0_{2\times 1} \end{bmatrix}, \quad \hat{\Upsilon}_{2ik} = \begin{bmatrix} 0 \\ \bar{\sigma}_k \mathcal{B}_k - \bar{\sigma}_i \mathcal{B}_i \\ 0_{2\times 1} \end{bmatrix},$$
$$\hat{\Gamma}_{1ik} = \begin{bmatrix} \mathcal{R}\left(A_i - A_j\right)^T \\ 0_{3\times 1} \end{bmatrix}, \quad \hat{\Gamma}_{2ik} = \begin{bmatrix} \mathcal{R}\left(C_{yi} - C_{yk}\right)^T \\ 0_{3\times 1} \end{bmatrix},$$
$$\hat{\Upsilon}_{3ik} = \begin{bmatrix} 0 \\ \mathcal{S}\left(B_i - B_k\right) \\ 0_{2\times 1} \end{bmatrix}, \quad \hat{\Gamma}_{3ik} = \begin{bmatrix} \bar{\sigma}_k \mathcal{C}_k^T - \bar{\sigma}_i \mathcal{C}_i^T \\ 0_{3\times 1} \end{bmatrix}.$$

It is easy to see that

$$\sum_{j=1}^{3} \left(\hat{\Upsilon}_{jik} \hat{\Upsilon}_{jik}^T + \hat{\Gamma}_{jik} \hat{\Gamma}_{jik}^T \right) \geq \sum_{j=1}^{3} \left(\hat{\Upsilon}_{jik} \hat{\Gamma}_{jik}^T + \hat{\Gamma}_{jik} \hat{\Upsilon}_{jik}^T \right),$$

which means

$$\begin{bmatrix} \hat{\Xi}_{1ik} & \hat{\Xi}_{2ik} & \hat{\Xi}_{3ik} \\ * & \hat{\Xi}_{4ik} & \hat{\Xi}_{5ik} \\ * & * & \hat{\Xi}_{6ik} \end{bmatrix} - \bar{\sigma}_k \bar{\Lambda}_i - \bar{\sigma}_i \bar{\Lambda}_k + \bar{\Lambda}_i + \bar{\Lambda}_k$$

$$+ \sum_{j=1}^{3} \left(\hat{\Upsilon}_{jik} \hat{\Gamma}_{jik}^T + \hat{\Gamma}_{jik} \hat{\Upsilon}_{jik}^T \right) < 0, \quad \forall i, k. \tag{3.42}$$

Similarly, for (3.38), one can see

$$\begin{bmatrix} \check{\Xi}_{1ik} & \check{\Xi}_{2ik} & \check{\Xi}_{3ik} \\ * & \check{\Xi}_{4ik} & \check{\Xi}_{5ik} \\ * & * & \check{\Xi}_{6ik} \end{bmatrix} - \bar{\Lambda}_i + \sum_{j=1}^{3} \left(\check{\Upsilon}_{jik} \check{\Gamma}_{jik}^T + \check{\Gamma}_{jik} \check{\Upsilon}_{jik}^T \right) < 0, \quad \forall i, k, \tag{3.43}$$

where

$$\check{\Upsilon}_{1ik} = \begin{bmatrix} 0 \\ \mathcal{S} \\ 0_{2\times 1} \end{bmatrix}, \quad \check{\Gamma}_{1ik} = \begin{bmatrix} \mathcal{R} (A_i - A_k)^T \\ 0_{3\times 1} \end{bmatrix},$$

$$\check{\Upsilon}_{2ik} = \begin{bmatrix} 0 \\ \mathcal{B}_k \\ 0_{2\times 1} \end{bmatrix}, \quad \check{\Gamma}_{2ik} = \begin{bmatrix} \mathcal{R} \left(C_{yi} - C_{yk} \right)^T \\ 0_{3\times 1} \end{bmatrix},$$

$$\check{\Upsilon}_{3ik} = \begin{bmatrix} 0 \\ \mathcal{S} (B_i - B_k) \\ 0_{2\times 1} \end{bmatrix}, \quad \check{\Gamma}_{3ik} = \begin{bmatrix} \mathcal{C}_k^T \\ 0_{3\times 1} \end{bmatrix}.$$

To solve the parameters of the IT2 fuzzy output-feedback controller, matrix P is partitioned and inverted as

$$P = \begin{bmatrix} \mathcal{S} & \mathcal{N} \\ \mathcal{N}^T & \mathcal{Y} \end{bmatrix}, \quad P^{-1} = \begin{bmatrix} \mathcal{R} & \mathcal{M} \\ \mathcal{M}^T & \mathcal{T} \end{bmatrix}.$$

Consider that $PP^{-1} = I$, inequality (3.41) holds. From (3.35), it is obvious that

$$\begin{bmatrix} -\mathcal{R} & -I \\ -I & -\mathcal{S} \end{bmatrix} < 0,$$

which shows that $\mathcal{R} - \mathcal{S}^{-1} > 0$, this is to say $I - \mathcal{R}\mathcal{S}$ is nonsingular. This ensures that there are always nonsingular matrices $\mathcal{M}$ and $\mathcal{N}$ such that (3.41) is satisfied. Setting

$$X_1 = \begin{bmatrix} \mathcal{R} & I \\ \mathcal{M}^T & 0 \end{bmatrix}, \quad X_2 = \begin{bmatrix} I & \mathcal{S} \\ 0 & \mathcal{N}^T \end{bmatrix}, \tag{3.44}$$

Then, it obtained from (3.44) that $PX_1 = X_2$. It follows that

$$X_1^T P X_1 = X_1^T X_2 = \begin{bmatrix} \mathcal{R} & I \\ I & \mathcal{S} \end{bmatrix},$$

which means that X_1 and X_2 are positive definite and P can be expressed as $P = X_2 X_1^{-1} > 0$. Consider the following equations:

$$\begin{aligned} &\mathcal{S}A_i\mathcal{R} + \mathcal{B}_k C_{yi}\mathcal{R} + \mathcal{S}B_i\mathcal{C}_k + \mathcal{N}A_{ck}\mathcal{M}^T \\ &\quad = \mathcal{S}A_k\mathcal{R} + \mathcal{B}_k C_{yk}\mathcal{R} + \mathcal{S}B_k\mathcal{C}_k + \mathcal{N}A_{ck}\mathcal{M}^T \\ &\qquad + \mathcal{S}\left(A_i - A_k\right)\mathcal{R} + \mathcal{B}_k\left(C_{yi} - C_{yk}\right)\mathcal{R} + \mathcal{S}\left(B_i - B_k\right)\mathcal{C}_k, \end{aligned}$$

and

$$\begin{aligned} &\bar{\sigma}_i\left(\mathcal{S}A_k\mathcal{R} + \mathcal{B}_i C_{yk}\mathcal{R} + \mathcal{S}B_k\mathcal{C}_i + \mathcal{N}A_{ci}\mathcal{M}^T\right) \\ &+ \bar{\sigma}_k\left(\mathcal{S}A_i\mathcal{R} + \mathcal{B}_k C_{yi}\mathcal{R} + \mathcal{S}B_i\mathcal{C}_k + \mathcal{N}A_{ck}\mathcal{M}^T\right) \\ &\quad = \bar{\sigma}_i\left(\mathcal{S}A_i\mathcal{R} + \mathcal{B}_i C_{yi}\mathcal{R} + \mathcal{S}B_i\mathcal{C}_i + \mathcal{N}A_{ci}\mathcal{M}^T\right) \\ &\qquad + \bar{\sigma}_k\left(\mathcal{S}A_k\mathcal{R} + \mathcal{B}_k C_{yk}\mathcal{R} + \mathcal{S}B_k\mathcal{C}_k + \mathcal{N}A_{ck}\mathcal{M}^T\right) \\ &\qquad + \left(\bar{\sigma}_k - \bar{\sigma}_i\right)\mathcal{S}\left(A_i - A_k\right)\mathcal{R} + \left(\bar{\sigma}_k\mathcal{B}_k - \bar{\sigma}_i\mathcal{B}_i\right)\left(C_{yi} - C_{yk}\right)\mathcal{R} \\ &\qquad + \mathcal{S}\left(B_i - B_k\right)\left(\mathcal{C}_k - \mathcal{C}_k\right). \end{aligned}$$

By performing congruence transformation by diag$\{X_1^{-1}, I, I\}$ to (3.42) and (3.43), we know that conditions in (3.34) and (3.32) hold. On the other hand, we perform congruence transformation to (3.36), (3.37) and (3.39) by X_1^{-1}, diag$\{X_1^{-1}, I\}$ and diag$\{X_1^{-1}, I, I\}$, respectively. We can see that the conditions in (3.30), (3.31) and (3.33) hold. Therefore, all the conditions in Theorem 3.4 are satisfied. The proof is completed. □

Remark 3.6 The main contributions of this chapter can be summarized below: (1) A new performance index, including the H_∞ performance, the L_2-L_∞ performance, the passivity performance and dissipativity performance is considered. (2) Based on the new performance index, a novel IT2 fuzzy state-feedback controller is designed for IT2 fuzzy systems with mismatched membership functions. (3) A new IT2 fuzzy output-feedback controller is also designed for IT2 fuzzy systems with mismatched membership functions under a unified frame.

3.4 Simulation Results

To validate the effectiveness and the practicality of the proposed control design schemes, a simulation example is provided in this section. In Example 3.7, the effectiveness of both the IT2 fuzzy state-feedback and output-feedback control schemes is testified.

Example 3.7 Consider the following 3-rule IT2 fuzzy system:

Plant Rule i: IF $x_1(t)$ is W_{i1}, THEN

$$\begin{cases} \dot{x}(t) = A_i x(t) + B_i u(t) + D_{1i} w(t), \\ z(t) = C_i x(t) + D_{2i} w(t), \quad i = 1, 2, 3, \end{cases} \tag{3.45}$$

where

$$\begin{aligned} A_1 &= \begin{bmatrix} 3 & -5 \\ 0.01 & 0.3 \end{bmatrix}, \quad A_2 = \begin{bmatrix} 0.2 & -3 \\ 0.3 & 0.1 \end{bmatrix}, \quad A_3 = \begin{bmatrix} -14 & 6 \\ 0.4 & 0.1 \end{bmatrix}, \\ B_1 &= \begin{bmatrix} 1 \\ -1 \end{bmatrix}, \quad B_2 = \begin{bmatrix} 10 \\ 0 \end{bmatrix}, \quad B_3 = \begin{bmatrix} -15 \\ -1 \end{bmatrix}, \\ D_{11ij} &= \begin{bmatrix} 0.1 \\ 0.1 \end{bmatrix}, \quad D_{12ij} = \begin{bmatrix} 0.1 \\ 0.1 \end{bmatrix}, \quad D_{13ij} = \begin{bmatrix} -0.1 \\ 0.1 \end{bmatrix}, \\ C_1 &= \begin{bmatrix} 0.1 & 0.1 \end{bmatrix}, \quad C_2 = \begin{bmatrix} 0.1 & 0.1 \end{bmatrix}, \quad C_3 = \begin{bmatrix} -0.2 & 0.2 \end{bmatrix}, \\ D_{21} &= -0.1, \quad D_{22ij} = -0.2, \quad D_{23ij} = -0.1. \end{aligned}$$

The LMFs and UMFs are given in Table 3.1.

It is assumed that the disturbance $w(t)$ is

$$w(t) = \begin{cases} 0.1 \sin(3t), & 0 \le t \le 5, \\ 0, & \text{else}. \end{cases} \tag{3.46}$$

Under the initial condition $x(0) = \begin{bmatrix} 5 & -5 \end{bmatrix}^T$, Fig. 3.1 depicts the state responses of the open-loop system, which indicates that the open-loop system (3.45) is not stable. In this case, we design the IT2 fuzzy state-feedback controller in (3.6) to stabilize this unstable system in (3.45).

Next, according to the description in (3.4) and (3.5), the LMFs and the UMFs in (3.5) of the IT2 fuzzy controller are defined in Table 3.2. By choosing $\underline{\nu}_j(x(t)) = 0.5$ and $\bar{\nu}_j(x(t)) = 0.5$ $(j = 1, 2, 3)$. In this control scheme, we consider the H_∞ performance index for the system in (3.45). Based on Definition 3.1, by setting $\Phi = 0$, $\Psi_1 = -I$, $\Psi_2 = 0$ and $\Psi_3 = 0.4I$, and according to Theorem 3.3, with the parameters $\bar{\sigma}_k$ $(k = 1, 2, 3)$ chosen as $\bar{\sigma}_1 = 0.1$, $\bar{\sigma}_2 = 0.9$, $\bar{\sigma}_3 = 0.1$, by solving the conditions

Table 3.1 The membership functions of the plant

LMFs	UMFs
$\underline{\theta}_1(x_1) = 0.95 - \frac{0.925}{1+e^{-\frac{(x_1+4.5)}{8}}}$	$\overline{\theta}_1(x_1) = 0.95 - \frac{0.925}{1+e^{-\frac{(x_1+3.5)}{8}}}$
$\underline{\theta}_2(x_1) = 0.025 + \frac{0.925}{1+e^{-\frac{(x_1-4.5)}{8}}}$	$\overline{\theta}_2(x_1) = 0.025 + \frac{0.925}{1+e^{-\frac{(x_1-3.5)}{8}}}$
$\underline{\theta}_3(x_1) = 1 - \overline{\theta}_1(x_1) - \overline{\theta}_2(x_1)$	$\overline{\theta}_3(x_1) = 1 - \underline{\theta}_1(x_1) - \underline{\theta}_2(x_1)$

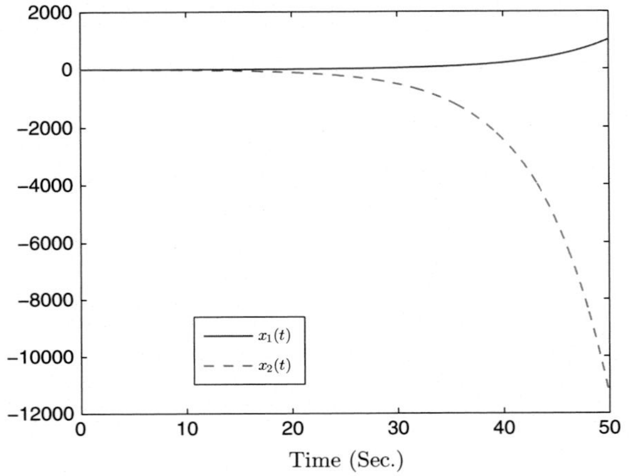

Fig. 3.1 State responses of the open-loop system

Table 3.2 The membership functions of the controller

LMFs	UMFs
$\underline{\eta}_1(x_1) = 1 - \frac{1}{1+e^{-\frac{x_1+5}{2}}}$	$\overline{\eta}_1(x_1) = 1 - \frac{1}{1+e^{-\frac{x_1+4}{2}}}$
$\underline{\eta}_2(x_1) = \frac{1}{1+e^{-\frac{x_1-5}{2}}}$	$\overline{\eta}_2(x_1) = \frac{1}{1+e^{-\frac{x_1-4}{2}}}$
$\underline{\eta}_3(x_1) = 1 - \overline{\eta}_1(x_1) - \overline{\eta}_2(x_1)$	$\overline{\eta}_3(x_1) = 1 - \underline{\eta}_1(x_1) - \underline{\eta}_2(x_1)$

in (3.15)–(3.19), we can obtain the H_∞ performance index is $\gamma = 0.6325$, and the controller gain matrices are obtained as follows:

$$K_1 = \begin{bmatrix} -0.7515 & -0.2119 \end{bmatrix}, \quad K_3 = \begin{bmatrix} -0.6212 & -0.1692 \end{bmatrix},$$
$$K_2 = \begin{bmatrix} -0.7148 & 0.0641 \end{bmatrix}.$$

Thus, under the same initial state condition, we can obtain the state responses of the closed-loop system in (3.45), which are plotted in Fig. 3.2. Obviously, the unstable system has been effectively stabilized by the designed IT2 fuzzy state-feedback controller. Therefore, the whole simulation in this control procedure has demonstrated the effectiveness of the designed IT2 fuzzy state-feedback control scheme.

We continue to consider that the state can not be measured. Then, the IT2 fuzzy dynamic output-feedback controller is designed to control the IT2 fuzzy system in (3.45). The measured output is given as $y(t) = C_{yi}x(t)$ $(i = 1, 2, 3)$, where

$$C_{y1} = \begin{bmatrix} 0.78 & 0.66 \end{bmatrix}, \quad C_{y2} = \begin{bmatrix} 0.33 & 0.75 \end{bmatrix}, \quad C_{y3} = \begin{bmatrix} 0.78 & 0.66 \end{bmatrix}.$$

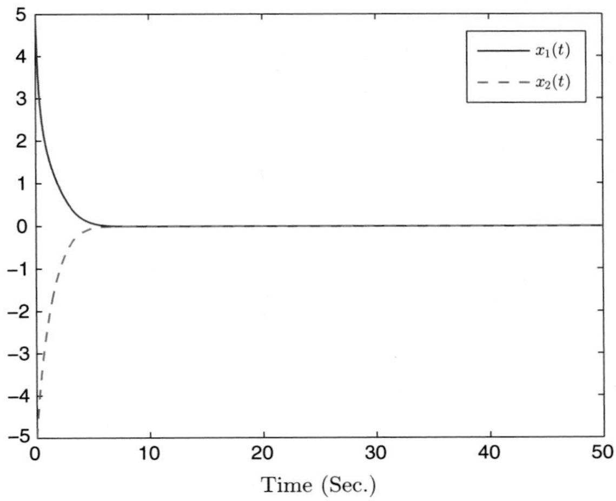

Fig. 3.2 States of the closed-loop system under IT2 fuzzy state-feedback controller

We consider the same membership functions in Table 3.2 for the IT2 fuzzy output-feedback controller design, i.e.,

$$\underline{\varpi}_1(x_1) = \underline{\eta}_1(x_1), \quad \overline{\varpi}_1(x_1) = \overline{\eta}_1(x_1),$$
$$\underline{\varpi}_2(x_1) = \underline{\eta}_2(x_1), \quad \overline{\varpi}_2(x_1) = \overline{\eta}_2(x_1),$$
$$\underline{\varpi}_3(x_1) = \underline{\eta}_3(x_1), \quad \overline{\varpi}_3(x_1) = \overline{\eta}_3(x_1).$$

In this control scheme, we consider the L_2-L_∞ performance index for the system in (3.45). From (3.10), we choose the constants $\underline{\nu}_j(x(t)) = 0.5$ and $\bar{\nu}_j(x(t)) = 0.5$ $(j = 1, 2, 3)$. Based on Definition 3.1, by setting $\Phi = I$, $\Psi_1 = 0$, $\Psi_2 = 0$, and $\Psi_3 = 0.1I$, and according to Theorem 3.5, by solving the conditions (3.35)–(3.40), with the parameters $\bar{\sigma}_k$ $(k = 1, 2, 3)$ chosen as $\bar{\sigma}_1 = 0.2$, $\bar{\sigma}_2 = 0.9$ and $\bar{\sigma}_3 = 0.3$, we can obtain the L_2-L_∞ performance index $\gamma = 1.3229$, and the controller gain matrices are obtained as follows:

$$A_{c1} = \begin{bmatrix} 0.5372 & -0.2595 \\ -56.8561 & 8.8242 \end{bmatrix}, \quad B_{c1} = 10^{-3} \times \begin{bmatrix} 0.1611 \\ -81.1136 \end{bmatrix},$$
$$A_{c2} = \begin{bmatrix} 0.0344 & 0.2058 \\ 1.3484 & -52.2277 \end{bmatrix}, \quad B_{c2} = 10^{-3} \times \begin{bmatrix} 5.5240 \\ 297.8980 \end{bmatrix},$$
$$A_{c3} = \begin{bmatrix} 0.2670 & 0.2830 \\ -35.5994 & 14.7994 \end{bmatrix}, \quad B_{c3} = 10^{-3} \times \begin{bmatrix} -0.1557 \\ 115.3068 \end{bmatrix},$$
$$C_{c1} = \begin{bmatrix} -0.3436 & 1.6988 \end{bmatrix}, \quad C_{c2} = \begin{bmatrix} 2.6330 & -3.9223 \end{bmatrix},$$
$$C_{c3} = \begin{bmatrix} -0.6807 & -2.2134 \end{bmatrix}.$$

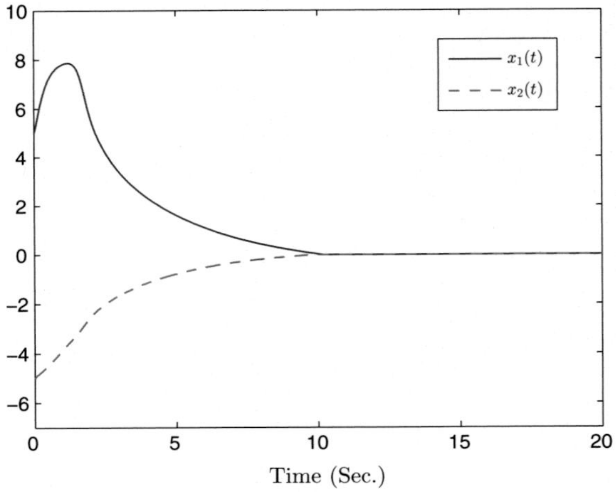

Fig. 3.3 States of the closed-loop system under IT2 fuzzy output-feedback controller

Thus, under the same initial state condition, we can obtain the state responses of the closed-loop system in (3.45), which are plotted in Fig. 3.3.

3.5 Conclusion

In this chapter, the problems of state and output-feedback controllers design have been solved for the IT2 T–S fuzzy system with mismatched membership functions. Under a unified framework, the IT2 fuzzy controllers have been designed for IT2 fuzzy systems based on a new performance index. In this new performance index, H_∞, L_2-L_∞, passive and dissipativity performances are included. By using Lyapunov stability theory and the convex optimization technique, the existence conditions of the state and output-feedback controllers have been expressed. A numerical example has illustrated the effectiveness of the proposed designed method. In future work, the actuator delay and fault will be considered in the IT2 fuzzy systems and the fault-tolerant controller will be designed for the systems with actuator delay and fault.

Chapter 4
Sampled-Data Control of Interval Type-2 Fuzzy-Model-Based Systems

4.1 Introduction

This chapter studies the problem of state-feedback sampled-data controller design for IT2 fuzzy systems with actuator fault. The IT2 fuzzy system can deal with uncertain grades of membership well if parameter uncertainties of nonlinear plants are considered, and the IT2 fuzzy controller can provide better performance. The IT2 fuzzy systems and the IT2 sampled-data controller do not share the same membership functions. When the actuator fault of IT2 fuzzy control system is considered, the IT2 fuzzy controller can stabilize the IT2 system well. Firstly, considering the mismatched membership functions, the IT2 fuzzy system and the IT2 state-feedback sampled-data controller are constructed. Secondly, based on Lyapunov stability theory, an IT2 state-feedback sampled-data controller is designed such that the closed-loop system is asymptotically stable for all possible actuator failures. The existence conditions of the IT2 fuzzy H_∞ sampled-data controller can be expressed by a convex optimization problem.

4.2 Problem Formulation

In this section, we consider the following IT2 fuzzy system:

♦ **Plant Form**:

Rule i: IF $f_1(x(t))$ is W_{i1} and ... and $f_b(x(t))$ is W_{ib}, THEN,

$$\begin{cases} \dot{x}(t) = A_i x(t) + B_i u^f(t) + B_{1i} w(t), \\ z(t) = C_i x(t) + D_i u^f(t) + D_{1i} w(t), \end{cases} \tag{4.1}$$

where W_{is} is the IT2 fuzzy set of the corresponding function $f_s(x(t))$, $i = 1, 2, \ldots, r$, $s = 1, 2, \ldots, b$ with r being the number of plant rules, b being the number of premise variables, $x(t) \in \mathbf{R}^n$ denotes the state vector, $u^f(t) \in \mathbf{R}^m$ stands for the input vector,

H. Li et al., *Analysis and Synthesis for Interval Type-2 Fuzzy-Model-Based Systems*, DOI 10.1007/978-981-10-0593-0_4

$z(t) \in \mathbf{R}^p$ is the controlled output and $w(t) \in \mathbf{R}^q$ is the disturbance input that belongs to $\mathcal{L}_2[0, \infty)$. $A_i, B_i, B_{1i}, C_i, D_i$ and D_{1i} are the known real constant matrices with appropriate dimensions. The following interval set expresses the firing strength of the ith rule.

$$\begin{aligned}\tilde{\theta}_i(x(t)) &= \left[\prod_{s=1}^{b} \underline{\mu}_{W_{is}}(f_s(x(t))), \prod_{s=1}^{b} \overline{\mu}_{W_{is}}(f_s(x(t)))\right] \\ &= \left[\underline{\theta}_i(x(t)), \overline{\theta}_i(x(t))\right], \quad i = 1, 2, \ldots, r,\end{aligned}$$

where $\overline{\theta}_i(x(t))$ and $\underline{\theta}_i(x(t))$ $\left(\overline{\theta}_i(x(t)) \geq \underline{\theta}_i(x(t)) \geq 0\right)$ denote the upper and lower grade of membership, respectively. $\underline{\mu}_{W_{is}}(f_s(x(t)))$ and $\overline{\mu}_{W_{is}}(f_s(x(t)))$ $\left(\overline{\mu}_{W_{is}}(f_s(x(t))) \geq \underline{\mu}_{W_{is}}(f_s(x(t))) \geq 0\right)$ stand for the LMF and UMF, respectively. The nonlinear system in (4.1) can be represented by the following IT2 T–S fuzzy systems:

$$\begin{cases}\dot{x}(t) = \sum_{i=1}^{r} \theta_i(x(t)) \left(A_i x(t) + B_i u^f(t) + B_{1i} w(t)\right), \\ z(t) = \sum_{i=1}^{r} \theta_i(x(t)) \left(C_i x(t) + D_i u^f(t) + D_{1i} w(t)\right),\end{cases} \tag{4.2}$$

where

$$\begin{aligned}\theta_i = h_i(x(t)) \underline{\theta}_i(x(t)) + (1 - h_i(x(t))) \overline{\theta}_i(x(t)) \geq 0, \quad &\forall i, \\ 0 \leq h_i(x(t)) \leq 1, \quad &\forall i, \\ \sum_{i=1}^{r} \theta_i(x(t)) = 1,\end{aligned}$$

and $h_i(x(t))$ is a nonlinear function, $\theta_i(x(t))$ stands for the grade of membership of the embedded membership function. Consider the actuator failure model with failure matrix M_a,

$$u^f(t) = M_a u(t). \tag{4.3}$$

The actuator fault matrix $M_a = \text{diag}\{m_{a1}, m_{a2}, \ldots, m_{am}\}$, where $0 \leq \underline{m}_{ai} \leq m_{ai} \leq \overline{m}_{ai} \leq 1$, and $\underline{m}_{ai}$ and $\overline{m}_{ai}$ are constant scalars, which denote the admissible failures of the actuator. We consider the following three different actuator failure cases:

(a) If $\underline{m}_{ai} = \overline{m}_{ai} = 0$, then, $m_{ai} = 0$, which represents that the corresponding actuator $u_i^f(t)$ has completely failed.
(b) When $\underline{m}_{ai} = \overline{m}_{ai} = 1$, we obtain $m_{ai} = 1$, which implies that there is no failure in the actuator $u_i^f(t)$.
(c) While $0 < \underline{m}_{ai} < \overline{m}_{ai} \leq 1$, it means that there exists a partial fault in the corresponding actuator $u_i^f(t)$.

Suppose that the updating signal successfully transmitted from the sampler to the controller and the Zero-Order-Hold (ZOH) at the instant t_k. It is assumed that the sampling intervals are bounded $t_{k+1} - t_k \leq h_M$, where h_M denotes the maximum time span between the time t_k and t_{k+1}. The initial conditions of $x(t)$ and $u(t)$ are given as $x(t) = \varphi(t)$ and $u(t) = 0$ for $t \in [t_0 - h_M, t_0]$, where $\varphi(t)$ is a differentiable function, t_0 is the initial time. The following IT2 sampled-date fuzzy control law is constructed:

Rule j: **IF** $g_1(x(t))$ is M_{j1} and ... and g_p is M_{jp}, **THEN**:

$$u(t) = K_{aj} x(t_k), \quad t_k \leq t < t_{k+1}, \tag{4.4}$$

where M_{js} denotes the fuzzy set of rule j corresponding to the function $g_s(x(t))$, $j = 1, 2, \ldots, r$, $s = 1, 2, \ldots, p$; r is a positive integer; $K_{aj} \in \mathbf{R}^{m \times n}$ stands for the state-feedback gain matrix of rule j. The following interval set expresses the firing strength of the jth rule.

$$\begin{aligned}\tilde{\eta}_j(x(t)) &= \left[\prod_{s=1}^{p} \underline{\mu}_{M_{js}}(g_s(x(t))), \prod_{s=1}^{p} \overline{\mu}_{M_{js}}(g_s(x(t)))\right] \\ &= \left[\underline{\eta}_j(x(t)), \overline{\eta}_j(x(t))\right], \quad j = 1, 2, \ldots, r,\end{aligned}$$

where $\underline{\eta}_j(x(t))$ and $\overline{\eta}_j(x(t))$ stand for the lower and upper grades of membership, respectively. The LMFs and UMFs are denoted as $\underline{\mu}_{M_{js}}(g_s(x(t)))$ and $\overline{\mu}_{M_{js}}(g_s(x(t)))$, respectively. Here, $\overline{\mu}_{M_{js}}(g_s(x(t))) \geq \underline{\mu}_{M_{js}}(g_s(x(t))) \geq 0$ and $\overline{\eta}_j(x(t)) \geq \underline{\eta}_j(x(t)) \geq 0$ for all j. The overall IT2 fuzzy control law is represented by

$$u(t) = \sum_{j=1}^{r} \eta_j(x(t)) K_{aj} x(t_k), \tag{4.5}$$

where t_k $(k = 1, 2, \ldots, r)$ denotes the kth sampling instant, $t_0 \geq 0$, and $\lim_{k \to \infty} t_k = \infty$. And

$$\eta_j(x(t)) = \frac{\alpha_j(x(t)) \underline{\eta}_j(x(t)) + \left(1 - \alpha_j(x(t))\right) \overline{\eta}_j(x(t))}{\sum_{l=1}^{r} \left(\alpha_l(x(t)) \underline{\eta}_l(x(t)) + (1 - \alpha_l(x(t))) \overline{\eta}_l(x(t))\right)} \geq 0, \quad \forall j,$$

$$\sum_{j=1}^{r} \eta_j(x(t)) = 1,$$

$$0 \leq \alpha_j(x(t)) \leq 1, \quad \forall j,$$

in which $\alpha_j(x(t))$ is a predefined function, $\eta_j(x(t))$ denotes the grade of membership of the embedded membership function.

Denote $h(t) = t - t_k$ for $t_k \leq t \leq t_{k+1}$. It is known that $0 \leq h(t) \leq t_{k+1} - t_k \leq h_M$. Then, it can be seen that $h(t)$ is piecewise-linear with derivative $\dot{h}(t) = 1$. Utilizing $t_k = t - h(t)$, we have

$$u(t) = \sum_{j=1}^{r} \eta_j(x(t)) K_{aj} x(t - h(t)). \tag{4.6}$$

In the following part, we use the notations $\theta_i(x(t)) = \theta_i$, and $\eta_j(x(t)) = \eta_j$ for $\forall i, j$. By substituting (4.6) into (4.2), and considering (4.3), the resulting closed-loop system can be described by

$$\begin{cases} \dot{x}(t) = \sum_{i=1}^{r} \sum_{j=1}^{r} \theta_i \eta_j \left(A_i x(t) + B_i M_a K_{aj} x(t - h(t)) + B_{1i} w(t) \right), \\ z(t) = \sum_{i=1}^{r} \sum_{j=1}^{r} \theta_i \eta_j \left(C_i x(t) + D_i M_a K_{aj} x(t - h(t)) + D_{1i} w(t) \right). \end{cases} \tag{4.7}$$

The main objective of this chapter is to design the fault-tolerant IT2 sampled-data fuzzy controller in (4.5) such that the system (4.1) with actuator faults is asymptotically stable and satisfies the H_∞ performance. Furthermore, when there exist no actuator failures in system (4.1), the standard IT2 sampled-data fuzzy controller is also designed in next section. The advantages of the fault-tolerant controller over the standard controller will be demonstrated in the simulation results.

Remark 4.1 It can be seen from (4.2) and (4.6), if the membership functions of the plant and controller are matched, the controller can be also designed based on PDC concept. In our study, the proposed IT2 fuzzy controller (4.6) does not need to share the same membership functions as those of the IT2 T–S fuzzy model (4.2). This offers a higher design flexibility to the IT2 fuzzy controller.

4.3 Main Results

In this section, we first introduce the following matrices, which will be used in the latter part.

$$\begin{aligned} M_{a0} &= \operatorname{diag}\{m_{a01}, m_{a02}, \ldots, m_{a0m}\}, \\ L_a &= \operatorname{diag}\{l_{a1}, l_{a2}, \ldots, l_{am}\}, \\ J_a &= \operatorname{diag}\{j_{a1}, j_{a2}, \ldots, j_{am}\}, \end{aligned}$$

where $m_{a0i} = \left(\underline{m}_{ai} + \overline{m}_{ai}\right)/2$, $l_{ai} = \left(m_{ai}(t) - m_{a0i}\right)/m_{a0i}$ and $j_{ai} = (\overline{m}_{ai} - \underline{m}_{ai})/(\underline{m}_{ai} + \overline{m}_{ai})$ with $i = 1, 2, \ldots, m$. Then, we have $M_a = M_{a0}(I + L_a)$ and $L_a^T L_a \leq J_a^T J_a \leq I$.

4.3.1 Stability Analysis

Then, we will solve the reliable fuzzy H_∞ state-feedback controller design problem for system (4.1) with actuator faults in this section. Firstly, we have the following theorem.

Theorem 4.2 *For a given scalar $h_M > 0$, and matrices K_{aj}, the closed-loop system (4.7) is asymptotically stable and satisfies $\|z(t)\|_2 \le \gamma \|w(t)\|_2$ for any nonzero $w(t) \in \mathcal{L}_2[0,\infty)$, if there exist matrices $P = P^T > 0$, $Q = Q^T$, $\Delta_i = \Delta_i^T$, $R = R^T$, N_1, N_2, L_1, L_2, and L_3 with appropriate dimensions and the condition $\eta_j - \lambda_j\theta_j > 0$ such that the following inequalities hold for all i, j:*

$$\lambda_j\Pi_{1ij} + \lambda_i\Pi_{1ji} + \lambda_j h_M\Pi_{2ij} + \lambda_i h_M\Pi_{2ji} + \Psi_{0ij} < 0, \quad i < j, \tag{4.8}$$

$$\lambda_i\Pi_{1ii} + \lambda_i h_M\Pi_{2ii} - \lambda_i\Delta_i + \Delta_i < 0, \tag{4.9}$$

$$\Pi_{1ij} + h_M\Pi_{2ij} - \Delta_i < 0, \tag{4.10}$$

$$\begin{bmatrix} \lambda_j\Pi_{1ij} + \lambda_i\Pi_{1ji} + \Psi_{0ij} & -\lambda_j h_M N - \lambda_i h_M N \\ * & -\lambda_j h_M R - \lambda_i h_M R \end{bmatrix} < 0, \quad i < j, \tag{4.11}$$

$$\begin{bmatrix} \lambda_i\Pi_{1ii} - \lambda_i\Delta_i + \Delta_i & -\lambda_i h_M N \\ * & -\lambda_i h_M R \end{bmatrix} < 0, \tag{4.12}$$

$$\begin{bmatrix} \Pi_{1ij} - \Delta_i & -h_M N \\ * & -h_M R \end{bmatrix} < 0, \tag{4.13}$$

where

$$\Pi_{1ij} = \begin{bmatrix} \Psi_{11ij} & \Psi_{12ij} & \Psi_{13ij} & L_1B_{1i} & C_i^T & \varepsilon L_1B_i & 0 \\ * & \Psi_{22ij} & \Psi_{23ij} & L_2B_{1i} & K_{aj}^T M_{a0}^T D_i^T & \varepsilon L_2B_i & K_{aj}^T M_{a0}^T \\ * & * & \Psi_{33ij} & L_3B_{1i} & 0 & \varepsilon L_3B_i & 0 \\ * & * & * & -\gamma^2 I & D_{1i}^T & 0 & 0 \\ * & * & * & * & -I & \varepsilon D_i & 0 \\ * & * & * & * & * & -\varepsilon J_a^{-1} & 0 \\ * & * & * & * & * & * & -\varepsilon J_a^{-1} \end{bmatrix},$$

$$\begin{aligned} \Psi_{0ij} &= -\lambda_j\Delta_i - \lambda_i\Delta_j + \Delta_i + \Delta_j, \quad \Psi_{13ij} = P + A_i^T L_3^T - L_1, \\ \Psi_{11ij} &= -Q_1 + \mathbf{He}(N_1 + L_1A_i), \quad \Psi_{23ij} = K_{aj}^T M_{a0}^T B_i^T L_3^T - L_2, \\ \Psi_{12ij} &= -Q_2 - N_1 + N_2^T + L_1B_iM_{a0}K_{aj} + A_i^T L_2^T, \\ \Psi_{22ij} &= -Q_3 + \mathbf{He}\left(L_2B_iM_{a0}K_{aj} - N_2\right), \quad \Psi_{33ij} = -L_3 - L_3^T, \end{aligned}$$

$$\Pi_{2ij} = \begin{bmatrix} 0 & 0 & Q_1 & 0 \\ * & 0 & Q_2^T & 0 \\ * & * & R & 0 \\ * & * & * & 0_{4\times4} \end{bmatrix}, \quad N = \begin{bmatrix} N_1 \\ N_2 \\ 0_{5\times1} \end{bmatrix}, \quad Q = \begin{bmatrix} Q_1 & Q_2 \\ * & Q_3 \end{bmatrix}.$$

Proof Consider the following Lyapunov–Krasovskii functional:

$$V(t) = x^T(t)Px(t) + (h_M - h(t)) \times \left[\bar{x}^T(t)Q\bar{x}(t) + \int_{t-h(t)}^{t} \dot{x}^T(s)R\dot{x}(s)ds\right], \tag{4.14}$$

where $\bar{x}(t) = \left[x^T(t)\ x^T(t-h(t))\right]^T$. Then, the time-derivative of $V(t)$ gives

$$\dot{V}(t) = 2x^T(t)P\dot{x}(t) - \bar{x}^T(t)Q\bar{x}(t) - \int_{t-h(t)}^{t} \dot{x}^T(s)R\dot{x}(s)ds + (h_M - h(t)) \times \left[2\bar{x}^T(t)Q\dot{\bar{x}}(t) + \dot{x}^T(t)R\dot{x}(t) - \dot{x}^T(t-h(t))R\dot{x}(t-h(t))\right].$$

According to $\dot{x}(t-h(t)) = 0$, one can get

$$\dot{V}(t) = 2x^T(t)P\dot{x}(t) - \bar{x}^T(t)Q\bar{x}(t) - \int_{t-h(t)}^{t} \dot{x}^T(s)R\dot{x}(s)ds + (h_M - h(t))\left(2\bar{x}^T(t)Q\begin{bmatrix}\dot{x}(t)\\ 0\end{bmatrix} + \dot{x}^T(t)R\dot{x}(t)\right). \tag{4.15}$$

For the matrices $\tilde{N} = \left[N_1^T\ N_2^T\ 0\ 0\right]^T, L = \left[L_1^T\ L_2^T\ L_3^T\ 0\right]^T$ with appropriate dimensions, it can be seen that the following equalities can be verified easily:

$$0 = 2\hat{x}^T(t)\tilde{N}\left[x(t) - x(t-h(t)) - \int_{t-h(t)}^{t} \dot{x}(s)ds\right], \tag{4.16}$$

$$0 = 2\hat{x}^T(t)L\left(-\dot{x}(t) + \sum_{i=1}^{r}\sum_{j=1}^{r}\theta_i\eta_j \times \left(A_ix(t) + B_iM_aK_{aj}x(t-h(t)) + B_{1i}w(t)\right)\right), \tag{4.17}$$

where $\hat{x}(t) = \left[x^T(t)\ x^T(t-h(t))\ \dot{x}^T(t)\ w^T(t)\right]^T$. Then, according to the definition of $z(t)$, one can have

$$z^T(t)z(t) \le \sum_{i=1}^{r}\sum_{j=1}^{r}\theta_i\eta_j\left[\left(C_ix(t) + D_iM_aK_{aj}(x(t-h(t))) + D_{1i}w(t)\right)^T \times \left(C_ix(t) + D_iM_aK_{aj}(x(t-h(t))) + D_{1i}w(t)\right)\right]. \tag{4.18}$$

Combining (4.15)–(4.18), the following inequality can be obtained:

$$\begin{aligned}
&\dot{V}(t) + z^T(t)z(t) - \gamma^2 w^T(t)w(t) \\
&= \sum_{i=1}^{r}\sum_{j=1}^{r}\theta_i\eta_j\Bigg(\hat{x}^T(t)\left(\Phi_{1ij} + \Phi_{3ij}^T\Phi_{3ij} + (h_M - h(t))\Phi_{2ij}\right)\hat{x}(t) \\
&\quad + \int_{t-h(t)}^{t}\begin{bmatrix}\hat{x}^T(t)\\ \dot{x}^T(s)\end{bmatrix}^T\begin{bmatrix}0 & -\tilde{N}\\ * & -R\end{bmatrix}\begin{bmatrix}\hat{x}^T(t)\\ \dot{x}^T(s)\end{bmatrix}ds\Bigg) \\
&= \sum_{i=1}^{r}\sum_{j=1}^{r}\theta_i\eta_j\left\{\frac{h_M - h(t)}{h_M}\hat{x}^T(t)\left(\Phi_{1ij} + \Phi_{3ij}^T\Phi_{3ij} + h_M\Phi_{2ij}\right)\hat{x}(t) + \frac{1}{h_M}\right. \\
&\quad \left.\times\int_{t-h(t)}^{t}\begin{bmatrix}\hat{x}^T(t)\\ \dot{x}^T(s)\end{bmatrix}^T\begin{bmatrix}\Phi_{1ij} + \Phi_{3ij}^T\Phi_{3ij} & -h_M\tilde{N}\\ * & -h_M R\end{bmatrix}\begin{bmatrix}\hat{x}^T(t)\\ \dot{x}^T(s)\end{bmatrix}ds\right\}, \qquad (4.19)
\end{aligned}$$

where

$$\Phi_{1ij} = \begin{bmatrix}\Psi_{11} & \bar{\Psi}_{12} & \Psi_{13ij} & L_1B_{1i}\\ * & \bar{\Psi}_{22} & \bar{\Psi}_{23} & L_2B_{1i}\\ * & * & \Psi_{33} & L_3B_{1i}\\ * & * & * & -\gamma^2 I\end{bmatrix},\quad \Phi_{2ij} = \begin{bmatrix}0 & 0 & Q_1 & 0\\ * & 0 & Q_2^T & 0\\ * & * & R & 0\\ * & * & * & 0\end{bmatrix},$$

$$\begin{aligned}
\bar{\Psi}_{12} &= -Q_2 - N_1 + N_2^T + L_1B_iM_aK_{aj} + A_i^TL_2^T,\\
\bar{\Psi}_{22} &= -Q_3 + \mathbf{He}\left(L_2B_iM_aK_{aj} - N_2\right),\quad \bar{\Psi}_{23} = K_{aj}^TM_a^TB_i^TL_3^T - L_2,\\
\bar{\Phi}_{3ij} &= \left[\, C_i \;\; D_iM_aK_{aj} \;\; 0 \;\; D_{1i}\,\right].
\end{aligned}$$

On the other hand, consider $\sum_{i=1}^{r}\sum_{j=1}^{r}\theta_i\left(\theta_j - \eta_j\right)\Delta_i = 0$. Then

$$\begin{aligned}
&\sum_{i=1}^{r}\sum_{j=1}^{r}\theta_i\eta_j\left(\Pi_{1ij} + h_M\Pi_{2ij}\right)\\
&= \sum_{i=1}^{r}\sum_{j=1}^{r}\theta_i\eta_j\left(\Pi_{1ij} + h_M\Pi_{2ij}\right) + \sum_{i=1}^{r}\sum_{j=1}^{r}\theta_i\left(\theta_j - \eta_j\right)\Delta_i\\
&= \sum_{i=1}^{r}\sum_{j=1}^{r}\theta_i\eta_j\left(\Pi_{1ij} + h_M\Pi_{2ij}\right) + \sum_{i=1}^{r}\sum_{j=1}^{r}\theta_i\left(\theta_j - \eta_j + \lambda_j\theta_j - \lambda_j\theta_j\right)\Delta_i\\
&= \sum_{i=1}^{r}\sum_{j=1}^{r}\theta_i\left(\eta_j + \lambda_j\theta_j - \lambda_j\theta_j\right)\left(\Pi_{1ij} + h_M\Pi_{2ij}\right) + \sum_{i=1}^{r}\sum_{j=1}^{r}\theta_i\\
&\quad \times\left(\theta_j - \lambda_j\theta_j\right)\Delta_i - \sum_{i=1}^{r}\sum_{j=1}^{r}\theta_i\left(\eta_j - \lambda_j\theta_j\right)\Delta_i\\
&= \sum_{i=1}^{r}\sum_{j=1}^{r}\theta_i\theta_j\left(\lambda_j\left(\Pi_{1ij} + h_M\Pi_{2ij}\right) - \lambda_j\Delta_i + \Delta_i\right)
\end{aligned}$$

$$+\sum_{i=1}^{r}\sum_{j=1}^{r}\theta_i\left(\eta_j-\lambda_j\theta_j\right)\left(\Pi_{1ij}+h_M\Pi_{2ij}-\Delta_i\right)$$

$$=\sum_{i=1}^{r}\sum_{j=1}^{r}\theta_i^2\left(\lambda_i\left(\Pi_{1ii}+h_M\Pi_{2ii}\right)-\lambda_i\Delta_i+\Delta_i\right)+\sum_{i=1}^{r-1}\sum_{j=i+1}^{r}\theta_i\theta_j\left(\lambda_j\right.$$

$$\times\left(\Pi_{1ij}+h_M\Pi_{2ij}\right)-\lambda_j\Delta_i+\Delta_i+\lambda_i\left(\Pi_{1ji}+h_M\Pi_{2ji}\right)$$

$$\left.-\lambda_i\Delta_j+\Delta_j\right)+\sum_{i=1}^{r}\sum_{j=1}^{r}\theta_i\left(\eta_j-\lambda_j\theta_j\right)\left(\Pi_{1ij}+h_M\Pi_{2ij}-\Delta_i\right). \quad (4.20)$$

Similarly, one can have

$$\sum_{i=1}^{r}\sum_{j=1}^{r}\theta_i\eta_j\left(\Pi_{1ij}+h_MNR^{-1}N^T\right)$$

$$=\sum_{i=1}^{r}\sum_{j=1}^{r}\theta_i^2\left(\lambda_i\left(\Pi_{1ij}+h_MNR^{-1}N^T\right)-\lambda_i\Delta_i+\Delta_i\right)$$

$$+\sum_{i=1}^{r-1}\sum_{j=i+1}^{r}\theta_i\theta_j\left(\lambda_j\left(\Pi_{1ij}+h_MNR^{-1}N^T\right)-\lambda_j\Delta_i+\Delta_i\right.$$

$$\left.+\lambda_i\left(\Pi_{1ij}+h_MNR^{-1}N^T\right)-\lambda_i\Delta_j+\Delta_j\right)$$

$$+\sum_{i=1}^{r}\sum_{j=1}^{r}\theta_i\left(\eta_j-\lambda_j\theta_j\right)\left(\Pi_{1ij}+h_MNR^{-1}N^T-\Delta_i\right).$$

From the inequalities (4.8)–(4.13), it can be seen that

$$\sum_{i=1}^{r}\sum_{j=1}^{r}\theta_i\eta_j\left(\Pi_{1ij}+h_M\Pi_{2ij}\right)<0, \quad (4.21)$$

$$\sum_{i=1}^{r}\sum_{j=1}^{r}\theta_i\eta_j\left(\Pi_{1ij}+h_MNR^{-1}N^T\right)<0. \quad (4.22)$$

For a given scalar $\varepsilon>0$, it holds that

$$\mathbf{He}\left(F_iJ_aE_j\right)\leq\varepsilon F_iJ_aF_i^T+\varepsilon^{-1}E_j^TJ_aE_j, \quad (4.23)$$

where

$$F=\begin{bmatrix}B_i^TL_1^T & B_i^TL_2^T & B_i^TL_3^T & 0 & D_i^T\end{bmatrix}^T,\quad E=\begin{bmatrix}0 & M_{a0}K_{aj} & 0_{1\times 3}\end{bmatrix}.$$

Using Schur complement to (4.21) and (4.22), respectively, based on the condition (4.23), it can be concluded that the following inequalities hold:

$$\sum_{i=1}^{r}\sum_{j=1}^{r}\theta_i\eta_j\left(\tilde{\Phi}_{1ij}+h_M\Phi_{2ij}+\mathbf{He}\left(F_iJ_aE_j\right)\right)<0, \tag{4.24}$$

$$\sum_{i=1}^{r}\sum_{j=1}^{r}\theta_i\eta_j\left(\tilde{\Phi}_{1ij}+h_M\breve{N}R^{-1}\breve{N}^T+\mathbf{He}\left(F_iJ_aE_j\right)\right)<0, \tag{4.25}$$

where

$$\tilde{\Phi}_{1ij}=\begin{bmatrix}\Psi_{11ij} & \Psi_{12ij} & \Psi_{13ij} & L_1B_{1i} & C_i^T\\ * & \Psi_{22ij} & \Psi_{23ij} & L_2B_{1i} & K_{aj}^TM_{a0}^TD_i^T\\ * & * & \Psi_{33ij} & L_3B_{1i} & 0\\ * & * & * & -\gamma^2I & D_{1i}^T\\ * & * & * & * & -I\end{bmatrix}.$$

Under the conditions $M_a = M_{a0}\left(I+L_a\right)$ and $L_a^TL_a \le J_a^TJ_a \le I$, we can obtain:

$$\sum_{i=1}^{r}\sum_{j=1}^{r}\theta_i\eta_j\left(\bar{\Phi}_{1ij}+h_M\bar{\Phi}_{2ij}\right)<0, \tag{4.26}$$

$$\sum_{i=1}^{r}\sum_{j=1}^{r}\theta_i\eta_j\left(\bar{\Phi}_{1ij}+h_M\breve{N}R^{-1}\breve{N}^T\right)<0, \tag{4.27}$$

where

$$\bar{\Phi}_{1ij}=\begin{bmatrix}\Phi_{1ij} & \Phi_{3ij}^T\\ * & -I\end{bmatrix},\quad \bar{\Phi}_{2ij}=\begin{bmatrix}\Phi_{2ij} & 0\\ 0 & 0\end{bmatrix},\quad \breve{N}=\begin{bmatrix}\tilde{N}\\ 0\end{bmatrix}.$$

Then, it is clear that the following two inequalities hold via Schur complement,

$$\sum_{i=1}^{r}\sum_{j=1}^{r}\theta_i\eta_j\left(\Phi_{1ij}+\Phi_{3ij}^T\Phi_{3ij}+h_M\Phi_{2ij}\right)<0, \tag{4.28}$$

$$\sum_{i=1}^{r}\sum_{j=1}^{r}\theta_i\eta_j\begin{bmatrix}\Phi_{1ij}+\Phi_{3ij}^T\Phi_{3ij} & -h_M\tilde{N}\\ * & -h_MR\end{bmatrix}<0, \tag{4.29}$$

which mean that the inequality $\dot{V}(t)+z^T(t)z(t)-\gamma^2w^T(t)w(t)<0$ holds in (4.19). The condition $\dot{V}(t)+z^T(t)z(t)-\gamma^2w^T(t)w(t)<0$ can guarantee that the closed-loop system (4.7) satisfies the H_∞ performance. In addition, when $w(t)=0$, from the conditions in Theorem 4.2, it also can be proved that the system (4.7) is asymptotically stable by following the above same line. The proof is completed. □

4.3.2 Sampled-Data Fault-Tolerant Control

In the following theorem, the control gain matrices K_{aj} in (4.7) can be obtained based on Theorem 4.2.

Theorem 4.3 *For a scalar $h_M > 0$, system (4.38) is asymptotically stable with H_∞ performance, if there exist matrices $\hat{P}^T = \hat{P}^T > 0$, $\hat{Q} = \hat{Q}^T$, $\hat{\Delta}_i = \hat{\Delta}_i^T$, any appropriate dimensioned matrices G, N_1^T, N_2^T, and Y_{aj}, the parameters a and b, and the condition $\eta_j - \lambda_j\theta_j > 0$ such that the following LMIs hold for all i, j:*

$$\lambda_j\hat{\Pi}_{1ij} + \lambda_i\hat{\Pi}_{1ji} + \lambda_j h_M\hat{\Pi}_{2ij} + \lambda_i h_M\hat{\Pi}_{2ji} + \hat{\Psi}_0 < 0, \quad i < j, \tag{4.30}$$

$$\lambda_i\hat{\Pi}_{1ii} + \lambda_i h_M\hat{\Pi}_{2ii} - \lambda_i\hat{\Delta}_i + \hat{\Delta}_i < 0, \tag{4.31}$$

$$\hat{\Pi}_{1ij} + h_M\hat{\Pi}_{2ij} - \hat{\Delta}_i < 0, \tag{4.32}$$

$$\begin{bmatrix} \lambda_j\hat{\Pi}_{1ij} + \lambda_i\hat{\Pi}_{1ji} + \hat{\Psi}_0 & -\lambda_j h_M\hat{N} - \lambda_i h_M\hat{N} \\ * & -\lambda_j h_M\hat{R} - \lambda_i h_M\hat{R} \end{bmatrix} < 0, \quad i < j, \tag{4.33}$$

$$\begin{bmatrix} \lambda_i\hat{\Pi}_{1ii} - \lambda_i\hat{\Delta}_i + \hat{\Delta}_i & -\lambda_i h_M\hat{N} \\ * & -\lambda_i h_M\hat{R} \end{bmatrix} < 0, \tag{4.34}$$

$$\begin{bmatrix} \hat{\Pi}_{1ij} - \hat{\Delta}_i & -h_M\hat{N} \\ * & -h_M\hat{R} \end{bmatrix} < 0, \tag{4.35}$$

where

$$\hat{\Pi}_{1ij} = \begin{bmatrix} \hat{\Psi}_{11ij} & \hat{\Psi}_{12ij} & \hat{\Psi}_{13ij} & B_{1i} & GC_i^T & \varepsilon B_i & 0 \\ * & \hat{\Psi}_{22} & \hat{\Psi}_{23ij} & aB_{1i} & Y_{aj}^T D_i^T & a\varepsilon B_i & Y_{aj}^T \\ * & * & \hat{\Psi}_{33ij} & bB_{1i} & 0 & b\varepsilon B_i & 0 \\ * & * & * & -\gamma^2 I & D_{1i}^T & 0 & 0 \\ * & * & * & * & -I & \varepsilon D_i^T & 0 \\ * & * & * & * & * & -\varepsilon J_a^{-1} & 0 \\ * & * & * & * & * & * & -\varepsilon J_a^{-1} \end{bmatrix},$$

$$\hat{\Pi}_{2ij} = \begin{bmatrix} 0 & 0 & \hat{Q}_1 & 0_{1\times 4} \\ * & 0 & \hat{Q}_2^T & 0_{1\times 4} \\ * & * & \hat{R} & 0_{1\times 4} \\ * & * & * & 0_{4\times 4} \end{bmatrix}, \quad \hat{Q} = \begin{bmatrix} \hat{Q}_1 & \hat{Q}_2 \\ * & \hat{Q}_3 \end{bmatrix},$$

$$\hat{\Psi}_0 = -\lambda_j\hat{\Delta}_i - \lambda_i\hat{\Delta}_j + \hat{\Delta}_i + \hat{\Delta}_j, \quad \hat{\Psi}_{11ij} = -\hat{Q}_1 + \mathbf{He}\left(\hat{N}_1 + A_i G^T\right),$$

$$\hat{\Psi}_{12ij} = -\hat{Q}_2 - \hat{N}_1 + \hat{N}_2^T + B_i Y_{aj} + aGA_i^T, \quad \hat{\Psi}_{13ij} = \hat{P} + bGA_i^T - G^T,$$

$$\hat{\Psi}_{22} = -\hat{Q}_3 + \mathbf{He}\left(aB_i Y_{aj} - \hat{N}_2\right), \quad \hat{\Psi}_{23ij} = bY_{aj}^T B_i^T - aG^T,$$

$$\hat{\Psi}_{33ij} = -b\left(G + G^T\right), \quad \hat{N} = \left[\hat{N}_1^T \; \hat{N}_2^T \; 0_{5\times 1}\right]^T,$$

and the IT2 fuzzy control gain matrices can be given by $K_{aj} = M_{a0}^{-1} Y_{aj} G^{-T}$.

Proof Firstly, we define the following new variables:

$$\begin{aligned} &L_1 = G^{-1}, \quad L_2 = aG^{-1}, \quad L_3 = bG^{-1}, \\ &\hat{P} = GPG^T, \quad \hat{N}_1 = GN_1G^T, \quad \hat{N}_2 = GN_2G^T, \quad \hat{R} = GRG^T, \\ &Y_{aj} = M_{a0}K_{aj}G^T, \quad \hat{\Delta}_i = E\text{diag}\{\Delta_i, 0, 0\}E^T, \\ &E_1 = \text{diag}\{G, G, G, I, I, I, I\}, \quad E_2 = \text{diag}\{G, G, G, I, I, I, I, G\}. \end{aligned}$$

Pre- and post-multiply (4.30)–(4.32) by E_1^{-1} and E_1^{-T}, and pre- and post-multiply (4.33)–(4.35) by E_2^{-1} and E_2^{-T}, respectively. Then the conditions in (4.8)–(4.13) hold. Therefore, all the conditions in Theorem 4.3 are satisfied. The proof is completed. □

Consider the system (4.1) with no actuator fault, the overall IT2 fuzzy model is inferred as follows:

$$\begin{cases} \dot{x}(t) = \sum_{i=1}^{r} \theta_i(x(t))(A_i x(t) + B_i u(t) + B_{1i} w(t)), \\ z(t) = \sum_{i=1}^{r} \theta_i(x(t))(C_i x(t) + D_i u(t) + D_{1i} w(t)). \end{cases} \tag{4.36}$$

Similar to the controller (4.6), the overall IT2 fuzzy sampled-data control law is described as

$$u(t) = \sum_{j=1}^{r} \eta_j(x) K_j x(t - h(t)). \tag{4.37}$$

The closed-loop system under the case of standard controller (4.37) is represented as

$$\begin{cases} \dot{x}(t) = \sum_{i=1}^{r}\sum_{j=1}^{r} \theta_i\eta_j \left(A_i x(t) + B_i K_j x(t - h(t)) + B_{1i} w(t)\right), \\ z(t) = \sum_{i=1}^{r}\sum_{j=1}^{r} \theta_i\eta_j \left(C_i x(t) + D_i K_j (x(t - h(t))) + D_{1i} w(t)\right). \end{cases} \tag{4.38}$$

In the following theorem, the standard controller gain matrices K_j for system (4.38) with no actuator fault can be solved. Considering the Lyapunov function defined in (4.27) and following Theorems 4.2 and 4.3, the following theorem can be presented directly.

Theorem 4.4 *For a given scalar $h_M > 0$, system (4.38) is asymptotically stable with H_∞ performance, and the IT2 fuzzy control gain matrix can be given by $K_j = X_j G^{-T}$ $(j = 1, 2, \ldots, r)$, if there exist matrices $\hat{P} = \hat{P}^T > 0$, $\hat{Q} = \hat{Q}^T$, $\hat{\Delta}_i = \hat{\Delta}_i^T$, and any appropriate dimensioned matrices G, $\hat{N} = \left[\hat{N}_1^T\ \hat{N}_2^T\ 0\ 0\ 0\right]^T$, and X_j, the parameters a and b, and the condition $\eta_j - \lambda_j\theta_j > 0$ such that the following LMIs hold:*

$$\lambda_j\hat{\Psi}_{1ij} + \lambda_i\hat{\Psi}_{1ji} + \lambda_j h_M\hat{\Psi}_{2ij} + \lambda_i h_M\hat{\Psi}_{2ji} + \hat{\Psi}_0 < 0, \quad i < j, \tag{4.39}$$

$$\lambda_i\hat{\Psi}_{1ii} + \lambda_i h_M\hat{\Psi}_{2ii} - \lambda_i\hat{\Delta}_i + \hat{\Delta}_i < 0, \tag{4.40}$$

$$\hat{\Psi}_{1ij} + h_M\hat{\Psi}_{2ij} - \hat{\Delta}_i < 0, \tag{4.41}$$

$$\begin{bmatrix} \lambda_j\hat{\Psi}_{1ij} + \lambda_i\hat{\Psi}_{1ji} + \hat{\Psi}_0 & -\lambda_j h_M\hat{N} - \lambda_i h_M\hat{N} \\ * & -\lambda_j h_M\hat{R} - \lambda_i h_M\hat{R} \end{bmatrix} < 0, \quad i < j, \tag{4.42}$$

$$\begin{bmatrix} \lambda_i\hat{\Psi}_{1ii} - \lambda_i\hat{\Delta}_i + \hat{\Delta}_i & -\lambda_i h_M\hat{N} \\ * & -\lambda_i h_M\hat{R} \end{bmatrix} < 0, \tag{4.43}$$

$$\begin{bmatrix} \hat{\Psi}_{1ij} - \hat{\Delta}_i & -h_M\hat{N} \\ * & -h_M\hat{R} \end{bmatrix} < 0, \tag{4.44}$$

where

$$\hat{\Psi}_{1ij} = \begin{bmatrix} \hat{\Psi}_{11ij} & \hat{\Psi}_{12ij} & \hat{\Psi}_{13ij} & B_{1i} & G_1C_i^T \\ * & \hat{\Psi}_{22} & \hat{\Psi}_{23ij} & aB_{1i} & X_j^TD_i^T \\ * & * & \hat{\Psi}_{33ij} & bB_{1i} & 0 \\ * & * & * & -\gamma^2 I & D_{1i}^T \\ * & * & * & * & -I \end{bmatrix}, \quad \hat{\Phi}_{2ij} = \begin{bmatrix} 0 & 0 & \hat{Q}_1 & 0 & 0 \\ * & 0 & \hat{Q}_2^T & 0 & 0 \\ * & * & \hat{R} & 0 & 0 \\ * & * & * & 0 & 0 \\ * & * & * & * & 0 \end{bmatrix},$$

$$\hat{\Psi}_0 = -\lambda_j\hat{\Delta}_i - \lambda_i\hat{\Delta}_j + \hat{\Delta}_i + \hat{\Delta}_j,$$

$$\hat{\Psi}_{11ij} = -\hat{Q}_1 + \mathbf{He}\left(\hat{N}_1 + A_iG_1^T\right),$$

$$\hat{\Psi}_{12ij} = -\hat{Q}_2 - \hat{N}_1 + \hat{N}_2^T + B_iX_j + aG_1A_i^T,$$

$$\hat{\Psi}_{13ij} = \hat{P} + bG_1A_i^T - G_1^T, \quad \hat{\Psi}_{22} = -\hat{Q}_3 + \mathbf{He}\left(aB_iX_j - \hat{N}_2\right),$$

$$\hat{\Psi}_{23ij} = bX_j^TB_i^T - aG_1^T, \quad \hat{\Psi}_{33ij} = \mathbf{He}\left(-bG_1\right), \quad \hat{Q} = \begin{bmatrix} \hat{Q}_1 & \hat{Q}_2 \\ * & \hat{Q}_3 \end{bmatrix}.$$

Remark 4.5 In this chapter, the sampled-data controller for IT2 fuzzy systems with actuator fault is designed to guarantee that the closed-loop system is asymptotically stable and satisfy the H_∞ performance when the actuator experiences failure. It should be mentioned that this chapter first presents the reliable control design method for IT2 fuzzy systems with actuator fault. Furthermore, the sampled-data controller is also the first time to be considered in the IT2 fuzzy systems.

4.4 Simulation Results

Example 4.6 This section gives an example to show the effectiveness of the proposed results. Consider an inverted pendulum shown in Fig. 2.1 subject to parameter uncertainties [94]. The dynamic equation for the inverted pendulum is given by (2.24). It is assumed that $m_{p\min} = 2\,\text{kg} \le m_p \le 5\,\text{kg} = m_{p\max}$ and $M_{c\min} = 8\,\text{kg} \le M_c \le 18\,\text{kg}$

$= M_{c\max}$, respectively. $a = 1/\left(m_p + M_c\right)$, $2L = 1$ m is the length of the pendulum. $u(t)$ is the force (N) applied to the cart. The operating domain of the inverted pendulum is given by $x_1(t) = \theta(t) \in \left[-\frac{5\pi}{12}, \frac{5\pi}{12}\right]$, and $x_2(t) = \dot{\theta}(t) \in [-5, 5]$. We can obtain a 4-rule IT2 fuzzy model to describe the inverted pendulum subject to parameter uncertainties in the following format:

Plant Rule i: IF $x_1(t)$ is M_{i1} and $x_2(t)$ is M_{i2}, THEN

$$\dot{x}(t) = A_i x(t) + B_i u(t), \quad i = 1, 2, 3, 4, \tag{4.45}$$

where

$$A_1 = A_2 = \begin{bmatrix} 0 & 1 \\ f_{1\min} & 0 \end{bmatrix}, \quad A_3 = A_4 = \begin{bmatrix} 0 & 1 \\ f_{1\max} & 0 \end{bmatrix},$$
$$B_1 = B_3 = \begin{bmatrix} 0 \\ f_{2\min} \end{bmatrix}, \quad B_2 = B_4 = \begin{bmatrix} 0 \\ f_{2\max} \end{bmatrix},$$
$$f_{1\min} = 10.0078, \quad f_{1\max} = 18.4800, \quad f_{2\min} = -0.1765, \quad f_{2\max} = -0.0261.$$

In order to utilize this system to demonstrate the effectiveness of the proposed design results, we shall give other parameters in system (4.1) as follows:

$$B_{11} = B_{13} = \begin{bmatrix} 0.5 \\ 0.1 \end{bmatrix}, \quad B_{12} = B_{14} = \begin{bmatrix} -0.5 \\ -0.1 \end{bmatrix}, \quad C_1 = C_3 = \begin{bmatrix} 0.1 & 0.1 \end{bmatrix},$$
$$C_2 = C_4 = \begin{bmatrix} -0.1 & 0.1 \end{bmatrix}, \quad D_1 = D_2 = 0.1, \quad D_3 = D_4 = 0.2,$$
$$D_{11} = D_{13} = 0.1, \quad D_{12} = D_{14} = 0.2.$$

The LMFs and UMFs are defined in Table 4.1.

Table 4.1 LMFs and UMFs of the IT2 T–S fuzzy model of inverted pendulum

LMFs	UMFs
$\underline{\mu}_{W_{11}}(x_1) = 1 - e^{-\frac{x_1^2}{1.2}}$	$\overline{\mu}_{W_{11}}(x_1) = 1 - 0.23e^{-\frac{x_1^2}{0.25}}$
$\underline{\mu}_{W_{12}}(x_1) = 1 - e^{-\frac{x_1^2}{1.2}}$	$\overline{\mu}_{W_{12}}(x_1) = 1 - 0.23e^{-\frac{x_1^2}{0.25}}$
$\underline{\mu}_{W_{13}}(x_1) = 0.23e^{-\frac{x_1^2}{0.25}}$	$\overline{\mu}_{W_{13}}(x_1) = e^{-\frac{x_1^2}{1.2}}$
$\underline{\mu}_{W_{14}}(x_1) = 0.23e^{-\frac{x_1^2}{0.25}}$	$\overline{\mu}_{W_{14}}(x_1) = e^{-\frac{x_1^2}{1.2}}$
$\underline{\mu}_{W_{21}}(x_1) = 0.5e^{-\frac{x_1^2}{0.25}}$	$\overline{\mu}_{W_{21}}(x_1) = e^{-\frac{x_1^2}{1.5}}$
$\underline{\mu}_{W_{22}}(x_1) = 1 - e^{-\frac{x_1^2}{1.5}}$	$\overline{\mu}_{W_{22}}(x_1) = 1 - 0.5e^{-\frac{x_1^2}{0.25}}$
$\underline{\mu}_{W_{23}}(x_1) = 0.5e^{-\frac{x_1^2}{0.25}}$	$\overline{\mu}_{W_{23}}(x_1) = e^{-\frac{x_1^2}{1.5}}$
$\underline{\mu}_{W_{24}}(x_1) = 1 - e^{-\frac{x_1^2}{1.5}}$	$\overline{\mu}_{W_{24}}(x_1) = 1 - 0.5e^{-\frac{x_1^2}{0.25}}$

Under the initial condition $x(0) = \left[\frac{5}{12}\pi \; 0\right]^T$, Fig. 4.1 indicates the state responses of the open-loop system and shows that this system is not stable. Then, the state-feedback controller will be designed to guarantee that the system (4.45) is asymptotically stable and satisfies the H_∞ performance in the following part.

Firstly, we consider the standard sampled-data fuzzy control design problem of this system. A 4-rule IT2 fuzzy controller is employed to stabilize the inverted pendulum with the LMFs and UMFs chosen as $\underline{\eta}_1 = \underline{\mu}_{M_{11}} = \overline{\eta}_1 = \overline{\mu}_{M_{11}} = e^{-\frac{3-x_1^2}{2.5}}$, $\underline{\eta}_2 = \underline{\mu}_{M_{12}} = \overline{\eta}_2 = \overline{\mu}_{M_{12}} = e^{-\frac{2-x_1^2}{2.5}}$, $\underline{\eta}_3 = \underline{\mu}_{M_{13}} = \overline{\eta}_3 = \overline{\mu}_{M_{13}} = e^{-\frac{1-x_1^2}{2.5}}$, $\underline{\eta}_4 = \underline{\mu}_{M_{14}} = \overline{\eta}_4 = \overline{\mu}_{M_{14}} = 1 - \overline{\mu}_{M_{11}} - \overline{\mu}_{M_{12}} - \overline{\mu}_{M_{13}}$ and $\alpha_j = 0.5$. The parameters $\bar{\sigma}_k$ $(k = 1, 2, 3)$ are chosen as $\lambda_1 = 0.1$, $\lambda_2 = 0.2$, $\lambda_3 = 0.3$, $\lambda_4 = 0.4$. Choosing $\gamma = 0.05$, $h_M = 0.05$ and applying Theorem 4.4 with $a = 0.1$ and $b = 0.2$, the state-feedback gain matrices are given as follows:

$$K_1 = \left[494.7782 \; 107.8609\right], \quad K_2 = \left[970.0731 \; 247.7838\right],$$
$$K_3 = \left[801.5091 \; 189.3779\right], \quad K_4 = \left[1114.4674 \; 295.7482\right].$$

Under the initial condition $x(0) = \left[\frac{5\pi}{12} \; 0\right]^T$, Fig. 4.2 plots the state trajectories of the closed-loop system and shows that the closed-loop system is asymptotically stable. This shows the effectiveness of the proposed sampled-data fuzzy control design method.

In the following part, it will be shown that the proposed reliable fuzzy H_∞ controller is effective for this system with actuator fault. If there exists an actuator fault M_a satisfying $\underline{m}_{a01} = 0.2$, and $\overline{m}_{a01} = 0.8$. It is concluded that $M_{a0} = 0.5$. The parameters $\bar{\sigma}_k$ $(k = 1, 2, 3)$ are chosen as $\lambda_1 = 0.1$, $\lambda_2 = 0.2$, $\lambda_3 = 0.3$, $\lambda_4 = 0.4$.

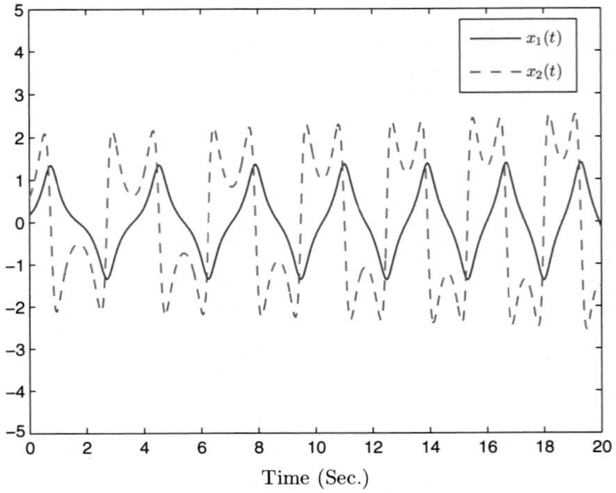

Fig. 4.1 States of the open-loop system

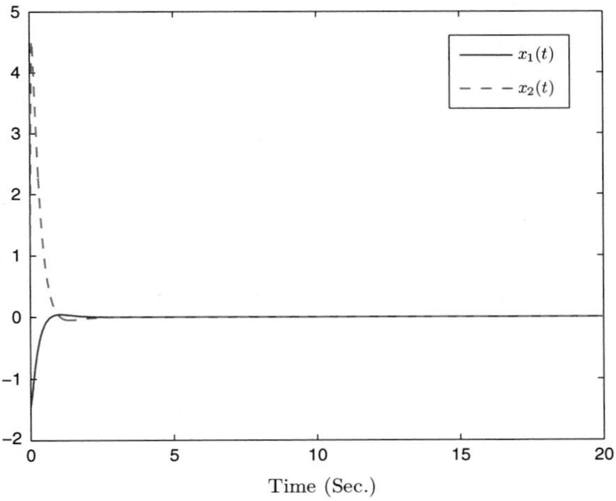

Fig. 4.2 States of the closed-loop system

Choosing $\gamma = 0.05, h_M = 0.05$ and applying Theorem 4.3 with $a = 0.1$ and $b = 0.2$, the reliable fuzzy controller gain matrices can be found as follows:

$$K_{a1} = \begin{bmatrix} 508.9578 & 132.7518 \end{bmatrix}, \quad K_{a2} = \begin{bmatrix} 1193.93658 & 304.9654 \end{bmatrix},$$
$$K_{a3} = \begin{bmatrix} 986.4728 & 233.0805 \end{bmatrix}, \quad K_{a4} = \begin{bmatrix} 1371.6507 & 363.9984 \end{bmatrix}.$$

Figures 4.3 and 4.4 show the state responses of closed-loop system with the actuator fault under the different controllers. From Figs. 4.2, 4.3 and 4.4, it is obtained that the designed controller for the IT2 fuzzy system with actuator fault can stabilize the system well. On the other hand, the proposed method in this chapter can deal with the uncertainties in membership functions well.

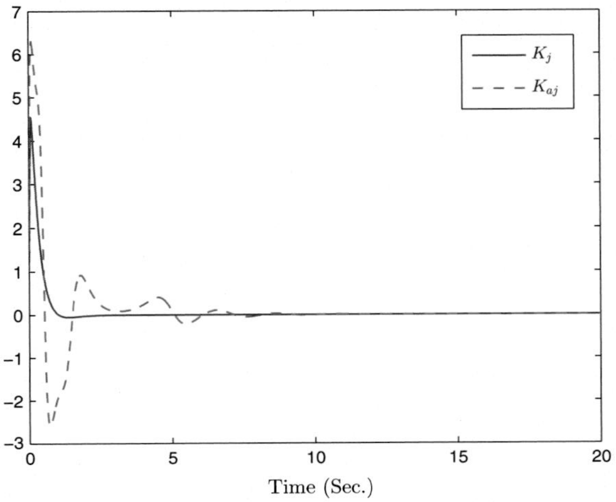

Fig. 4.3 State $x_1(t)$ under the standard controller and the reliable controller

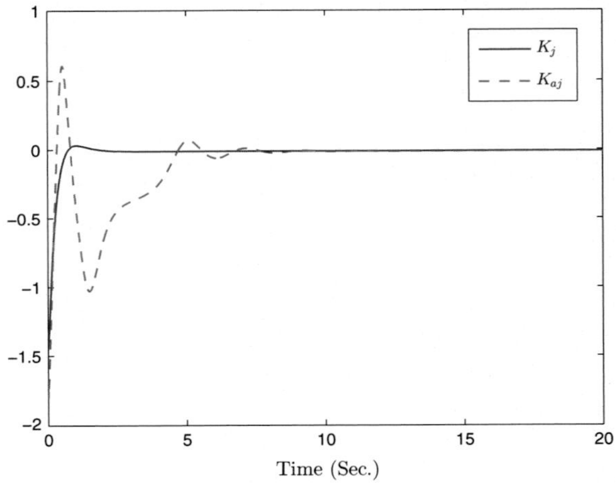

Fig. 4.4 State $x_2(t)$ under the standard controller and the reliable controller

4.5 Conclusion

In this chapter, the problem of state-feedback sampled-data controller design has been investigated for IT2 fuzzy systems with actuator fault. The IT2 fuzzy system and the IT2 sampled-data controller do not need to share the same membership functions. By considering the mismatched membership functions, the IT2 fuzzy model and the IT2 state-feedback sampled-data controller have been firstly constructed. Secondly, using some new techniques, a novel reliable IT2 state-feedback sampled-data controller has been designed such that the closed-loop system is asymptotically stable for all possible actuator failures. The existence condition of the IT2 fuzzy H_∞ sampled-data controller has been expressed by a convex optimization problem. Furthermore, the standard IT2 fuzzy H_∞ sampled-data controller has been designed. Finally, an inverted pendulum model has been used to demonstrate the effectiveness of the proposed results.

Chapter 5
Output Tracking Control of Interval Type-2 Fuzzy-Model-Based Systems

5.1 Introduction

This chapter is concerned with the output tracking control problem for continuous-time nonlinear systems via IT2 fuzzy model approach. The IT2 fuzzy system can deal with uncertain grades of membership well if parameter uncertainties of nonlinear plants are considered, and the IT2 fuzzy controller can provide better performance. The IT2 fuzzy systems and the IT2 output tracking controller do not share the same membership functions. Firstly, an IT2 fuzzy system and output tracking controller are constructed respectively. The IT2 FMB control system is subject to imperfect premise membership functions, the LMFs and UMFs characterizing the FOU are chosen to be a favorable representation. This favorable representation allows the LMFs and UMFs to be taken into the stability analysis. Therefore, the stability conditions are membership functions dependent. The result in [93] provides the technical support to the work in this chapter. Then, we divide the FOU into a number of sub-FOUs to further relax the stability conditions. The information of the sub-FOUs along with those of LMFs and UMFs are brought to the stability analysis. Secondly, based on Lyapunov stability theory, an IT2 fuzzy controller is designed to guarantee that the output of the system can track the output of a given reference model in the H_∞ sense. The existence condition of the IT2 fuzzy controller can be expressed by a convex optimization problem.

5.2 System Description and Preliminaries

Consider the following continuous-time IT2 fuzzy system:

♦ **Plant Form**:

Rule i: IF $f_1(x(t))$ is F_1^i and ... and $f_\psi(x(t))$ is F_ψ^i, THEN,

H. Li et al., *Analysis and Synthesis for Interval Type-2 Fuzzy-Model-Based Systems*, DOI 10.1007/978-981-10-0593-0_5

$$\begin{cases} \dot{x}(t) = A_i x(t) + B_{wi} w(t) + B_{ui} u^f(t), \\ y(t) = C_i x(t) + D_{ui} u^f(t), \end{cases} \tag{5.1}$$

where $x(t) \in \mathbf{R}^{n_x}$, $y(t) \in \mathbf{R}^{n_y}$, $w(t) \in \mathbf{R}^{n_w}$ and $u^f(t) \in \mathbf{R}^{n_u}$ stand for the state, the output, the bounded external disturbance and the fault control input vector, respectively; A_i, B_{wi}, B_{ui}, C_i and D_{ui} are system matrices with appropriate dimensions. $f_\nu(x(t))$ and F_ν^i are the premise variables and the IT2 fuzzy sets ($i = 1, 2, \ldots, p, \nu = 1, 2, \ldots, \psi$), respectively, where p is the number of IT2 IF-THEN rules and ψ is the number of premise variables. The firing strength of the ith rule is expressed by the following interval set,

$$\begin{aligned} \tilde{\theta}_i x(t) &= \left[\prod_{\nu=1}^{\psi} \underline{\mu}_{F_\nu^i}(f_\nu(x(t))), \prod_{\nu=1}^{\psi} \overline{\mu}_{F_\nu^i}(f_\nu(x(t))) \right] \\ &= \left[\underline{\theta}_i(x(t)), \overline{\theta}_i(x(t)) \right], \end{aligned}$$

where $\underline{\theta}_i(x(t))$ and $\overline{\theta}_i(x(t))$ are the lower and upper grades of membership, the LMFs and UMFs are denoted as $\underline{\mu}_{F_\nu^i}(f_\nu(x(t)))$ and $\overline{\mu}_{F_\nu^i}(f_\nu(x(t)))$, respectively. Hence, the overall fuzzy model is inferred as

$$\begin{cases} \dot{x}(t) = \sum_{i=1}^{p} (\theta_i(x(t))) \left(A_i x(t) + B_{wi} w(t) + B_{ui} u^f(t)\right), \\ y(t) = \sum_{i=1}^{p} (\theta_i(x(t))) \left(C_i x(t) + D_{ui} u^f(t)\right), \end{cases} \tag{5.2}$$

where $\theta_i(x(t)) = \underline{\alpha}_i(x(t)) \underline{\theta}_i(x(t)) + \overline{\alpha}_i(x(t)) \overline{\theta}_i(x(t))$. $\underline{\alpha}_i(x(t))$, $\overline{\alpha}_i(x(t)) \in [0, 1]$. $\underline{\alpha}_i(x(t)) + \overline{\alpha}_i(x(t)) = 1$, $\underline{\alpha}_i(x(t))$ and $\overline{\alpha}_i(x(t))$ are nonlinear functions. It is obvious that $\sum_{i=1}^{p} \theta_i(x(t)) = 1$ with $0 \le \theta_i(x(t)) \le 1$. Consider the actuator failure model with failure matrix M_a,

$$u^f(t) = M_a u(t). \tag{5.3}$$

The actuator fault matrix $M_a = \text{diag}\{m_{a1}, m_{a2}, \ldots, m_{an_u}\}$, where $0 \le \underline{m}_{ai} \le m_{ai} \le \overline{m}_{ai} \le 1$, and $\underline{m}_{ai}$ and $\overline{m}_{ai}$ ($i = 1, 2, \ldots, n_u$) are the constant scalars which denote the admissible failures of the actuator. m_{ai} represents the possible fault of the corresponding actuator $u_i^f(t)$. We consider the following three different cases of actuator failure.

1. When $\underline{m}_{ai} = \overline{m}_{ai} = 0$, we obtain $m_{ai} = 0$, which indicates that the corresponding actuator $u_i^f(t)$ has completely failed.
2. If $\underline{m}_{ai} = \overline{m}_{ai} = 1$, then, $m_{ai} = 1$, which implies that there is no failure in the actuator $u_i^f(t)$.
3. When $0 \le \underline{m}_{ai} \le \overline{m}_{ai} \le 1$, it means that there exists a partial fault in the corresponding actuator $u_i^f(t)$.

In this chapter, the purpose is to design a control scheme for the IT2 fuzzy system with actuator fault in (5.2) such that the output tracks a reference signal to meet a desired tracking performance. Suppose the reference signal $y_d(t)$ is generated by

$$\begin{cases} \dot{x}_d(t) = Gx_d(t) + r(t), \\ y_d(t) = Hx_d(t), \end{cases} \tag{5.4}$$

where y_d has the same dimension as y, x_d, $r \in \mathbf{R}^n$ are, respectively, the reference state and the bounded reference input, G and H are appropriately dimensional constant matrices with G Hurwitz. Then, the state-feedback controller is constructed as follows:

Rule j: IF $g_1(x(t))$ is W_1^j and ... and $g_\varphi(x(t))$ is W_φ^j, THEN:

$$u(t) = K_j x(t) + K_{dj} x_d(t), \tag{5.5}$$

where W_s^j denotes the fuzzy set of rule j corresponding to the function $g_s(x(t))$, $j = 1, 2, \ldots, q, s = 1, 2, \ldots, \varphi$ and q is a positive integer. K_j and K_{dj} are the state-feedback control gains of rule j. The following interval set is expressed the firing strength of the jth rule.

$$\begin{aligned} \tilde{\eta}_j(x(t)) &= \left[\prod_{s=1}^{\varphi} \underline{\mu}_{W_s^j}(g_s(x(t))), \prod_{s=1}^{\varphi} \overline{\mu}_{W_s^j}(g_s(x(t))) \right] \\ &= \left[\underline{\eta}_j(x(t)), \overline{\eta}_j(x(t)) \right], \end{aligned}$$

where $\underline{\eta}_j(x(t))$ and $\overline{\eta}_j(x(t))$ are the lower and upper grades of membership, respectively. The LMFs and UMFs are denoted as $\underline{\mu}_{W_s^j}(g_s(x(t)))$ and $\overline{\mu}_{W_s^j}(g_s(x(t)))$. The overall IT2 fuzzy control law is represented by

$$u(t) = \sum_{j=1}^{q} \eta_j(x(t)) \left(K_j x(t) + K_{dj} x_r(t) \right), \tag{5.6}$$

where

$$\eta_j(x(t)) = \frac{\underline{\beta}_j(x(t))\underline{\eta}_j(x(t)) + \overline{\beta}_j(x(t))\overline{\eta}_j(x(t))}{\sum_{l=1}^{q} \left(\underline{\beta}_j(x(t))\underline{\eta}_j(x(t)) + \overline{\beta}_j(x(t))\overline{\eta}_j(x(t)) \right)},$$

with $\underline{\beta}_j(x(t))$, $\overline{\beta}_j(x(t)) \in [0, 1]$, and $\underline{\beta}_j(x(t)) + \overline{\beta}_j(x(t)) = 1$, where $\underline{\beta}_j(x(t))$ and $\overline{\beta}_j(x(t))$ are nonlinear functions. Obviously, $\sum_{j=1}^{q} \eta_j(x(t)) = 1$ with $1 \geq \eta_j(x(t)) \geq 0$.

Therefore, from (5.1)–(5.4), the following augmented closed-loop system can be obtained:

$$\begin{cases} \dot{\xi}(t) = \sum_{i=1}^{p} \sum_{j=1}^{q} \theta_i(x(t)) \eta_j(x(t)) \left(\bar{A}_{ij} \xi(t) + \bar{B}_{ij} v(t) \right), \\ e(t) = \sum_{i=1}^{p} \sum_{j=1}^{q} \theta_i(x(t)) \eta_j(x(t)) \bar{C}_{ij} \xi(t), \end{cases} \tag{5.7}$$

where

$$\begin{aligned} \xi(t) &= \left[x^T(t) \; x_d^T(t) \right]^T, \quad e(t) = y(t) - y_d(t), \\ \bar{A}_{ij} &= \begin{bmatrix} A_i + B_{ui} M_a K_j & B_{ui} M_a K_{dj} \\ 0 & G \end{bmatrix}, \quad \bar{B}_{ij} = \begin{bmatrix} B_{wi} & 0 \\ 0 & I \end{bmatrix}, \\ \bar{C}_{ij} &= \left[C_i + D_{ui} M_a K_j \; D_{ui} M_a K_{dj} - H \right], \quad v(t) = \left[w^T(t) \; r^T(t) \right]^T. \end{aligned}$$

Next, following tracking requirements are given.

1. The closed-loop system in (5.7) with $v(t) \equiv 0$ is asymptotically stable;
2. The attenuation for the effect of $v(t)$ on the tracking error $e(t)$ is below the following desire level

$$\int_0^\infty e^T(t) e(t) \, \mathrm{dt} \le \gamma^2 \int_0^\infty v^T(t) v(t) \, \mathrm{dt}, \tag{5.8}$$

for all nonzero $v \in \mathcal{L}_2[0, \infty)$, where $\gamma > 0$ is a performance index. More specifically, it is required that

$$\|e(t)\|_2 < \gamma \|v(t)\|_2 . \tag{5.9}$$

Then, if the above two requirements are satisfied, then we can say that the H_∞ output tracking performance index γ is achieved.

To facilitate the stability analysis and controller synthesis of the IT2 control system (5.7), the state space of interest denoted as Φ is divided into c connected sub-state spaces denoted as $\Phi_k, k = 1, 2, \ldots, c$ such that $\Phi = \bigcup_{k=1}^{c} \Phi_k$. Furthermore, to consider more information of the IT2 membership functions, local LMFs and UMFs within the FOU are introduced. Considering the FOU divided into $\tau + 1$ sub-FOUs, in the lth sub-FOU, $l = 1, 2, \ldots, \tau + 1$, the LMFs and UMFs are defined as follows ($\forall i, j, k, l$):

$$\underline{\omega}_{ijl}(x(t)) = \sum_{k=1}^{c} \sum_{i_1=1}^{2} \cdots \sum_{i_n=1}^{2} \prod_{r=1}^{n} \rho_{ri_r kl}(x_r(t)) \underline{\varrho}_{iji_1 i_2 \ldots i_n kl}, \tag{5.10}$$

$$\overline{\omega}_{ijl}(x(t)) = \sum_{k=1}^{c} \sum_{i_1=1}^{2} \cdots \sum_{i_n=1}^{2} \prod_{r=1}^{n} \rho_{ri_r kl}(x_r(t)) \overline{\varrho}_{iji_1 i_2 \ldots i_n kl}, \tag{5.11}$$

$$0 \le \underline{\omega}_{ijl}(x(t)) \le \overline{\omega}_{ijl}(x(t)) \le 1,$$
$$0 \le \underline{\varrho}_{iji_1i_2\ldots i_nkl} \le \overline{\varrho}_{iji_1i_2\ldots i_nkl} \le 1,$$

where $\underline{\varrho}_{iji_1i_2\ldots i_nkl}$ and $\overline{\varrho}_{iji_1i_2\ldots i_nkl}$ are constant scalars to be determined, $0 \le \rho_{ri_skl}(x_r(t)) \le 1$ and $\rho_{r1kl}(x_r(t)) + \rho_{r2kl}(x_r(t)) = 1$ for $r, s = 1, 2, \ldots, n$, l=1, 2, ..., $\tau + 1; i_r = 1, 2; x(t) \in \Phi_k$, otherwise, $\rho_{ri_skl}(x_r(t)) = 0$. It follows that $\sum_{k=1}^{c}\sum_{i_1=1}^{2}\cdots\sum_{i_n=1}^{2}\prod_{r=1}^{n}\rho_{ri_rkl}(x_r(t)) = 1$ for all l. Then, the closed-loop IT2 system (5.7) is redescribed as the following favorable form:

$$\begin{cases} \dot{\xi}(t) = \sum_{i=1}^{p}\sum_{j=1}^{q}\omega_{ij}(x(t))\left(\bar{A}_{ij}\xi(t) + \bar{B}_{ij}v(t)\right), \\ e(t) = \sum_{i=1}^{p}\sum_{j=1}^{q}\omega_{ij}(x(t))\bar{C}_{ij}\xi(t), \end{cases}$$

where

$$\begin{aligned} \omega_{ij}(x(t)) &\equiv \theta_i(x(t))\eta_j(x(t)) \\ &= \sum_{l=1}^{\tau+1}\varsigma_{ijl}(x(t))\left[\underline{\gamma}_{ijl}(x(t))\underline{\omega}_{ijl}(x(t)) + \overline{\gamma}_{ijl}(x(t))\overline{\omega}_{ijl}(x(t))\right], \end{aligned} \tag{5.12}$$

with $\sum_{i=1}^{p}\sum_{j=1}^{q}\omega_{ij}(x(t)) = 1$, and $0 \le \underline{\gamma}_{ijl}(x(t)) \le \overline{\gamma}_{ijl}(x(t)) \le 1$ are two functions, which are not necessary to be known, exhibiting the property that $\underline{\gamma}_{ijl}(x(t)) + \overline{\gamma}_{ijl}(x(t)) = 1$ for all i, j and l; $\varsigma_{ijl}(x(t)) = 1$ if the membership function $\omega_{ijl}(x(t))$ is within the sub-FOU l, otherwise, $\varsigma_{ijl}(x(t)) = 0$. For brevity, in the following part, the variables $\underline{\theta}_i(x(t))$, $\overline{\theta}_i(x(t))$, $\underline{\eta}_i(x(t))$, $\overline{\eta}_j(x(t))$, $\omega_{ij}(x(t))$, $\underline{\omega}_{ijl}(x(t))$, $\overline{\omega}_{ijl}(x(t))$, $\underline{\gamma}_{ijl}(x(t))$, $\overline{\gamma}_{ijl}(x(t))$, $\rho_{1i_1kl}(x_1(t))$, $\rho_{2i_2kl}(x_2(t))$, ..., $\rho_{ni_nkl}(x_n(t))$ and $\varsigma_{ijl}(x(t))$ are denoted by $\underline{\theta}_i, \overline{\theta}_i, \underline{\eta}_i, \overline{\eta}_j, \omega_{ij}, \underline{\omega}_{ijl}, \overline{\omega}_{ijl}, \underline{\gamma}_{ijl}, \overline{\gamma}_{ijl}, \rho_{1i_1kl}, \rho_{2i_2kl}, \ldots, \rho_{ni_nkl}$ and ς_{ijl}, respectively.

5.3 Main Results

In this section, we first introduce the following matrices, which will be used in the later part.

$$M_{a0} = \mathrm{diag}\left\{m_{a01}, m_{a02}, \ldots, m_{a0n_u}\right\}, \tag{5.13}$$
$$L_a = \mathrm{diag}\left\{l_{a1}, l_{a2}, \ldots, l_{an_u}\right\}, \tag{5.14}$$
$$J_a = \mathrm{diag}\left\{j_{a1}, j_{a2}, \ldots, j_{an_u}\right\}, \tag{5.15}$$

where $m_{a0i} = \left(\underline{m}_{ai} + \overline{m}_{ai}\right)/2$, $l_{ai} = (m_{ai}(t) - m_{a0i})/m_{a0i}$ and $j_{ai} = (\overline{m}_{ai} - \underline{m}_{ai})/(\underline{m}_{ai} + \overline{m}_{ai})$ with $i = 1, 2, \ldots, n_u$. Then, we have $M_a = M_{a0}(I + L_a)$ and $L_a^T L_a \leq J_a^T J_a \leq I$.

5.3.1 Stability Analysis

Then, the H_∞ output tracking controller design problem for the systems with actuator faults in (5.1) will be solved in this section. Next, an important theorem in the controller design problem is given in the following part.

Theorem 5.1 *Consider the closed-loop system in (5.12). For the given matrices A_i, B_{ui}, B_{wi}, C_i, D_{ui}, A, C, M_a and the controller gains K_j and K_{dj}, the closed-loop system (5.12) achieves the H_∞ output tracking performance γ if there exist matrices $P > 0$, $W_{ijl} = W_{ijl}^T$ and $M = M^T$, $i = 1, 2, \ldots, p$; $j = 1, 2, \ldots, q$; $l = 1, 2, \ldots, \tau + 1$, such that the following LMIs are satisfied.*

$$W_{ijl} > 0, \quad \forall i, j, l, \tag{5.16}$$

$$\Psi_{ij} + W_{ijl} + M > 0, \quad \forall i, j, l, \tag{5.17}$$

$$\sum_{i=1}^{p}\sum_{j=1}^{q}\Big[\overline{\varrho}_{iji_1i_2\ldots i_nkl}\Psi_{ij} - \left(\underline{\varrho}_{iji_1i_2\ldots i_nkl} - \overline{\varrho}_{iji_1i_2\ldots i_nkl}\right)W_{ijl} + \overline{\varrho}_{iji_1i_2\ldots i_nkl}M\Big] - M < 0, \quad \forall i_1, i_2, \ldots, i_n, k, l, \tag{5.18}$$

where $\underline{\varrho}_{iji_1i_2\ldots i_nkl}$, $\overline{\varrho}_{iji_1i_2\ldots i_nkl}$, $i = 1, 2, \ldots, p$, $j = 1, 2, \ldots, q$, $i_1, i_2, \ldots, i_n = 1, 2$, $k = 1, 2, \ldots, c$, $l = 1, 2, \ldots, \tau + 1$ are pre-defined constant scalars satisfying (5.10) and (5.11), and

$$\Psi_{ij} = \begin{bmatrix} \mathbf{He}(P\bar{A}_{ij}) & P\bar{B}_{ij} & \bar{C}_{ij}^T \\ * & -\gamma^2 I & 0 \\ * & * & -I \end{bmatrix}.$$

Proof Choose the Lyapunov–Krasovskii functional as follows:

$$V(t) = \xi^T(t) P \xi(t).$$

Then, the time derivative of $V(t)$ is given by

$$\begin{aligned} \dot{V}(t) &= 2\xi^T(t) P \dot{\xi}(t) \\ &= 2\sum_{i=1}^{p}\sum_{j=1}^{q}\omega_{ij}\xi^T(t) P \left(\bar{A}_{ij}\xi(t) + \bar{B}_{ij}v(t)\right). \end{aligned} \tag{5.19}$$

Next, consider the following index:

$$J = \int_0^\infty \left[e^T(t) e(t) - \gamma^2 v^T(t) v(t) \right] dt. \tag{5.20}$$

It is obvious that $V(0) = 0$ and $V(\infty) \geq 0$ under the zero-initial condition. Then, it is obtained that

$$\begin{aligned} J &= \int_0^\infty \left[e^T(t) e(t) - \gamma^2 v^T(t) v(t) + \dot{V}(t) \right] dt + V(\infty) \\ &\leq \int_0^\infty \left[e^T(t) e(t) - \gamma^2 v^T(t) v(t) + \dot{V}(t) \right] dt. \end{aligned} \tag{5.21}$$

Submitting (5.19) into (5.21) and considering the equation in (5.12), it is obtained that

$$\begin{aligned} &\dot{V}(t) + e^T(t) e(t) - \gamma^2 v^T(t) v(t) \\ &\leq \sum_{i=1}^{p} \sum_{j=1}^{q} \omega_{ij} \left[2\xi^T(t) P \left(\bar{A}_{ij} \xi(t) + \bar{B}_{ij} v(t) \right) \right. \\ &\qquad \left. + \xi^T(t) \bar{C}_{ij}^T \bar{C}_{ij} \xi(t) - \gamma^2 v^T(t) v(t) \right] \\ &= \sum_{i=1}^{p} \sum_{j=1}^{q} \omega_{ij} \tilde{\xi}^T(t) \Theta_{ij} \tilde{\xi}(t) \\ &= \sum_{i=1}^{p} \sum_{j=1}^{q} \sum_{l=1}^{\tau+1} \varsigma_{ijl} \left(\underline{\gamma}_{ijl} \underline{\omega}_{ijl} + \overline{\gamma}_{ijl} \overline{\omega}_{ijl} \right) \tilde{\xi}^T(t) \Theta_{ij} \tilde{\xi}(t), \end{aligned} \tag{5.22}$$

where

$$\Theta_{ij} = \begin{bmatrix} \mathbf{He}(P\bar{A}_{ij}) + \bar{C}_{ij}^T \bar{C}_{ij} & P\bar{B}_{ij} \\ * & -\gamma^2 I \end{bmatrix}, \quad \tilde{\xi}(t) = \begin{bmatrix} \xi(t) \\ v(t) \end{bmatrix}.$$

From (5.22), we know that if

$$\sum_{i=1}^{p} \sum_{j=1}^{q} \sum_{l=1}^{\tau+1} \varsigma_{ijl} \left(\underline{\gamma}_{ijl} \underline{\omega}_{ijl} + \overline{\gamma}_{ijl} \overline{\omega}_{ijl} \right) \Theta_{ij} < 0,$$

then the closed-loop system (5.7) satisfies the H_∞ output tracking performance. By Schur complement in [210], it is shown that

$$\sum_{i=1}^{p} \sum_{j=1}^{q} \sum_{l=1}^{\tau+1} \varsigma_{ijl} \left(\underline{\gamma}_{ijl} \underline{\omega}_{ijl} + \overline{\gamma}_{ijl} \overline{\omega}_{ijl} \right) \Psi_{ij} < 0.$$

Noting that $0 \le \underline{\omega}_{ijl} \le \overline{\omega}_{ijl} \le 1$, $0 \le \underline{\gamma}_{ijl} \le 1$, $0 \le \overline{\gamma}_{ijl} \le 1$ and $\underline{\gamma}_{ijl} + \overline{\gamma}_{ijl} = 1$ for all i, j and l. Then, the following inequalities are introduced,

$$\left(\sum_{i=1}^{p}\sum_{j=1}^{q}\sum_{l=1}^{\tau+1}\varsigma_{ijl}\left(\underline{\gamma}_{ijl}\underline{\omega}_{ijl} + \overline{\gamma}_{ijl}\overline{\omega}_{ijl}\right) - 1\right) M = 0, \tag{5.23}$$

$$-\sum_{i=1}^{p}\sum_{j=1}^{q}\left(1 - \underline{\gamma}_{ijl}\right)\left(\underline{\omega}_{ijl} - \overline{\omega}_{ijl}\right) W_{ijl} \ge 0, \tag{5.24}$$

where $M = M^T$ and $W_{ijl} = W_{ijl}^T \ge 0$ are arbitrary matrices with appropriate dimensions.

From (5.12), (5.22)–(5.24), one can have

$$\begin{aligned}
&\sum_{i=1}^{p}\sum_{j=1}^{q}\sum_{l=1}^{\tau+1}\varsigma_{ijl}\left(\underline{\gamma}_{ijl}\underline{\omega}_{ijl} + \overline{\gamma}_{ijl}\overline{\omega}_{ijl}\right)\Psi_{ij} \\
&\le \sum_{i=1}^{p}\sum_{j=1}^{q}\sum_{l=1}^{\tau+1}\varsigma_{ijl}\left(\underline{\gamma}_{ijl}\underline{\omega}_{ijl} + \left(1 - \underline{\gamma}_{ijl}\right)\overline{\omega}_{ijl}\right)\Psi_{ij} \\
&-\sum_{i=1}^{p}\sum_{j=1}^{q}\sum_{l=1}^{\tau+1}\varsigma_{ijl}\left(1 - \underline{\gamma}_{ijl}\right)\left(\underline{\omega}_{ijl} - \overline{\omega}_{ijl}\right) W_{ijl} \\
&+\left[\sum_{i=1}^{p}\sum_{j=1}^{q}\sum_{l=1}^{\tau+1}\varsigma_{ijl}\left(\underline{\gamma}_{ijl}\underline{\omega}_{ijl} + \left(1 - \underline{\gamma}_{ijl}\right)\overline{\omega}_{ijl}\right) - 1\right] M \\
&= \sum_{i=1}^{p}\sum_{j=1}^{q}\sum_{l=1}^{\tau+1}\varsigma_{ijl}\left(\overline{\omega}_{ijl}\Psi_{ij} - \left(\underline{\omega}_{ijl} - \overline{\omega}_{ijl}\right) W_{ijl} + \overline{\omega}_{ijl} M\right) - M \\
&+\sum_{i=1}^{p}\sum_{j=1}^{q}\sum_{l=1}^{\tau+1}\varsigma_{ijl}\underline{\gamma}_{ijl}\left(\underline{\omega}_{ijl} - \overline{\omega}_{ijl}\right)\left(\Psi_{ij} + W_{ijl} + M\right).
\end{aligned} \tag{5.25}$$

From (5.25), the inequality $\dot{V}(t) + e^T(t)\, e(t) - \gamma^2 v^T(t)\, v(t) < 0$ will be satisfied if the inequality in (5.17) and the following inequality holds for all i, j and l

$$\sum_{i=1}^{p}\sum_{j=1}^{q}\sum_{l=1}^{\tau+1}\varsigma_{ijl}\left[\overline{\omega}_{ijl}\Psi_{ij} - \left(\underline{\omega}_{ijl} - \overline{\omega}_{ijl}\right) W_{ijl} + \overline{\omega}_{ijl} M\right] - M < 0. \tag{5.26}$$

Recalling that only one $\varsigma_{ijl} = 1$ for each fixed value of i and j such that $\sum_{l=1}^{\tau+1}\varsigma_{ijl} = 1$, the inequality (5.26) will be satisfied if the following inequalities hold for all $i_1, i_2, \ldots, i_n, i, j, k, l$,

$$\sum_{k=1}^{c}\sum_{i_1=1}^{2}\sum_{i_2=1}^{2}\cdots\sum_{i_n=1}^{2}\prod_{r=1}^{n}\rho_{ri_rkl}\left\{\sum_{i=1}^{p}\sum_{j=1}^{q}\left[\overline{\varrho}_{iji_1i_2\ldots i_nkl}\Psi_{ij}\right.\right.$$
$$\left.\left.-\left(\underline{\varrho}_{iji_1i_2\ldots i_nkl}-\overline{\varrho}_{iji_1i_2\ldots i_nkl}\right)W_{ijl}+\overline{\varrho}_{iji_1i_2\ldots i_nkl}M\right]-M\right\}<0. \quad (5.27)$$

From (5.18), we can obtain

$$\sum_{i=1}^{p}\sum_{j=1}^{q}\sum_{l=1}^{\tau+1}\varsigma_{ijl}\left[\overline{\omega}_{ijl}\Psi_{ij}-\left(\underline{\omega}_{ijl}-\overline{\omega}_{ijl}\right)W_{ijl}+\overline{\omega}_{ijl}M\right]-M<0,$$

which is equivalent to (5.27). Therefore, we have that $\dot{V}(t)+e^T(t)e(t)-\gamma^2 v^T(t)v(t)<0$ for all nonzero $v(t)\in\mathcal{L}_2[0,\infty)$, which means $J<0$, that is, $\|e(t)\|_2<\gamma\|v(t)\|_2$. The proof is completed. □

5.3.2 Output Tracking Control

Next, the fuzzy controller existence condition is presented and the gain matrices K_j and K_{dj} of a desired controller in the form of (5.3) will be obtained base on Theorem 5.1.

Theorem 5.2 *The closed-loop system in (5.12) achieves the H_∞ output tracking performance γ if there exist matrices $U=U^T$, $V=V^T$, X_j, Y_j, $\check{W}_{ijl}=\check{W}_{ijl}^T$ and $\check{M}=\check{M}^T$, $(i=1,2,\ldots,p,\ j=1,2,\ldots,q,\ l=1,2,\ldots,\tau+1)$, such that following LMIs hold:*

$$\check{W}_{ijl}>0,\quad \forall i,j,l, \quad (5.28)$$
$$\tilde{\Psi}_{1ij}+\tilde{W}_{ijl}+\tilde{M}>0,\quad \forall i,j,l, \quad (5.29)$$
$$\sum_{i=1}^{p}\sum_{j=1}^{q}\left[\overline{\varrho}_{iji_1i_2\ldots i_nkl}\tilde{\Psi}_{2ij}-\left(\underline{\varrho}_{iji_1i_2\ldots i_nkl}-\overline{\varrho}_{iji_1i_2\ldots i_nkl}\right)\tilde{W}_{ijl}\right.$$
$$\left.+\overline{\varrho}_{iji_1i_2\ldots i_nkl}\tilde{M}\right]-\tilde{M}<0,\quad \forall i_1,i_2,\ldots,i_n,k,l, \quad (5.30)$$

where $\underline{\varrho}_{iji_1i_2\ldots i_nkl}$, $\overline{\varrho}_{iji_1i_2\ldots i_nkl}$, $i=1,2,\ldots,p$, $j=1,2,\ldots,q$, $i_1,i_2,\ldots,i_n=1,2$, $k=1,2,\ldots,c$, $l=1,2,\ldots,\tau+1$ are pre-defined constant scalars satisfying (5.10) and (5.11), and

$$\tilde{\Psi}_{1ij} = \begin{bmatrix} \Theta_{aij} & B_{ui}Y_j^T & B_{wi} & 0 & \Theta_{bij} & -X_j \\ * & \mathbf{He}\left(GV^T\right) & 0 & I & \Theta_{3ij} & -Y_j \\ * & * & -\gamma^2 I & 0 & 0 & 0 \\ * & * & * & -\gamma^2 I & 0 & 0 \\ * & * & * & * & \Theta_{cij} & 0 \\ * & * & * & * & * & \varepsilon_{aij} I \end{bmatrix},$$

$$\tilde{\Psi}_{2ij} = \begin{bmatrix} \Theta_{1ij} & B_{ui}Y_j^T & B_{wi} & 0 & \Theta_{2ij} & X_j \\ * & \mathbf{He}\left(GV^T\right) & 0 & I & \Theta_{3ij} & Y_j \\ * & * & -\gamma^2 I & 0 & 0 & 0 \\ * & * & * & -\gamma^2 I & 0 & 0 \\ * & * & * & * & \Theta_{4ij} & 0 \\ * & * & * & * & * & -\varepsilon_{ij} J_a^{-1} \end{bmatrix},$$

$$\begin{aligned} \Theta_{aij} &= \mathbf{He}\left(A_i U^T + B_{ui} X_j^T\right) - \varepsilon_{aij} B_{ui} B_{ui}^T, \quad \tilde{M} = \text{diag}\left\{\check{M}, 0\right\}, \\ \Theta_{1ij} &= \mathbf{He}\left(A_i U^T + B_{ui} X_j^T\right) + \varepsilon_{ij} B_{ui} J_a B_{ui}^T, \\ \Theta_{2ij} &= U C_i^T + X_j D_{ui}^T + \varepsilon_{ij} B_{ui} J_a D_{ui}^T, \quad \Theta_{3ij} = -V H^T + Y_j D_{ui}^T, \\ \Theta_{4ij} &= -I + \varepsilon_{ij} D_{ui} J_a D_{ui}^T, \quad \Theta_{cij} = -I - \varepsilon_{aij} D_{ui} D_{ui}^T, \\ \tilde{W}_{ijl} &= \text{diag}\left\{\check{W}_{ijl}, 0\right\}, \quad \Theta_{bij} = U C_i^T + X_j D_{ui}^T - \varepsilon_{aij} B_{ui} D_{ui}^T. \end{aligned}$$

Then, the control gains K_j and K_{dj} in (5.5) are given by $K_j = U^{-T} X_j^T M_{a0}^{-T}$ and $K_{dj} = V^{-T} Y_j^T M_{a0}^{-T}$.

Proof By defining $\bar{P} = P^{-1} = \text{diag}\{U, V\}$, $\check{W}_{ijl} = \bar{P} W_{ijl} \bar{P}^T$ and $\check{M} = \bar{P} M \bar{P}^T$, performing a congruence transformation with diag $\left\{\bar{P}, I, I\right\}$ and noting the conditions (5.13)–(5.15), (5.17) and (5.18) are respectively rewritten as

$$\check{\Psi}_{ij} + \check{W}_{ijl} + \check{M}_{ijl} > 0, \tag{5.31}$$

$$\sum_{i=1}^{p} \sum_{j=1}^{q} \left(\overline{\varrho}_{iji_1 i_2 \ldots i_n kl} \check{\Psi}_{ij} - \check{\varrho} \check{W}_{ijl} + \overline{\varrho}_{iji_1 i_2 \ldots i_n kl} \check{M} \right) - \check{M} < 0, \tag{5.32}$$

where

$$\check{\Psi}_{ij} = \begin{bmatrix} \Phi_{1ij} & B_{ui}Y_j^T & B_{wi} & 0 & \Phi_{2ij} \\ * & \mathbf{He}\left(GV^T\right) & 0 & I & \Phi_{3ij} \\ * & * & -\gamma^2 I & 0 & 0 \\ * & * & * & -\gamma^2 I & 0 \\ * & * & * & * & -I \end{bmatrix} + \mathbf{He}\left(E L_a F\right)$$

$$\check{\varrho} = \underline{\varrho}_{iji_1 i_2 \ldots i_n kl} - \overline{\varrho}_{iji_1 i_2 \ldots i_n kl},$$

$$E = \begin{bmatrix} B_{ui}^T & 0 & 0 & 0 & D_{ui}^T \end{bmatrix}^T, \quad F = \begin{bmatrix} X_j^T & Y_j^T & 0 & 0 & 0 \end{bmatrix},$$
$$\Phi_{1ij} = \mathbf{He}\left(A_i U^T + B_{ui} X_j^T\right), \quad \Phi_{2ij} = U C_i^T + X_j D_{ui}^T,$$
$$\Phi_{3ij} = V H^T + Y_j D_{ui}^T, \quad X_j = U K_j^T M_{a0}^T, \quad Y_j = V K_{dj}^T M_{a0}^T.$$

Recalling that $L_a^T L_a \le J_a^T J_a \le I$ and Schur complement, the condition (5.30) is rewritten as

$$\sum_{i=1}^{p}\sum_{j=1}^{q}\Big[\overline{\varrho}_{iji_1i_2\ldots i_nkl}\left(\check{\Psi}_{ij} + \varepsilon_{ij}^{-1} F^T J_a F + \varepsilon_{ij} E J_a E^T\right)$$
$$-\left(\underline{\varrho}_{iji_1i_2\ldots i_nkl} - \overline{\varrho}_{iji_1i_2\ldots i_nkl}\right)\check{W}_{ijl} + \overline{\varrho}_{iji_1i_2\ldots i_nkl}\check{M}\Big] - \check{M} < 0.$$

Form Lemma 1.3, one can have

$$\sum_{i=1}^{p}\sum_{j=1}^{q}\Big[\overline{\varrho}_{iji_1i_2\ldots i_nkl}\left(\check{\Psi}_{ij} + \mathbf{He}\left(E L_a F\right)\right)$$
$$-\left(\underline{\varrho}_{iji_1i_2\ldots i_nkl} - \overline{\varrho}_{iji_1i_2\ldots i_nkl}\right)\check{W}_{ijl} + \overline{\varrho}_{iji_1i_2\ldots i_nkl}\check{M}\Big] - \check{M} < 0. \quad (5.33)$$

From (5.32) and (5.33), the condition (5.18) in Theorem 5.1 is obtained. Similarly, one can obtain that (5.29) is equivalent to (5.17). Otherwise, $\tilde{W}_{ijl} > 0$ is equivalent to $W_{ijl} > 0$. Therefore, all the conditions in Theorem 5.1 are satisfied. Hence, the proof is completed. □

Noting that, if the system (5.1) has no actuator fault, the overall IT2 fuzzy model is inferred as follows:

$$\begin{cases} \dot{x}(t) = \sum_{i=1}^{p} \theta_i\left(x(t)\right)\left(A_i x(t) + B_{wi} w(t) + B_{ui} u(t)\right), \\ y(t) = \sum_{i=1}^{p} \theta_i\left(x(t)\right)\left(C_i x(t) + D_{ui} u(t)\right). \end{cases} \quad (5.34)$$

Moreover, the overall IT2 fuzzy control law is described as follows:

$$u(t) = \sum_{j=1}^{q} \eta_j\left(x(t)\right)\left(K_{aj} x(t) + K_{daj} x_d(t)\right). \quad (5.35)$$

Considering the standard controller in (5.35), the augmented closed-loop system with the reference signal $y_d(t)$ is represented as follows:

$$
\begin{cases}
\dot{\xi}(t) = \sum_{i=1}^{p} \sum_{j=1}^{q} \omega_{ij}(x(t)) \left(\tilde{A}_{ij} \xi(t) + \tilde{B}_{ij} \upsilon(t) \right), \\
e(t) = \sum_{i=1}^{p} \sum_{j=1}^{q} \omega_{ij}(x(t)) \tilde{C}_{ij} \xi(t),
\end{cases}
\tag{5.36}
$$

where

$$
\tilde{A}_{ij} = \begin{bmatrix} A_i + B_{ui} K_{aj} & B_{ui} K_{daj} \\ 0 & G \end{bmatrix}, \quad \tilde{B}_{ij} = \begin{bmatrix} B_{wi} & 0 \\ 0 & I \end{bmatrix},
$$
$$
\tilde{C}_{ij} = \begin{bmatrix} C_i + D_{ui} K_{aj} & D_{ui} K_{daj} - H \end{bmatrix}.
$$

Using the similar approach in Theorems 5.1 and 5.2, the following theorem is obtained for the H_∞ output tracking controller synthesis of system (5.36) with no actuator fault.

Theorem 5.3 *The closed-loop system in (5.36) achieves the H_∞ output tracking performance γ, and the fuzzy control gain matrices can be given by $K_{aj} = U^{-T} X_j^T$ and $K_{daj} = V^{-T} Y_j^T$, if there exist matrices $U = U^T$, $V = V^T$, X_j, Y_j, $\check{W}_{ijl} = \check{W}_{ijl}^T$ and $\check{M} = \check{M}^T$ such that following LMIs hold.*

$$
\check{W}_{ijl} > 0, \quad \forall i, j, l, \tag{5.37}
$$
$$
\Upsilon_{ij} + \check{W}_{ijl} + \check{M} > 0, \quad \forall i, j, l, \tag{5.38}
$$
$$
\sum_{i=1}^{p} \sum_{j=1}^{q} \left[\overline{\varrho}_{ij i_1 i_2 \ldots i_n k l} \Upsilon_{ij} - \left(\underline{\varrho}_{ij i_1 i_2 \ldots i_n k l} - \overline{\varrho}_{ij i_1 i_2 \ldots i_n k l} \right) \check{W}_{ijl} \right.
$$
$$
\left. + \overline{\varrho}_{ij i_1 i_2 \ldots i_n k l} \check{M} \right] - \check{M} < 0, \quad \forall i_1, i_2, \ldots, i_n, k, l, \tag{5.39}
$$

where $\underline{\varrho}_{ij i_1 i_2 \ldots i_n k l}$, $\overline{\varrho}_{ij i_1 i_2 \ldots i_n k l}$, $i = 1, 2, \ldots, p$, $j = 1, 2, \ldots, q$, $i_1, i_2, \ldots, i_n = 1, 2$, $k = 1, 2, \ldots, c$, $l = 1, 2, \ldots, \tau + 1$ are pre-defined constant scalars satisfying (5.10) and (5.11), and

$$
\Upsilon_{ij} = \begin{bmatrix}
\Upsilon_{1ij} & B_{ui} Y_j^T & B_{wi} & 0 & \Upsilon_{2ij} \\
* & \mathbf{He}\left(GV^T\right) & 0 & I & \Upsilon_{3ij} \\
* & * & -\gamma^2 I & 0 & 0 \\
* & * & * & -\gamma^2 I & 0 \\
* & * & * & * & -I
\end{bmatrix},
$$

with $\Upsilon_{1ij} = \mathbf{He}(A_i U^T + B_{ui} X_j^T)$, $\Upsilon_{2ij} = U C_i^T + X_j D_{ui}^T$ and $\Upsilon_{3ij} = -V H^T + Y_j D_{ui}^T$.

Remark 5.4 In this study, the IT2 tracking controller is the first time to be designed to guarantee the plant with actuator fault is asymptotically stable and satisfies the H_∞ output tracking performance. Meanwhile, the faults occurred in the actuator can be tolerant.

5.4 Simulation Results

To illustrate the effectiveness of our results, a practical example is given in this section.

Example 5.5 Consider the well-studied continuous stirred tank reactor (CSTR). The following nonlinear system is taken from [238].

$$\begin{cases} \dot{z}_1(t) = -z_1(t) + \bar{D}_a(1-z_1)e^{-\frac{1}{z_2+\bar{\gamma}}}, \\ \dot{z}_2(t) = -\left(1+\bar{\beta}\right)z_2 + \bar{H}D_a(1-z_1)e^{-\frac{1}{z_2+\bar{\gamma}}} + \bar{\beta}\left(\bar{u}(t)+\bar{h}\right), \end{cases}$$

where $z = [z_1 \quad z_2]^T$, $\bar{u}$ is associated with the inputs, $\bar{h}$ is related to external disturbance which may be caused by an uncontrollable change in the ambient temperature, and the parameters are taken from [21] as $\bar{D}_a = 0.072$, $\bar{H} = 8$, $\bar{\gamma} = 20$, and $\bar{\beta} = 0.3$. For detailed explanations about these parameters, please refer to [21, 238]. By using the method in [181], the following three IF-THEN rules are given:

Plant Rule 1: IF the temperature is low (i.e., $x_2(t)$ is about -1), THEN,

$$x(t) = A_1x(t) + B_{u1}\left(u(t)+\bar{h}\right),$$

Plant Rule 2: IF the temperature is low (i.e., $x_2(t)$ is about 0), THEN,

$$x(t) = A_2x(t) + B_{u2}\left(u(t)+\bar{h}\right),$$

Plant Rule 3: IF the temperature is low (i.e., $x_2(t)$ is about 1), THEN,

$$x(t) = A_3x(t) + B_{u3}\left(u(t)+\bar{h}\right),$$

where $x(t) = [x_1 \quad x_2]^T = z - z_e$, $u = \bar{u} - u_e$, and

$$A_1 = \begin{bmatrix} -1.0684 & 0.0002 \\ -0.5471 & 1.2987 \end{bmatrix}, \quad A_2 = \begin{bmatrix} -1.0686 & 0.0002 \\ -0.5484 & -1.2988 \end{bmatrix},$$
$$A_3 = \begin{bmatrix} -1.0687 & 0.0002 \\ -0.5497 & 1.2988 \end{bmatrix}, \quad B_{u1} = B_{u2} = B_{u3} = \begin{bmatrix} 0 \\ 0.3 \end{bmatrix}.$$

In these fuzzy rules, select the Gaussian functions as the membership functions

$$\theta_1 = \frac{e^{\left(-\frac{(x_2+1)^2}{\sigma^2}\right)}}{e^{\left(-\frac{(x_2+1)^2}{\sigma^2}\right)} + e^{\left(-\frac{x_2^2}{\sigma^2}\right)} + e^{\left(-\frac{(x_2-1)^2}{\sigma^2}\right)}},$$

$$\theta_2 = \frac{e^{\left(-\frac{x_2^2}{\sigma^2}\right)}}{e^{\left(-\frac{(x_2+1)^2}{\sigma^2}\right)} + e^{\left(-\frac{x_2^2}{\sigma^2}\right)} + e^{\left(-\frac{(x_2-1)^2}{\sigma^2}\right)}},$$

$$\theta_3 = \frac{e^{\left(-\frac{(x_2-1)^2}{\sigma^2}\right)}}{e^{\left(-\frac{(x_2+1)^2}{\sigma^2}\right)} + e^{\left(-\frac{x_2^2}{\sigma^2}\right)} + e^{\left(-\frac{(x_2-1)^2}{\sigma^2}\right)}},$$

where $\sigma \in [0.5, 0.8]$ is regarded as the parameter uncertainties. In this example, the state $x(t)$ is uniformly globally bounded, the reference model is considered as follows

$$\begin{cases} \dot{x}_d(t) = -x_d(t) + r(t), \\ y_d(t) = 0.5x_d(t), \end{cases}$$

and our purpose is to control the temperature follow the reference signal $y_d(t)$ to meet H_∞ sense. The overall fuzzy model is inferred as

$$\begin{cases} \dot{x}(t) = \sum_{i=1}^{3} \theta_i \left(A_i x(t) + B_{wi} w(t) + B_{ui} u(t)\right), \\ y(t) = \sum_{i=1}^{3} \theta_i \left(C_i x(t) + D_{ui} u(t)\right), \end{cases}$$

where $C_1 = C_2 = C_3 = [0 \quad 1]$, $D_{u1} = D_{u2} = D_{u3} = 0.2$ and $B_{w1} = B_{w2} = B_{w3} = [1 \quad 3]^T$. The LMFs and UMFs are selected as Table 5.1.

Next, we select $w(t) = \sin(t)$ and $r(t) = 0.4\sin(t+1)$ to show the simulations. Choosing the initial conditions $x(0) = \begin{bmatrix} 1 & 0 \end{bmatrix}^T$ and $x_d(0) = -0.3$. Then, there exists an actuator fault M_a satisfying $\underline{m}_{a01} = 0.2$, and $\overline{m}_{a01} = 0.8$. It is implied that $M_{a0} = 0.5$, let $\gamma = 0.9$, by using MATLAB LMI Toolbox, the controller gain of (5.6) are computed as follows:

$$K_1 = \begin{bmatrix} 0.6462 & -6.3887 \end{bmatrix}, \quad K_2 = \begin{bmatrix} 0.6462 & -6.3908 \end{bmatrix},$$
$$K_{d1} = 1.6682, \quad K_{d2} = 1.6685.$$

Table 5.1 LMFs and UMFs of the plant and the controller

LMFs of the pant
$\underline{\mu}_{W_{11}}(x_1) = \frac{e^{\left(-\frac{(x_2+1)^2}{0.5^2}\right)}}{e^{\left(-\frac{(x_2+1)^2}{0.5^2}\right)}+e^{\left(-\frac{x_2^2}{0.5^2}\right)}+e^{\left(-\frac{(x_2-1)^2}{0.5^2}\right)}}$
$\underline{\mu}_{W_{12}}(x_1) = \frac{e^{\left(-\frac{x_2^2}{0.8^2}\right)}}{e^{\left(-\frac{(x_2+1)^2}{0.8^2}\right)}+e^{\left(-\frac{x_2^2}{0.8^2}\right)}+e^{\left(-\frac{(x_2-1)^2}{0.8^2}\right)}}$
$\underline{\mu}_{W_{13}}(x_1) = \frac{e^{\left(-\frac{(x_2-1)^2}{0.5^2}\right)}}{e^{\left(-\frac{(x_2+1)^2}{0.5^2}\right)}+e^{\left(-\frac{x_2^2}{0.5^2}\right)}+e^{\left(-\frac{(x_2-1)^2}{0.5^2}\right)}}$
UMFs of the plant
$\overline{\mu}_{W_{11}}(x_1) = \frac{e^{\left(-\frac{(x_2+1)^2}{0.8^2}\right)}}{e^{\left(-\frac{(x_2+1)^2}{0.8^2}\right)}+e^{\left(-\frac{x_2^2}{0.8^2}\right)}+e^{\left(-\frac{(x_2-1)^2}{0.8^2}\right)}}$
$\overline{\mu}_{W_{12}}(x_1) = \frac{e^{\left(-\frac{x_2^2}{0.5^2}\right)}}{e^{\left(-\frac{(x_2+1)^2}{0.5^2}\right)}+e^{\left(-\frac{x_2^2}{0.5^2}\right)}+e^{\left(-\frac{(x_2-1)^2}{0.5^2}\right)}}$
$\overline{\mu}_{W_{13}}(x_1) = \frac{e^{\left(-\frac{(x_2-1)^2}{0.8^2}\right)}}{e^{\left(-\frac{(x_2+1)^2}{0.8^2}\right)}+e^{\left(-\frac{x_2^2}{0.8^2}\right)}+e^{\left(-\frac{(x_2-1)^2}{0.8^2}\right)}}$
LMFs of the controller
$\underline{\mu}_{M_{11}}(x_1) = e^{\left(-\frac{x_2^2}{0.5}\right)}$
$\underline{\mu}_{M_{12}}(x_1) = 1 - e^{\left(-\frac{x_2^2}{0.5}\right)}$
UMFs of the controller
$\overline{\mu}_{M_{11}}(x_1) = e^{\left(-\frac{x_2^2}{0.5}\right)}$
$\overline{\mu}_{M_{12}}(x_1) = 1 - e^{\left(-\frac{x_2^2}{0.5}\right)}$

On the other hand, if there is no actuator fault, the controller gains of (5.35) are given:

$$K_{a1} = \left[0.3590\ -3.5489\right],\quad K_{a2} = \left[0.4308\ -4.2601\right],$$
$$K_{da1} = 2.0018,\quad K_{da2} = 2.1691.$$

Based on the above controller gains, Figs. 5.1, 5.2 and 5.3 show the responses of the system output $y(t)$ with actuator fault and the desire signal $y_d(t)$, the state $x(t)$ and the system output $y_a(t)$ with no actuator fault and the reference signal $y_d(t)$,

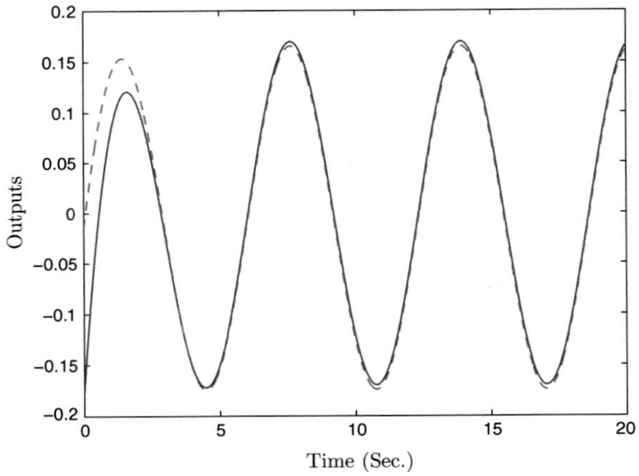

Fig. 5.1 Responses of $y(t)$ and $y_d(t)$

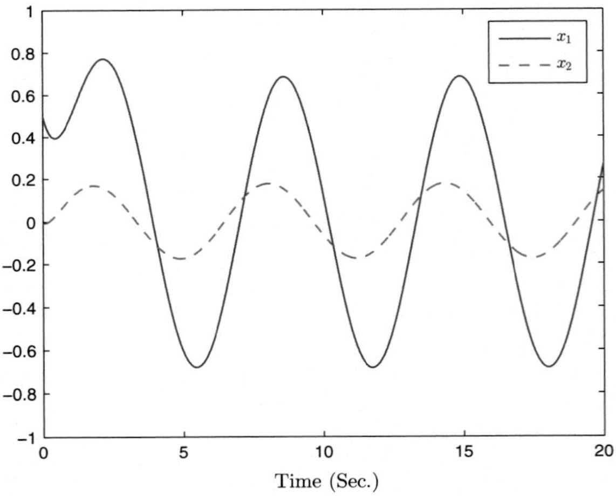

Fig. 5.2 Responses of $x(t)$

respectively. From Figs. 5.1 and 5.3, we can obtain that the designed controller can guarantee the system output track the desired output well even there exists actuator fault.

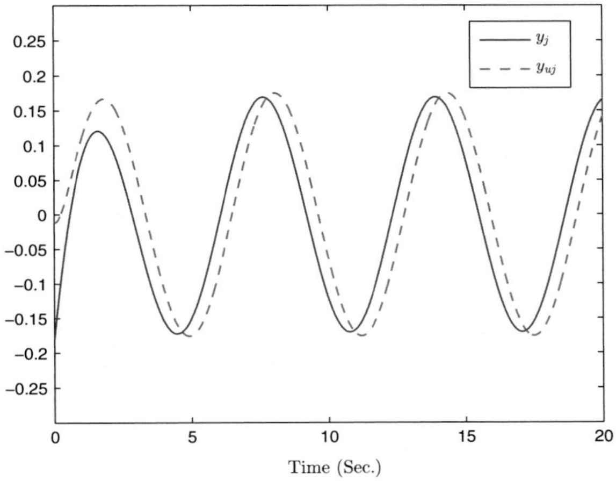

Fig. 5.3 Responses of $y_a(t)$ and $y_d(t)$

5.5 Conclusion

In this chapter, the problem of H_∞ output tracking controller design has been solved for IT2 fuzzy systems with actuator fault. Firstly, considering the actuator fault, the IT2 fuzzy model and the IT2 state-feedback tracking controller has been constructed. Secondly, based on Lyapunov stability theory and LMI techniques, the reliable fuzzy H_∞ IT2 output tracing controller has been designed, of which the existence condition can be expressed by a convex optimization problem. Furthermore, the standard IT2 fuzzy H_∞ tracking controller has been also designed. Finally, simulation results have clearly illustrated that the designed reliable fuzzy controller has the capability of guaranteeing a better tracking performance under actuator fault case.

Chapter 6
Switched Control of Interval Type-2 Fuzzy-Model-Based Systems

6.1 Introduction

Recently, Dong and Yang introduced a switched dynamic output-feedback controller approach for continuous-time T–S fuzzy systems in [46]. The switched control approach can obtain better performance. However, it should be pointed out that the aforementioned results are under the condition that the grades of membership are certain. This chapter investigates the dynamic output-feedback H_∞ control problem for IT2 fuzzy systems. The parameter uncertainties can be expressed by the LMF and UMF, respectively. Based on Lyapunov stability theory, a novel switched controller is proposed to ensure that the closed-loop system is asymptotically stable with an H_∞ performance. In the design procedure, the gains of the IT2 switched controller can be solved by the standard software. A practical example is given to show the feasibility and advantage of the proposed scheme over the existing ones. The main contributions of this chapter can be summarized as follows: (1) The switched controller is first time designed for nonlinear systems subject to parameter uncertainties on the basis of IT2 T–S fuzzy model; (2) The parameter uncertainties of the plants can be captured by using the LMF and UMF, respectively. It should be mentioned that the existing type-1 switched controller for T–S fuzzy system can not solve the uncertain parameters problem; (3) To obtain the controller parameters, the methods proposed in this chapter can reduce the number of the controller gain matrices required.

6.2 Problem Formulation

Consider the following IT2 fuzzy model that represents a continuous-time nonlinear system:

♦ **Plant Form:**

Rule i: IF $f_1(x(t))$ is Q_1^i and ... and $f_p(x(t))$ is Q_p^i, THEN,

H. Li et al., *Analysis and Synthesis for Interval Type-2 Fuzzy-Model-Based Systems*, DOI 10.1007/978-981-10-0593-0_6

$$\begin{cases} \dot{x}(t) = A_i x(t) + B_{1i} w(t) + B_{2i} u(t), \\ z(t) = C_{1i} x(t) + D_{1i} w(t) + D_{2i} u(t), \\ y(t) = C_{2i} x(t) + D_{3i} w(t), \end{cases} \tag{6.1}$$

where $f_a(x(t))$ denotes the premise variables and Q_a^i is an IT2 fuzzy set, $i = 1, 2, \ldots, r$, $a = 1, 2, \ldots, p$. p is a positive integer. $x(t) \in \mathbf{R}^n$ stands for the system state vector, $w(t) \in \mathbf{R}^q$ denotes the disturbance input and $u(t) \in \mathbf{R}^m$ stands for the control input vector, $z(t) \in \mathbf{R}^v$ is the controller output vector, $y(t) \in \mathbf{R}^l$ is the measure output vector. $A_i, B_{1i}, B_{2i}, C_{1i}, D_{1i}, D_{2i}, C_{2i}$, and D_{3i} are the known matrices with appropriate dimensions. The following interval sets denote firing strength of the ith rule:

$$W_i(x(t)) = \left[\underline{\delta}_i(x(t)), \overline{\delta}_i(x(t))\right], \quad i = 1, 2, \ldots, r,$$

where $\underline{\delta}_i(x(t)) = \prod_{a=1}^{p} \underline{\vartheta}_{Q_a^i}(f_a(x(t))) \geq 0$ and $\overline{\delta}_i(x(t)) = \prod_{a=1}^{p} \bar{\vartheta}_{Q_a^i}(f_a(x(t))) \geq 0$ are lower and upper grades of membership, respectively. $\underline{\vartheta}_{Q_a^i}(f_a(x(t))) \geq 0$ and $\bar{\vartheta}_{Q_a^i}(f_a(x(t))) \geq 0$ stand for the LMFs and UMFs, respectively. Therefore, it can be found that $\bar{\vartheta}_{Q_a^i}(f_a(x(t))) \geq \underline{\vartheta}_{Q_a^i}(f_a(x(t)))$ and $\overline{\delta}_i(x(t)) \geq \underline{\delta}_i(x(t))$ for all i. Then, the IT2 T–S fuzzy system is obtained as:

Rule i: IF $f_1(x(t))$ is Q_1^i and ... and $f_p(x(t))$ is Q_p^i, THEN,

$$\begin{cases} \dot{x}(t) = \sum_{i=1}^{r} \tilde{\delta}_i(x(t)) \left[A_i x(t) + B_{1i} w(t) + B_{2i} u(t)\right], \\ z(t) = \sum_{i=1}^{r} \tilde{\delta}_i(x(t)) \left[C_{1i} x(t) + D_{1i} w(t) + D_{2i} u(t)\right], \\ y(t) = \sum_{i=1}^{r} \tilde{\delta}_i(x(t)) \left[C_{2i} x(t) + D_{3i} w(t)\right], \end{cases} \tag{6.2}$$

where

$$\tilde{\delta}_i(x(t)) = \frac{\underline{\varsigma}_i(x(t))\,\underline{\delta}_i(x(t)) + \bar{\varsigma}_i(x(t))\,\overline{\delta}_i(x(t))}{\sum_{j=1}^{r} \underline{\varsigma}_j(x(t))\,\underline{\delta}_j(x(t)) + \bar{\varsigma}_j(x(t))\,\overline{\delta}_j(x(t))} \geq 0,$$

$$0 \leq \tilde{\delta}_i(x(t)) \leq 1, \quad \sum_{i=1}^{r} \tilde{\delta}_i(x(t)) = 1, \quad \forall i,$$

in which $0 \leq \underline{\varsigma}_i(x(t)) \leq 1$ and $0 \leq \bar{\varsigma}_i(x(t)) \leq 1$ are nonlinear functions and possess the trait of $\underline{\varsigma}_i(x(t)) + \bar{\varsigma}_i(x(t)) = 1$ for all i. $\tilde{\delta}_i(x(t))$ stands for the normalized membership functions. For a simple description, $\tilde{\delta}_i(x(t))$ is denoted as $\tilde{\delta}_i$ and the vector $\tilde{\delta}(x(t)) = \left[\tilde{\delta}_1(x(t)), \ldots, \tilde{\delta}_r(x(t))\right]^T$ is denoted as $\tilde{\delta}$.

Define

$$\begin{aligned}
\Xi &= \left\{ \tilde{\delta} : 0 \le \tilde{\delta}_i \le 1, 1 \le i \le r, \sum_{i=1}^{r} \tilde{\delta}_i = 1 \right\}, \\
\Xi_l &= \left\{ \tilde{\delta} : 0 \le \tilde{\delta}_i \le \tilde{\delta}_l, 1 \le i \le r, \sum_{i=1}^{r} \tilde{\delta}_i = 1, \tilde{\delta} \in \Xi \right\}, \\
\partial \Xi_l &= \{ \tilde{\delta} : \exists i \ne l \text{ such that } \tilde{\delta}_i = \tilde{\delta}_l \text{ and } \tilde{\delta} \in \Xi_l \}, \\
\partial \Xi &= \cup_{l=1}^{r} \partial \Xi_l, \quad \Xi = \cup_{l=1}^{r} \Xi_l, \quad 1 \le l \le r.
\end{aligned}$$

At any moment t, there must exist one l, such that the vector $\tilde{\delta} \in \Xi_l$ or $\tilde{\delta} \in \partial \Xi_l$, which reveals that the lth subsystem is more important than other subsystems or as important as other subsystems. The values of premise variables $\tilde{\delta}_i$ can be obtained at any moment t. Then, a switched controller can be achieved. The following steps should be taken into account: (1) when $\tilde{\delta} \in \partial \Xi_l$, the controller gains do not switch. (2) when $\tilde{\delta} \in \Xi_l$, the controller gains should be switched.

From above discussion, if $\tilde{\delta} \in \Xi_l$, $1 \le l \le r$, a switched controller is given as follows:

$$\begin{cases}
\dot{\hat{x}}(t) = \sum_{j=1}^{r} \tilde{\delta}_j (\hat{x}(t)) \left[A_{Kjl} \hat{x}(t) + B_{Kjl} y(t) \right], \\
u(t) = \sum_{j=1}^{r} \tilde{\delta}_j (\hat{x}(t)) \left[C_{Kjl} \hat{x}(t) + D_{Kl} y(t) \right],
\end{cases} \tag{6.3}$$

where $\hat{x}(t) \in \mathbf{R}^n$ stands for the estimated state vector. A_{Kjl}, B_{Kjl}, C_{Kjl} and D_{Kl} are the controller gains to be designed.

Remark 6.1 The difference between type-1 switched controller [46] and IT2 switched controller is that the IT2 switched controller contains parameter uncertainties. The parameter uncertainties can result in the uncertainties of the membership functions. In this chapter, the uncertain parameter can be expressed by the LMFs and UMFs. With the upper and LMFs and relevant weighting functions, the values of membership functions for IT2 fuzzy system can be obtained at any time or moment t. The authors in [16] used an example to demonstrate the process. Then, the parameter uncertainties can be obtained and the IT2 switched controller can be achieved for the IT2 T–S fuzzy model.

By (6.2) and (6.3), we obtain the closed-loop system as follows:

$$\begin{cases}
\dot{\bar{x}}(t) = \sum_{i=1}^{r} \sum_{j=1}^{r} \tilde{\delta}_i (x(t)) \tilde{\delta}_j (\hat{x}(t)) \left[\bar{A}_{ijl} \bar{x}(t) + \bar{B}_{ijl} w(t) \right], \\
z(t) = \sum_{i=1}^{r} \sum_{j=1}^{r} \tilde{\delta}_i (x(t)) \tilde{\delta}_j (\hat{x}(t)) \left[\bar{C}_{ijl} \bar{x}(t) + \bar{D}_{ijl} w(t) \right],
\end{cases} \tag{6.4}$$

where $\tilde{\delta} \in \Xi_l$, $1 \leq l \leq r$, $\bar{x}(t) = \left[x^T(t)\ \hat{x}^T(t) \right]^T$,

$$\bar{A}_{ijl} = \begin{bmatrix} A_i + B_{2i} D_{Kl} C_{2i} & B_{2i} C_{Kjl} \\ B_{Kjl} C_{2i} & A_{Kjl} \end{bmatrix}, \quad \bar{B}_{ijl} = \begin{bmatrix} B_{1i} + B_{2i} D_{Kl} D_{3i} \\ B_{Kjl} D_{3i} \end{bmatrix},$$
$$\bar{C}_{ijl} = \left[C_{1i} + D_{2i} D_{Kl} C_{2i}\ \ D_{2i} C_{Kjl} \right], \quad \bar{D}_{ijl} = D_{1i} + D_{2i} D_{Kl} D_{3i}.$$

Therefore, the switched control problem to be investigated in this chapter can be summarized as follows: (1) The closed-loop system (6.4) is asymptotically stable. (2) Under the assumption of zero initial condition, the controlled output $z(t)$ satisfies $\|z(t)\|_2 < \gamma \|w(t)\|_2$ for all nonzero $w(t) \in \mathcal{L}_2[0, \infty)$.

6.3 Main Results

In this section, the stability conditions for the closed-loop system (6.4) with the H_∞ performance are first proposed in Theorem 6.2. Lemma 1.4 considers some matrix properties. Then, based on Theorem 6.2 and Lemma 1.4, the convex sufficient conditions of switched controller for IT2 fuzzy systems with an H_∞ performance are proposed.

Theorem 6.2 *The closed-loop system (6.4) is asymptotically stable with an H_∞ performance level γ, if there exists matrix $P > 0$ with appropriate dimensions, such that the following LMIs hold for $i, j = 1, 2, \ldots, r$:*

$$\Omega_{iil} < 0, \tag{6.5}$$
$$\Omega_{ijl} + \Omega_{jil} < 0, \quad 1 \leq i < j \leq r, \tag{6.6}$$

where

$$\Omega_{ijl} = \begin{bmatrix} \mathbf{He}(P\bar{A}_{ijl}) & P\bar{B}_{ijl} & \bar{C}_{ijl}^T \\ * & -\gamma^2 I & \bar{D}_{ijl}^T \\ * & * & -I \end{bmatrix}, \quad 1 \leq l \leq r, \ \tilde{\delta} \in \Xi_l.$$

Proof Consider the Lyapunov function as follows:

$$V(t) = \bar{x}^T(t) P \bar{x}(t).$$

Then the time derivative of $V(t)$ is expressed as:

$$\dot{V}(t) = 2\bar{x}^T(t) P \dot{\bar{x}}(t) = \sum_{i=1}^{r} \sum_{j=1}^{r} \tilde{\delta}_i \tilde{\delta}_j 2\bar{x}^T(t) P \left[\bar{A}_{ijl} \bar{x}(t) + \bar{B}_{ijl} w(t) \right].$$

Thus, we have

$$\begin{aligned}
&\dot{V}(t) + z^T(t)z(t) - \gamma^2 w^T(t)w(t) \\
&\le \sum_{i=1}^{r}\sum_{j=1}^{r} \tilde{\delta}_i\tilde{\delta}_j \left[\bar{x}^T(t)\left(P\bar{A}_{ijl} + \bar{A}_{ijl}^T P + \bar{C}_{ijl}^T\bar{C}_{ijl}\right)\bar{x}(t)\right. \\
&\quad \left. + 2\bar{x}^T(t)\left(P\bar{B}_{ijl} + \bar{C}_{ijl}^T\bar{D}_{ijl}\right)w(t) + w^T(t)\left(-\gamma^2 I + \bar{D}_{ijl}^T\bar{D}_{ijl}\right)w(t)\right] \\
&= \sum_{i=1}^{r}\sum_{j=1}^{r} \tilde{\delta}_i\tilde{\delta}_j \xi^T(t)\bar{\Omega}_{ijl}\xi(t),
\end{aligned}$$

where $\xi(t) = \left[\bar{x}^T(t)\ w^T(t)\right]^T$, and

$$\bar{\Omega}_{ijl} = \begin{bmatrix} \mathbf{He}\left(P\bar{A}_{ijl}\right) + \bar{C}_{ijl}^T\bar{C}_{ijl} & P\bar{B}_{ijl} + \bar{C}_{ijl}^T\bar{D}_{ijl} \\ * & -\gamma^2 I + \bar{D}_{ijl}^T\bar{D}_{ijl} \end{bmatrix}.$$

Moreover, we know that

$$\sum_{i=1}^{r}\sum_{j=1}^{r} \tilde{\delta}_i\tilde{\delta}_j\bar{\Omega}_{ijl} = \sum_{i=1}^{r}\sum_{j=i}^{r} \tilde{\delta}_i\tilde{\delta}_i\bar{\Omega}_{iil} + \sum_{i=1}^{r-1}\sum_{j=i+1}^{r} \tilde{\delta}_i\tilde{\delta}_j\left(\bar{\Omega}_{ijl} + \bar{\Omega}_{jil}\right).$$

Then, applying Schur complement to (6.5) and (6.6), we can obtain

$$\sum_{i=1}^{r}\sum_{j=1}^{r} \tilde{\delta}_i\tilde{\delta}_j\bar{\Omega}_{ijl} < 0.$$

Then, we can have

$$\dot{V}(t) + z^T(t)z(t) - \gamma^2 w^T(t)w(t) < 0. \tag{6.7}$$

Under zero initial conditions, integrating both sides of (6.7) yields $\|z(t)\|_2 < \gamma\|w(t)\|_2$ for all nonzero $w(t) \in \mathcal{L}_2[0, \infty)$. In addition, when $w(t) \equiv 0$, it can be seen from (6.7) that $\dot{V}(t) < 0$, which means that the system (6.4) is asymptotically stable. The proof is completed. □

6.3.1 Switched Output-Feedback Control

Based on Lemma 1.4, the following conditions for switched controller for the IT2 fuzzy system with an H_∞ performance can be obtained.

Theorem 6.3 *The closed-loop system (6.4) is asymptotically stable with an H_∞ performance level γ, if there exist matrices $X = X^T$, $Y = Y^T$, $G_{ijl} = G_{jil}^T$, $H_{ijl} \geq 0$, $\hat{A}_{Kjl}$, $\hat{B}_{Kjl}$, $\hat{C}_{Kjl}$, and $\hat{D}_{Kl}$ with appropriate dimensions such that the following LMIs hold for $i, j, l = 1, 2, \ldots, r$:*

$$\begin{bmatrix} X & I \\ I & Y \end{bmatrix} > 0, \tag{6.8}$$

$$\Psi_{iil} - G_{iil} < 0, \tag{6.9}$$

$$\hat{\Psi}_{ijl} + \hat{\Psi}_{jil} < 0, \quad 1 \leq i < j \leq r, \tag{6.10}$$

$$\left[G_{ijl}\right]_{r\times r} + \mathbf{He}\left(F_l^T \left[H_{ijl}\right]_{(r-1)\times r}\right) < 0, \quad 1 \leq l \leq r, \tag{6.11}$$

where

$$\Psi_{iil} = \begin{bmatrix} \Theta_{1iil} & \Theta_{2iil} & \Theta_{3iil} \\ * & -\gamma^2 I & D_{1i}^T + D_{3i}^T \hat{D}_{Kl}^T D_{2i}^T \\ * & * & -I \end{bmatrix},$$

$$\hat{\Psi}_{ijl} + \hat{\Psi}_{jil} = \begin{bmatrix} \Theta_{1ijl} + \Theta_{1jil} & \Theta_{2ijl} + \Theta_{2jil} & \Theta_{3ijl} + \Theta_{3jil} & \Theta_{4ijl} & \Theta_{5ijl} \\ * & -2\gamma^2 I & \Theta_{6ijl} + \Theta_{6jil} & 0 & 0 \\ * & * & -2I & 0 & 0 \\ * & * & * & \Theta_7 & 0 \\ * & * & * & * & \Theta_7 \end{bmatrix},$$

$$F_l = \begin{bmatrix} -I & 0 & \ldots & 0 & I & 0 & \ldots & 0 \\ 0 & -I & \ldots & 0 & I & 0 & \ldots & 0 \\ \vdots & \vdots & \ddots & \vdots & \vdots & \vdots & \ddots & \vdots \\ 0 & 0 & \ldots & -I & I & 0 & \ldots & 0 \\ 0 & 0 & \ldots & 0 & I & -I & \ldots & 0 \\ \vdots & \vdots & \ddots & \vdots & \vdots & \vdots & \ddots & \vdots \\ 0 & 0 & \ldots & 0 & I & 0 & \ldots & -I \end{bmatrix}_{(r-1)\times r},$$

$$\Theta_{1ijl} = \begin{bmatrix} \mathbf{He}\left(A_i X + B_{2i}\hat{C}_{Kjl}\right) & \hat{A}_{Kjl}^T + A_i + B_{2i}\hat{D}_{Kl}C_{2j} \\ * & \mathbf{He}\left(YA_i + \hat{B}_{Kjl}C_{2i}\right) \end{bmatrix},$$

$$\Theta_{3ijl} = \begin{bmatrix} XC_{1i}^T + \hat{C}_{Kjl}^T D_{2i}^T \\ C_{1i}^T + C_{2j}^T \hat{D}_{Kl}^T D_{2i}^T \end{bmatrix}, \quad \Theta_{4ijl} = \begin{bmatrix} 0 & X\left(C_{2i} - C_{2j}\right)^T \\ \hat{B}_{Kjl} - \hat{B}_{Kil} & 0 \end{bmatrix},$$

$$\Theta_{5ijl} = \begin{bmatrix} 0 & (\hat{C}_{Kjl} - \hat{C}_{Kil})^T \\ Y(B_{2i} - B_{2j}) & 0 \end{bmatrix}, \quad \Theta_{6ijl} = D_{1i}^T + D_{3j}^T \hat{D}_{Kl}^T D_{2i}^T,$$

$$\Theta_{2ijl} = \begin{bmatrix} B_{1i} + B_{2i}\hat{D}_{Kl}D_{3j} \\ YB_{1i} + \hat{B}_{Kjl}D_{3i} \end{bmatrix}, \quad \Theta_7 = \begin{bmatrix} -I & 0 \\ 0 & -I \end{bmatrix}. \tag{6.12}$$

The controller gain matrices can be computed by

$$\begin{aligned}
D_{Kl} &= \hat{D}_{Kl},\\
C_{Kjl} &= \left(\hat{C}_{Kjl} - \hat{D}_{Kl}C_{2j}X\right)U^{-T},\\
B_{Kjl} &= V^{-1}\left(\hat{B}_{Kjl} - YB_{2j}\hat{D}_{Kl}\right),\\
A_{Kjl} &= V^{-1}\left(\hat{A}_{Kjl} - YA_jX - \hat{B}_{Kjl}C_{2j}X\right.\\
&\qquad \left. - YB_{2j}\hat{C}_{Kjl} + YB_{2j}\hat{D}_{Kl}C_{2j}X\right)U^{-T},
\end{aligned} \tag{6.13}$$

where the matrices U and V satisfy the conditions of Lemma 1.4.

Proof Let

$$P = \begin{bmatrix} Y & V \\ V^T & -U^{-1}XV \end{bmatrix}, \quad P^{-1} = \begin{bmatrix} X & U \\ U^T & -V^{-1}YU \end{bmatrix},$$

where X, Y, U and V satisfy the conditions of Lemma 1.4, and partition M as $M = \begin{bmatrix} X & I \\ U^T & 0 \end{bmatrix}$, then we have

$$\begin{aligned}
M^T P M &= \begin{bmatrix} X & I \\ U^T & 0 \end{bmatrix}^T \begin{bmatrix} Y & V \\ V^T & -U^{-1}XV \end{bmatrix} \begin{bmatrix} X & I \\ U^T & 0 \end{bmatrix}\\
&= \begin{bmatrix} I & 0 \\ Y & V \end{bmatrix} \begin{bmatrix} X & I \\ U^T & 0 \end{bmatrix} = \begin{bmatrix} X & I \\ I & Y \end{bmatrix}.
\end{aligned}$$

Since U is nonsingular, we can see that M is also a nonsingular matrix. From the condition of (6.8), one can obtain $P > 0$. According to the scheme discussed in [46], we consider

$$\begin{aligned}
&\begin{bmatrix} \tilde{\delta}_1 I \\ \vdots \\ \tilde{\delta}_r I \end{bmatrix}^T \mathbf{He}\left(F_l^T \left[H_{ijl}\right]_{(r-1)\times r}\right) \begin{bmatrix} \tilde{\delta}_1 I \\ \vdots \\ \tilde{\delta}_r I \end{bmatrix}\\
&= \mathbf{He}\left(\begin{bmatrix} \left(\tilde{\delta}_l - \tilde{\delta}_1\right) I \\ \vdots \\ \left(\tilde{\delta}_l - \tilde{\delta}_{l-1}\right) I \\ \left(\tilde{\delta}_l - \tilde{\delta}_{l+1}\right) I \\ \vdots \\ \left(\tilde{\delta}_l - \tilde{\delta}_r\right) I \end{bmatrix}^T \left[H_{ijl}\right]_{(r-1)\times r} \begin{bmatrix} \tilde{\delta}_1 I \\ \vdots \\ \tilde{\delta}_r I \end{bmatrix} \right).
\end{aligned} \tag{6.14}$$

If $\tilde{\delta} \in \Xi_l$, then one can have $\tilde{\delta}_l - \tilde{\delta}_i \geq 0,\ 1 \leq i \neq l \leq r$. From (6.14) and $H_{ijl} \geq 0$, we can obtain

$$\begin{bmatrix} \tilde{\delta}_1 I \\ \vdots \\ \tilde{\delta}_r I \end{bmatrix}^T \mathbf{He}\left(F_l^T \left[H_{ijl}\right]_{(r-1)\times r}\right) \begin{bmatrix} \tilde{\delta}_1 I \\ \vdots \\ \tilde{\delta}_r I \end{bmatrix} \geq 0, \quad \tilde{\delta} \in \Xi_l, \quad 1 \leq l \leq r. \quad (6.15)$$

Then, pre- and post-multiplying (6.11) by $\left[\tilde{\delta}_1 I, \ldots, \tilde{\delta}_r I\right]$ and its transpose, respectively, we can obtain

$$\begin{bmatrix} \tilde{\delta}_1 I \\ \vdots \\ \tilde{\delta}_r I \end{bmatrix}^T \left[G_{ijl}\right]_{r\times r} \begin{bmatrix} \tilde{\delta}_1 I \\ \vdots \\ \tilde{\delta}_r I \end{bmatrix} + \begin{bmatrix} \tilde{\delta}_1 I \\ \vdots \\ \tilde{\delta}_r I \end{bmatrix}^T \mathbf{He}\left(F_l^T \left[H_{ijl}\right]_{(r-1)\times r}\right) \begin{bmatrix} \tilde{\delta}_1 I \\ \vdots \\ \tilde{\delta}_r I \end{bmatrix} < 0.$$

From (6.15), we can obtain

$$\begin{bmatrix} \tilde{\delta}_1 I \\ \vdots \\ \tilde{\delta}_r I \end{bmatrix}^T \left[G_{ijl}\right]_{r\times r} \begin{bmatrix} \tilde{\delta}_1 I \\ \vdots \\ \tilde{\delta}_r I \end{bmatrix} < 0, \quad 1 \leq l \leq r,$$

which means

$$G_{iil} < 0.$$

Then, it can be found from (6.9) that

$$\Psi_{iil} < 0. \quad (6.16)$$

Using Schur complement to $\hat{\Psi}_{ijl} + \hat{\Psi}_{jil} < 0$, it can be seen:

$$\begin{bmatrix} \Theta_{1ijl} + \Theta_{1jil} & \Theta_{2ijl} + \Theta_{2jil} & \Theta_{3ijl} + \Theta_{3jil} \\ * & -2\gamma^2 I & \Theta_{6ijl} + \Theta_{6jil} \\ * & * & -2I \end{bmatrix} + \sum_{j=1}^{2} \left(\phi_j \phi_j^T + \psi_j \psi_j^T\right) < 0,$$

where

$$\phi_1 = \begin{bmatrix} 0 \\ \hat{B}_{Kjl} - \hat{B}_{Kil} \\ 0_{2\times 1} \end{bmatrix}, \quad \phi_2 = \begin{bmatrix} 0 \\ Y\left(B_{2i} - B_{2j}\right) \\ 0_{2\times 1} \end{bmatrix},$$

$$\psi_1 = \begin{bmatrix} X\left(C_{2i} - C_{2j}\right)^T \\ 0_{3\times 1} \end{bmatrix}, \quad \psi_2 = \begin{bmatrix} \left(\hat{C}_{Kjl} - \hat{C}_{Kil}\right)^T \\ 0_{3\times 1} \end{bmatrix}.$$

Then, one can have

$$\sum_{j=1}^{2}\left(\phi_j\phi_j^T+\psi_j\psi_j^T\right)\geq\sum_{j=1}^{2}\left(\phi_j\psi_j^T+\psi_j\phi_j^T\right),$$

which concludes

$$\begin{bmatrix}\Theta_{1ijl}+\Theta_{1jil} & \Theta_{2ijl}+\Theta_{2jil} & \Theta_{3ijl}+\Theta_{3jil}\\ * & -2\gamma^2 I & \Theta_{6ijl}+\Theta_{6jil}\\ * & * & -2I\end{bmatrix}+\sum_{j=1}^{2}\left(\phi_j\psi_j^T+\psi_j\phi_j^T\right)<0. \quad (6.17)$$

Consider the following equation:

$$\begin{aligned}&YA_iX+VA_{Kil}U^T+\hat{B}_{Kil}C_{2i}X+YB_{2i}\hat{C}_{Kil}-YB_{2i}\hat{D}_{Kl}C_{2i}X\\&\quad+YA_jX+VA_{Kjl}U^T+\hat{B}_{Kjl}C_{2j}X+YB_{2j}\hat{C}_{Kjl}-YB_{2j}\hat{D}_{Kl}C_{2j}X\\&\quad+\left(\hat{B}_{Kjl}-\hat{B}_{Kil}\right)\left(C_{2i}-C_{2j}\right)X+Y\left(B_{2i}-B_{2j}\right)\left(\hat{C}_{Kjl}-\hat{C}_{Kil}\right)\\&\qquad=YA_iX+VA_{Kjl}U^T+\hat{B}_{Kjl}C_{2i}X+YB_{2i}\hat{C}_{Kjl}-YB_{2i}\hat{D}_{Kl}C_{2i}X\\&\qquad\quad+YA_jX+VA_{Kil}U^T+\hat{B}_{Kil}C_{2j}X\\&\qquad\quad+YB_{2j}\hat{C}_{Kil}-YB_{2j}\hat{D}_{Kl}C_{2j}X.\end{aligned} \quad (6.18)$$

From (6.13), we can have:

$$\begin{aligned}\hat{A}_{Kjl}&=YA_jX+VA_{Kjl}U^T+\hat{B}_{Kjl}C_{2j}X+YB_{2j}\hat{C}_{Kjl}-YB_{2j}\hat{D}_{Kl}C_{2j}X,\\\hat{B}_{Kjl}&=VB_{Kjl}+YB_{2j}\hat{D}_{Kl},\\\hat{C}_{Kjl}&=C_{Kjl}U^T+\hat{D}_{Kl}C_{2j}X,\\\hat{D}_{Kl}&=D_{Kl}.\end{aligned} \quad (6.19)$$

Combining (6.17)–(6.19), we can obtain

$$\Psi_{ijl}+\Psi_{jil}<0. \quad (6.20)$$

Then, pre- and post-multiplying Ω_{iil} and $\Omega_{ijl}+\Omega_{jil}$ by diag$\{M^T, I, I\}$ and its transpose respectively, we can have

$$\begin{aligned}\text{diag}\{M^T,I,I\}\Omega_{iil}\text{diag}\{M,I,I\}&=\Psi_{iil},\\\text{diag}\{M^T,I,I\}\left(\Omega_{ijl}+\Omega_{jil}\right)\text{diag}\{M,I,I\}&=\Psi_{ijl}+\Psi_{jil}.\end{aligned}$$

Because diag$\{M^T, I, I\}$ is nonsingular and based on (6.16) and (6.20), we can obtain $\Omega_{iil} < 0$ and $\Omega_{ijl} + \Omega_{jil} < 0$. According to Theorem 6.2, the closed-loop system (6.4) is asymptotically stable with an H_∞ performance. The proof is completed. □

6.3.2 Switched State-Feedback Control

In the following part, we will consider a switched state-feedback controller for the IT2 fuzzy system (6.2). First, the specific switched state-feedback controller is proposed as follows:

$$u(t) = \sum_{j=1}^{r} \tilde{\delta}_j(x(t)) K_{jl} x(t), \quad \tilde{\delta} \in \Xi_l, \quad 1 \le l \le r, \tag{6.21}$$

where K_{jl} is the controller parameter to be designed.

It can be seen from (6.2) and (6.21) that we can obtain closed-loop system as follows:

$$\begin{cases} \dot{x}(t) = \sum_{i=1}^{r}\sum_{j=1}^{r} \tilde{\delta}_i(x(t))\tilde{\delta}_j(\hat{x}(t)) \left[\left(A_i + B_{2i}K_{jl}\right)x(t) + B_{1i}w(t) \right], \\ z(t) = \sum_{i=1}^{r}\sum_{j=1}^{r} \tilde{\delta}_i(x(t))\tilde{\delta}_j(\hat{x}(t)) \left[\left(C_{1i} + D_{2i}K_{jl}\right)x(t) + D_{1i}w(t) \right], \end{cases} \tag{6.22}$$

where $\tilde{\delta} \in \Xi_l$, $1 \le l \le r$. Then, the stability condition with H_∞ performance of the closed-loop system (6.22) is presented in the following corollary.

Corollary 6.4 *The closed-loop system (6.22) is asymptotically stable with an H_∞ performance level γ, if there exists matrix $\hat{P} > 0$ with appropriate dimensions, such that the following LMIs hold for $i, j, l = 1, 2, \ldots, r$:*

$$\sum_{i=1}^{r}\sum_{j=1}^{r} \tilde{\delta}_i\tilde{\delta}_j \begin{bmatrix} \mathbf{He}\left(\hat{P}A_i + \hat{P}B_{2i}K_{jl}\right) & \hat{P}B_{1i} & \left(C_{1i} + D_{2i}K_{jl}\right)^T \\ * & -\gamma^2 I & D_{1i}^T \\ * & * & -I \end{bmatrix} < 0.$$

Proof This proof is similar to that of Theorem 6.2, thus it is omitted. □

Based on Corollary 6.4, the existence conditions of the H_∞ switched state-feedback controller for the IT2 fuzzy system are given in Corollary 6.5.

Corollary 6.5 *The closed-loop system (6.22) is asymptotically stable with an* H_∞ *performance level* γ*, if there exist matrices* $\bar{P} > 0$, $\bar{G}_{ijl} = \bar{G}^T_{jil}$, $\bar{H}_{ijl} \geq 0$, M_{jl} *with appropriate dimensions such that the following LMIs hold for* $i, j, l = 1, 2, \ldots, r$*:*

$$\begin{aligned} \bar{\Psi}_{iil} &< \bar{G}_{iil}, \\ \bar{\Psi}_{ijl} + \bar{\Psi}_{jil} &< \bar{G}_{ijl} + \bar{G}^T_{ijl}, \quad 1 \leq i < j \leq r, \\ \left[\bar{G}_{ijl}\right]_{r\times r} + \mathbf{He}\left(\bar{F}_l^T \left[\bar{H}_{ijl}\right]_{(r-1)\times r}\right) &< 0, \end{aligned}$$

where

$$\bar{F}_l = \begin{bmatrix} -I & 0 & \ldots & 0 & I & 0 & \ldots & 0 \\ 0 & -I & \ldots & 0 & I & 0 & \ldots & 0 \\ \vdots & \vdots & \ddots & \vdots & \vdots & \vdots & \ddots & \vdots \\ 0 & 0 & \ldots & -I & I & 0 & \ldots & 0 \\ 0 & 0 & \ldots & 0 & I & -I & \ldots & 0 \\ \vdots & \vdots & \ddots & \vdots & \vdots & \vdots & \ddots & \vdots \\ 0 & 0 & \ldots & 0 & I & 0 & \ldots & -I \end{bmatrix}_{(r-1)\times r},$$

$$\bar{\Psi}_{ijl} = \begin{bmatrix} \mathbf{He}\left(A_i\bar{P} + B_{2i}M_{jl}\right) & B_{1i} & \left(C_{1i}\bar{P} + D_{2i}M_{jl}\right)^T \\ * & -\gamma^2 I & D^T_{1i} \\ * & * & -I \end{bmatrix}.$$

Moreover, the controller parameter is given by

$$K_{jl} = M_{jl}\bar{P}^{-1}.$$

Proof The desired result can be carried out by employing the same techniques used as those in Theorem 1 discussed in [46] thus the proof is omitted. □

6.4 Simulation Results

In this section, a practical example is used to show the effectiveness and the merit of the proposed method.

Example 6.6 Consider the mass-spring-damping system shown in Fig. 6.1. According to Newton's law, we can obtain:

$$m\ddot{x} + F_f + F_s = u\left(t\right),$$

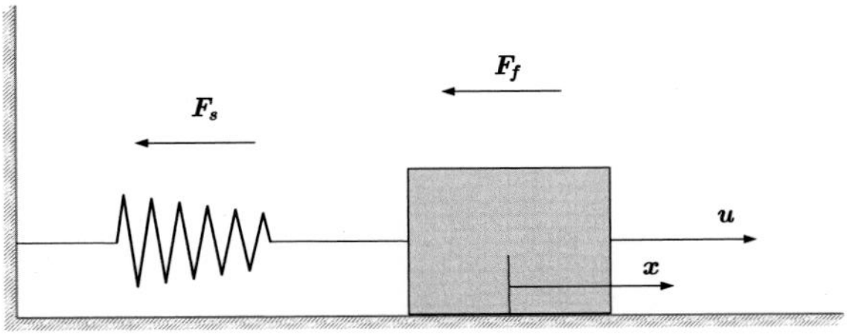

Fig. 6.1 Mass-spring-damping system. © [2015] IEEE. Reprinted, with permission, from ref. [5]

where m stands for the mass, F_f is the friction force, F_s and $u(t)$ denote the restoring force of the spring and the external control input, respectively. The friction force $F_f = c\dot{x}$ with $c > 0$ and the hardening spring force $F_s = \hat{k}\left(1 + a^2x^2\right)x$ with constants $\hat{k}$ and a. Then, the dynamic equation can be written as:

$$m\ddot{x} + c\dot{x} + \hat{k}x + \hat{k}a^2x^3 = u(t),$$

where x stands for the displacement from a reference point. Define

$$x(t) = \begin{bmatrix} x_1(t) \\ x_2(t) \end{bmatrix} = \begin{bmatrix} x \\ \dot{x} \end{bmatrix}, \quad \bar{f}(t) = \frac{-\hat{k} - \hat{k}a^2x_1^2(t)}{m}.$$

Let $x_1(t) \in [-2, 2]$, $m = 1\,\text{kg}$, $c = 2\,\text{N·m/s}$, and $a = 0.3\,\text{m}^{-1}$. Assume $\hat{k} \in [5, 8]\,\text{N/m}$, then $\bar{f}_{\max} = -5$ with $\hat{k} = 5$ and $x_1(t) = 0$. $\bar{f}_{\min} = -10.88$ with $\hat{k} = 8$ and $x_1(t) = \pm 2$. The matrices for the IT2 T–S fuzzy system (6.1) can be obtained:

$$A_1 = \begin{bmatrix} 0 & 1 \\ \bar{f}_{\min} & -\frac{c}{m} \end{bmatrix}, \quad A_2 = \begin{bmatrix} 0 & 1 \\ \bar{f}_{\max} & -\frac{c}{m} \end{bmatrix}, \quad B_{11} = \begin{bmatrix} -0.02 \\ -0.01 \end{bmatrix},$$
$$B_{12} = \begin{bmatrix} 0.01 \\ 0.02 \end{bmatrix}, \quad B_{21} = \begin{bmatrix} 0 \\ \frac{1}{m} \end{bmatrix}, \quad B_{22} = \begin{bmatrix} 0 \\ \frac{1}{m} \end{bmatrix}, \quad C_{11} = \begin{bmatrix} 0.1 & 0.1 \end{bmatrix},$$
$$C_{12} = \begin{bmatrix} 0.1 & 0.1 \end{bmatrix}, \quad D_{11} = -0.6, \quad D_{12} = 0.3, \quad D_{21} = 0.8, \quad D_{22} = 1.1,$$
$$C_{21} = \begin{bmatrix} 1 & 0 \end{bmatrix}, \quad C_{22} = \begin{bmatrix} 1 & 0 \end{bmatrix}, \quad D_{31} = -0.007, \quad D_{32} = 0.003.$$

According to the uncertain parameter $\hat{k}$, the LMFs and UMFs for IT2 T–S fuzzy system can be achieved respectively as follows:

$$\underline{\delta}_1(x_1(t)) = \frac{-\bar{f}\left(t, \hat{k} = 5\right) + \bar{f}_{\max}}{\bar{f}_{\max} - \bar{f}_{\min}}, \quad \bar{\delta}_1(x_1(t)) = \frac{-\bar{f}\left(t, \hat{k} = 8\right) + \bar{f}_{\max}}{\bar{f}_{\max} - \bar{f}_{\min}},$$
$$\underline{\delta}_2(x_1(t)) = \frac{\bar{f}\left(t, \hat{k} = 8\right) - \bar{f}_{\min}}{\bar{f}_{\max} - \bar{f}_{\min}}, \quad \bar{\delta}_2(x_1(t)) = \frac{\bar{f}\left(t, \hat{k} = 5\right) - \bar{f}_{\min}}{\bar{f}_{\max} - \bar{f}_{\min}}.$$

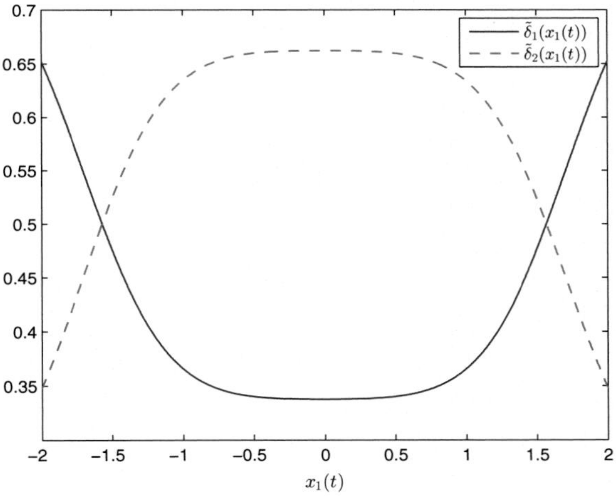

Fig. 6.2 Membership functions of the IT2 fuzzy system

Weighting functions are defined as:

$$\underline{\varsigma}_i\,(x_1\,(t)) = 0.6\sin^2\,(x_1\,(t))\,, \quad \bar{\varsigma}_i\,(x_1\,(t)) = 1 - \underline{\varsigma}_i\,(x_1\,(t))\,.$$

The membership functions for IT2 fuzzy system are shown in Fig. 6.2. According to the conditions of Theorem 6.3, we can obtain the controller parameters:

$$A_{K11} = \begin{bmatrix} -2.8998 & -0.7108 \\ 8.2562 & -0.7437 \end{bmatrix}, \quad A_{K12} = \begin{bmatrix} -2.2575 & -0.8618 \\ 9.9344 & -1.0806 \end{bmatrix},$$
$$A_{K21} = \begin{bmatrix} -3.3982 & -0.5610 \\ 6.6149 & 0.5408 \end{bmatrix}, \quad A_{K22} = \begin{bmatrix} -3.1949 & -0.6860 \\ 8.3397 & -0.2379 \end{bmatrix},$$
$$B_{K11} = \begin{bmatrix} -0.4813 \\ -0.2940 \end{bmatrix}, \quad B_{K21} = \begin{bmatrix} -0.4039 \\ -0.4038 \end{bmatrix},$$
$$B_{K12} = \begin{bmatrix} -0.2641 \\ 0.2081 \end{bmatrix}, \quad B_{K22} = \begin{bmatrix} -0.2771 \\ 0.4909 \end{bmatrix},$$
$$C_{K11} = \begin{bmatrix} -2.3728 & -0.2464 \end{bmatrix}, \quad C_{K21} = \begin{bmatrix} 1.0238 & 0.0980 \end{bmatrix},$$
$$C_{K12} = \begin{bmatrix} 0.2607 & -0.0904 \end{bmatrix}, \quad C_{K22} = \begin{bmatrix} -0.6078 & -0.5003 \end{bmatrix},$$
$$D_{K1} = -0.1151, \quad D_{K2} = -0.1029.$$

Moreover, if the value of parameter $\hat{k} = 8\,\text{N/m}$ rather than $\hat{k} \in [5, 8]\,\text{N/m}$, the type-1 switched dynamic output-feedback controller method [46] can be designed for the system under the same example. The matrices can be obtained:

$$A_1 = \begin{bmatrix} 0 & 1 \\ \frac{-\hat{k}-4\hat{k}a^2}{m} & -\frac{c}{m} \end{bmatrix}, \quad A_2 = \begin{bmatrix} 0 & 1 \\ \frac{-\hat{k}}{m} & -\frac{c}{m} \end{bmatrix}, \quad B_{21} = \begin{bmatrix} 0 \\ \frac{1}{m} \end{bmatrix}, \quad B_{22} = \begin{bmatrix} 0 \\ \frac{1}{m} \end{bmatrix}.$$

Other parameters are same as the IT2 switched controller. The membership functions for type-1 T–S fuzzy system can be defined as $a_1(x_1(t)) = \frac{x_1^2(t)}{4}$, $a_2(x_1(t)) = 1 - a_1(x_1(t))$.

Remark 6.7 It can be seen that the type-1 T–S fuzzy system can model the plant when no uncertain parameters exist in the plant. However, if the value of parameter $\hat{k}$ changes in a range rather than a fixed value, it is clear that the type-1 T–S fuzzy system can not handle the plant with fixed membership functions. Thus, on handling parameter uncertainties, the type-2 switched controller proposed in this chapter is better than the type-1 switched controller [46]. Moreover, the following figures illuminate that the type-2 switched controller can obtain better performance than the type-1 switched controller.

In order to show the advantages of the type-2 switched controller proposed in this chapter over the type-1 switched controller in [46], it is assumed that the switched controller proposed in Theorem 6.3 and [46] use the same $w(t)$:

$$w(t) = \begin{cases} -2, & 0.4 \le t \le 0.8, \\ 2, & 1.2 \le t \le 1.6, \\ 0, & \text{else.} \end{cases}$$

Fig. 6.3 Responses of the state $x_1(t)$

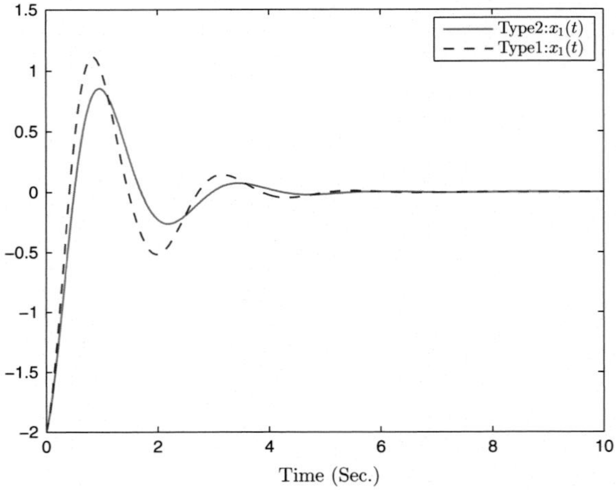

It can be found from Theorem 6.3 that the obtained type-2 switched controller optimal H_∞ performance index $\gamma = 3.2455$. Moreover, for [46], it can be found that the type-1 switched controller optimal H_∞ performance index $\gamma = 3.8983$. From the two indexes, it can be seen that the IT2 fuzzy system switched controller proposed in this chapter can obtain better performance than the type-1 switched control scheme [46]. To further show the advantages of the proposed IT2 switched control approach over the type-1 switched control scheme [46], we consider the following pictures. Figures 6.3 and 6.4 plot the response of the states $x_1(t)$ and $x_2(t)$ and Figs. 6.5 and 6.6 plot the response of the states $\hat{x}_1(t)$ and $\hat{x}_2(t)$, under the initial conditions of $x(t) = \left[-2\ 2\right]^T$ and $\hat{x}(t) = \left[-2.8\ 2.8\right]^T$, respectively. The control input $u(t)$ and output $z(t)$ are depicted in Figs. 6.7 and 6.8, respectively. These results illuminate

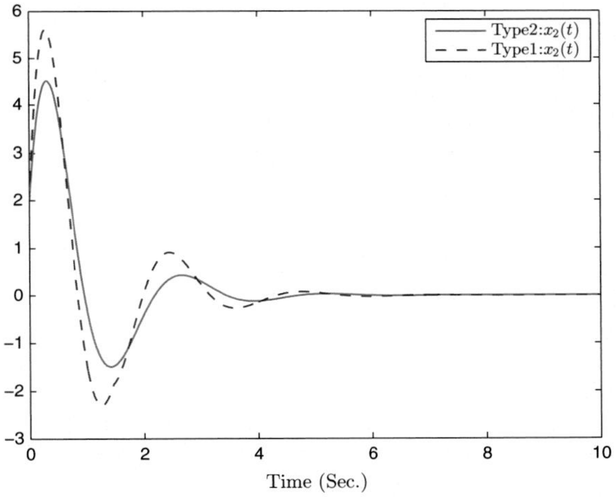

Fig. 6.4 Responses of the state $x_2(t)$

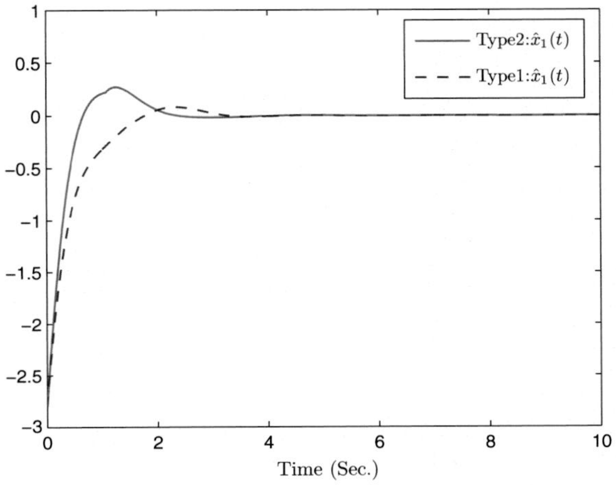

Fig. 6.5 Responses of the state $\hat{x}_1(t)$

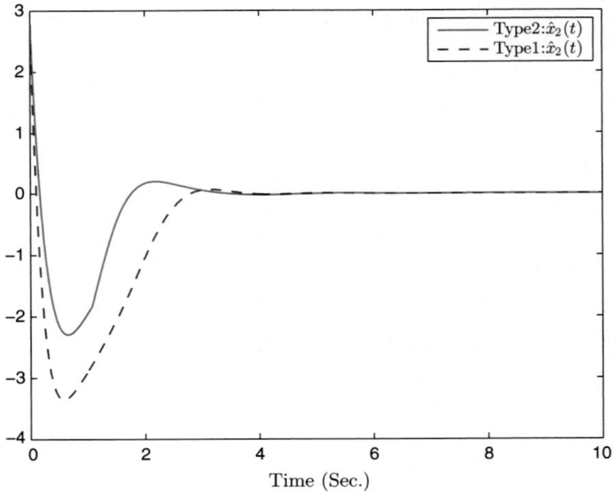

Fig. 6.6 Responses of the state $\hat{x}_2\ (t)$

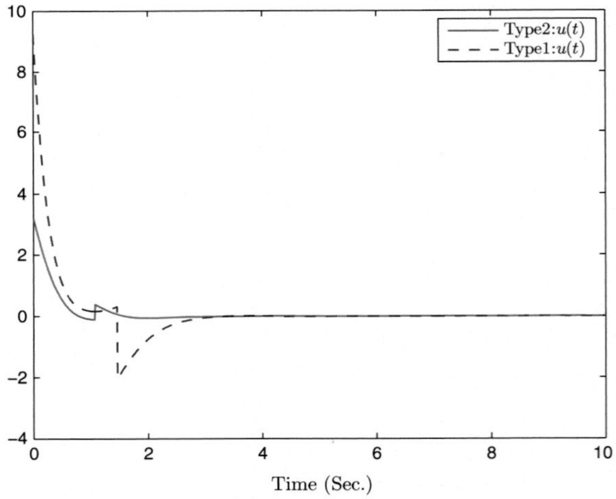

Fig. 6.7 Control input $u\ (t)$

the merits of the proposed IT2 switched controller approach, and it is clear that the IT2 fuzzy system switched controller can obtain better performance than the type-1 switched control scheme.

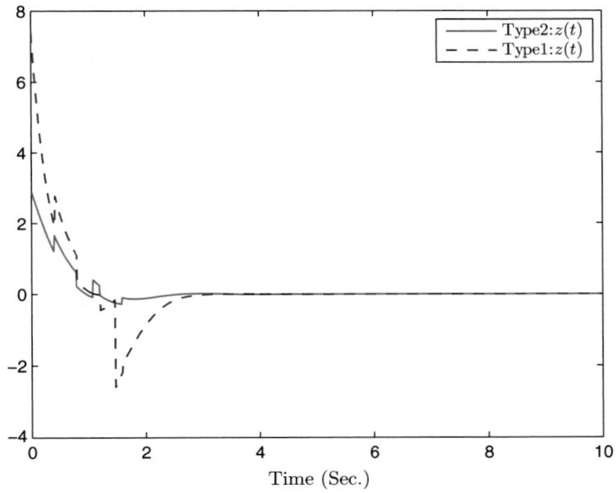

Fig. 6.8 Control output $z(t)$

6.5 Conclusion

In this chapter, the H_∞ switched control problem has been considered for a class of IT2 fuzzy systems. Firstly, the IT2 fuzzy system and the switched controller have been constructed. The switched controller has been proposed to guarantee that the closed-loop system is asymptotically stable with an H_∞ performance. In the design procedure, the parameters of the switched controller can be obtained by the standard software. Finally, a practical example has been given to show the merits of the proposed approach. The advantages of the switched controller for IT2 fuzzy systems have been clearly confirmed over type-1 results.

Chapter 7
Filter Design of Interval Type-2 Fuzzy-Model-Based Systems

7.1 Introduction

This chapter considers the problem of filter design for IT2 fuzzy systems with $\mathcal{D}$-stability constraints based on a new performance index. Firstly, IT2 fuzzy model and IT2 fuzzy filter are established and they do not share the same LMFs and UMFs. Secondly, the new performance index is introduced and covers H_∞, L_2-L_∞, passive and dissipativity performances. Under a unified frame, a novel type of IT2 filter is designed such that the filtering error system guarantees the prescribed H_∞, L_2-L_∞, passive and dissipativity performance levels with $\mathcal{D}$-stability constraints. The existence condition of the IT2 filter is expressed as the convex optimization problem and the filter parameters in the condition can be solved by the standard software.

7.2 Problem Formulation

Consider the following IT2 fuzzy model that represents a continuous-time nonlinear system:

♦ **Plant Form**:

Rule i: IF $f_1(x(t))$ is $\tilde{M}_1^i$ and ... and $f_p(x(t))$ is $\tilde{M}_p^i$, THEN,

$$\begin{cases} \dot{x}(t) = A_i x(t) + D_{1i} w(t), \\ z(t) = C_i x(t) + D_{2i} w(t), \\ y(t) = E_i x(t) + D_{3i} w(t), \end{cases} \tag{7.1}$$

where $\tilde{M}_a^i$ is an IT2 fuzzy set of rule i corresponding to the function $f_a(x(t))$, $i = 1, 2, \ldots, r$; $a = 1, 2, \ldots, p$; p is a positive integer; $x(t) \in \mathbf{R}^n$ is the system state vector and $w(t) \in \mathbf{R}^q$ is the disturbance input; $z(t) \in \mathbf{R}^v$ is the control output; $y(t) \in \mathbf{R}^m$ is the measure output; A_i, D_{1i} , C_i, D_{2i}, E_i and D_{3i} are the known matrices with

H. Li et al., *Analysis and Synthesis for Interval Type-2 Fuzzy-Model-Based Systems*, DOI 10.1007/978-981-10-0593-0_7

appropriate dimensions. The firing strength of the ith rule is the following interval set:

$$W_i(x(t)) = \left[\underline{w}_i(x(t)), \overline{w}_i(x(t))\right], \quad i = 1, 2, \ldots, r,$$

where $\underline{w}_i(x(t)) = \prod_{a=1}^{p} \underline{\mu}_{\tilde{M}_a^i}(f_a(x(t))) \geq 0$, $\overline{w}_i(x(t)) = \prod_{a=1}^{p} \bar{\mu}_{\tilde{M}_a^i}(f_a(x(t))) \geq 0$, $\underline{w}_i(x(t))$ and $\overline{w}_i(x(t))$ stand for the lower and upper grades of membership, respectively. $\underline{\mu}_{\tilde{M}_a^i}(f_a(x(t))) \geq 0$ and $\bar{\mu}_{\tilde{M}_a^i}(f_a(x(t))) \geq 0$ stand for the LMFs and UMFs, respectively. Therefore, it can be found that $\bar{\mu}_{\tilde{M}_a^i}(f_a(x(t))) \geq \underline{\mu}_{\tilde{M}_a^i}(f_a(x(t)))$ and $\overline{w}_i(x(t)) \geq \underline{w}_i(x(t))$ for all i. Then, the IT2 T–S fuzzy system is described as follows:

$$\begin{cases} \dot{x}(t) = \sum_{i=1}^{r} \tilde{w}_i(x(t)) \left[A_i x(t) + D_{1i} w(t)\right], \\ z(t) = \sum_{i=1}^{r} \tilde{w}_i(x(t)) \left[C_i x(t) + D_{2i} w(t)\right], \\ y(t) = \sum_{i=1}^{r} \tilde{w}_i(x(t)) \left[E_i(t) + D_{3i} w(t)\right], \end{cases} \tag{7.2}$$

where

$$\tilde{w}_i(x(t)) = \underline{a}_i(x(t)) \underline{w}_i(x(t)) + \bar{a}_i(x(t)) \overline{w}_i(x(t)) \geq 0, \quad \forall i,$$
$$\sum_{i=1}^{r} \tilde{w}_i(x(t)) = 1,$$

and $0 \leq \underline{a}_i(x(t)) \leq 1$ and $0 \leq \bar{a}_i(x(t)) \leq 1$ denote nonlinear functions and possess the trait of $\underline{a}_i(x(t)) + \bar{a}_i(x(t)) = 1$ for all i.

An IT2 fuzzy filter with r rules is constructed as follows:

$$\begin{cases} \dot{\hat{x}}(t) = A_{fj}\hat{x}(t) + B_{fj} y(t), \\ z_f(t) = C_{fj}\hat{x}(t), \end{cases} \tag{7.3}$$

where $\tilde{N}_\beta^j$ is an IT2 fuzzy set of rule j corresponding to the function $g_\beta(x(t))$, $j = 1, 2, \ldots, r$; $\beta = 1, 2, \ldots, l$; l is a positive integer; A_{fj}, B_{fj} and C_{fj} are the filter parameters to be designed. The firing strength of the jth rule is the following interval set:

$$M_j(x(t)) = \left[\underline{m}_j(x(t)), \overline{m}_j(x(t))\right], \quad j = 1, 2, \ldots, r,$$

where $\underline{m}_j(x(t)) = \prod_{\beta=1}^{l} \underline{\mu}_{\tilde{N}_\beta^j}\left(g_\beta(x(t))\right) \geq 0$ and $\overline{m}_j(x(t)) = \prod_{\beta=1}^{l} \bar{\mu}_{\tilde{N}_\beta^j}\left(g_\beta(x(t))\right) \geq 0$ stand for the lower and upper grades of membership, respectively.

$\underline{\mu}_{\tilde{N}_\beta^j}(g_\beta(x(t))) \geq 0$ and $\bar{\mu}_{\tilde{N}_\beta^j}(g_\beta(x(t))) \geq 0$ denote the LMFs and UMFs, respectively. Here, $\bar{\mu}_{\tilde{N}_\beta^j}(g_\beta(x(t))) \geq \underline{\mu}_{\tilde{N}_\beta^j}(g_\beta(x(t)))$ leading to $\overline{m}_j(x(t)) \geq \underline{m}_j(x(t))$ for all j. Assume that $x(t)$ is available for filter. The overall IT2 fuzzy filter is proposed as follows:

$$\begin{cases} \dot{\hat{x}}(t) = \sum_{j=1}^{r} \tilde{m}_j(x(t)) \left[A_{fj}\hat{x}(t) + B_{fj}y(t)\right], \\ z_f(t) = \sum_{j=1}^{r} \tilde{m}_j(x(t)) C_{fj}\hat{x}(t), \end{cases} \tag{7.4}$$

where

$$\tilde{m}_j(x(t)) = \frac{\underline{\beta}_j(x(t))\underline{m}_j(x(t)) + \bar{\beta}_j(x(t))\overline{m}_j(x(t))}{\sum_{k=1}^{r}\left(\underline{\beta}_k(x(t))\underline{m}_k(x(t)) + \bar{\beta}_k(x(t))\overline{m}_k(x(t))\right)} \geq 0, \quad \forall j,$$

$$\sum_{j=1}^{r} \tilde{m}_j(x(t)) = 1,$$

in which $0 \leq \underline{\beta}_j(x(t)) \leq 1$ and $0 \leq \bar{\beta}_j(x(t)) \leq 1$ are predefined functions and possess the trait of $\underline{\beta}_j(x(t)) + \bar{\beta}_j(x(t)) = 1$ for all j.

In order to have a simple description, we denote $\tilde{w}_i = \tilde{w}_i(x(t))$ and $\tilde{m}_j(x(t)) = \tilde{m}_j$, $i, j = 1, 2, \ldots, r$.

It can be seen from (7.2) and (7.4) that the IT2 filtering error system can be written in the following form:

$$\begin{cases} \dot{\bar{x}}(t) = \sum_{i=1}^{r}\sum_{j=1}^{r} \tilde{w}_i\tilde{m}_j \left[\bar{A}_{ij}\bar{x}(t) + \bar{D}_{1ij}w(t)\right], \\ e(t) = \sum_{i=1}^{r}\sum_{j=1}^{r} \tilde{w}_i\tilde{m}_j \left[\bar{C}_{ij}\bar{x}(t) + \bar{D}_{2i}w(t)\right], \end{cases} \tag{7.5}$$

where $\sum_{i=1}^{r} \tilde{w}_i = \sum_{j=1}^{r} \tilde{m}_j = \sum_{i=1}^{r}\sum_{j=1}^{r} \tilde{w}_i\tilde{m}_j = 1$ and

$$\bar{A}_{ij} = \begin{bmatrix} A_i & 0 \\ B_{fj}E_i & A_{fj} \end{bmatrix}, \quad \bar{D}_{1ij} = \begin{bmatrix} D_{1i} \\ B_{fj}D_{3i} \end{bmatrix}, \quad \bar{x}(t) = \begin{bmatrix} x(t) \\ \hat{x}(t) \end{bmatrix},$$
$$\bar{C}_{ij} = \begin{bmatrix} C_i & -C_{fj} \end{bmatrix}, \quad \bar{D}_{2i} = D_{2i}, \quad e(t) = z(t) - z_f(t).$$

Definition 7.1 ([230]) For given matrices Φ, Ψ_1, Ψ_2 and Ψ_3 satisfying Assumption 3.1, system (7.5) is said to be extended dissipative if there exists a scalar δ such that the following inequality holds for any $t > 0$ and all $w(t) \in \mathcal{L}_2[0, \infty)$:

$$\int_0^t J(s)ds - e^T(t)\Phi e(t) \geq \delta, \tag{7.6}$$

where $J(t) = e^T(t)\Psi_1 e(t) + 2e^T(t)\Psi_2 w(t) + w^T(t)\Psi_3 w(t)$.

Remark 7.2 The authors in [230] introduced this definition and pointed out that the performance index covers a few of well-known performance indexes. For example,

(a) Let $\Phi = 0$, $\Psi_1 = -I$, $\Psi_2 = 0$, $\Psi_3 = \gamma^2 I$ and $\delta = 0$, the inequality (7.6) reduces to the H_∞ performance [214];
(b) Let $\Phi = I$, $\Psi_1 = 0$, $\Psi_2 = 0$, $\Psi_3 = \gamma^2 I$ and $\delta = 0$, the inequality (7.6) becomes the L_2-L_∞ (energy-to-peak) performance [55];
(c) If the dimension of output $e(t)$ is the same as that of disturbance $w(t)$, then the inequality (7.6) with $\Phi = 0$, $\Psi_1 = 0$, $\Psi_2 = I$, $\Psi_3 = \gamma I$ and $\delta = 0$ becomes the passivity performance [216];
(d) Let $\Phi = 0$, $\Psi_1 = Q$, $\Psi_2 = S$, $\Psi_3 = R - \alpha I$ and $\delta = 0$, the inequality (7.6) reduces to the strict (Q, S, R)-dissipativity [116];
(e) When $\Phi = 0$, $\Psi_1 = -\epsilon I$, $\Psi_2 = I$, $\Psi_3 = -\sigma I$ with $\epsilon > 0$ and $\sigma > 0$, the inequality (7.6) becomes the very-strict passivity performance.

In the definition of the very-strict passivity performance, the scalar δ is not required to be zero. It was shown in [134] that δ should be a non-positive scalar. This fact can also be seen from Assumption 3.1 and Definition 7.1. Indeed, when $w = 0$, it follows from (7.6) that

$$\delta \leq \int_0^t e^T(s)\Psi_1 e(s)ds - e^T(t)\Phi e(t). \tag{7.7}$$

Note from Assumption 3.1 that $\Phi \geq 0$ and $\Psi_1 \leq 0$. Thus, the above inequality implies that $\delta \leq 0$.

Our purpose is to design the filter of the form (7.3) such that

(1) the filtering error system (7.5) is asymptotically stable.
(2) the filtering error system (7.5) guarantees the new performance index (7.6).
(3) the filtering error system (7.5) satisfy the $\mathcal{D}$-stability constraints (1.16).

7.3 Main Results

7.3.1 Filtering Performance Analysis

In this section, the filter design problem is considered. The following theorem presents a performance criterion for the filtering error system (7.5).

Theorem 7.3 *Given matrices Φ, Ψ_1, Ψ_2 and Ψ_3 satisfying Assumption 3.1 and the membership functions satisfying $\tilde{m}_j - \rho_j \tilde{w}_j \geq 0$ $(0 < \rho_j \leq 1)$, the IT2 filtering error system (7.5) is asymptotically stable and satisfy the index performance in Definition 7.1 and the poles lie in the disc $\mho(q, r)$, if there exist matrices $\hat{G} > 0$, $\hat{P} > 0$ and $\Lambda_i > 0$, $i = 1, 2, \ldots, r$, such that the following LMIs hold:*

$$\hat{G} - \hat{P} < 0, \tag{7.8}$$

$$\bar{C}_{ij}^T \Phi \bar{C}_{ij} - \hat{G} < 0, \tag{7.9}$$

$$\Xi_{ij} - \Lambda_i < 0, \tag{7.10}$$

$$\rho_i \Xi_{ii} - \rho_i \Lambda_i + \Lambda_i < 0, \tag{7.11}$$

$$\rho_j \Xi_{ij} + \rho_i \Xi_{ji} - \rho_j \Lambda_i - \rho_i \Lambda_j + \Lambda_i + \Lambda_j < 0, \quad i < j, \tag{7.12}$$

$$\begin{bmatrix} -\hat{P} & \hat{P}\left(\bar{A}_{ij} - qI\right) \\ * & -r^2 \hat{P} \end{bmatrix} < 0, \tag{7.13}$$

where

$$\begin{aligned}
\Xi_{ij} &= \begin{bmatrix} \Omega_{ij}^{11} & \Omega_{ij}^{12} \\ * & \Omega_{ij}^{13} \end{bmatrix}, \\
\Omega_{ij}^{11} &= \mathbf{He}\left(\hat{P}\bar{A}_{ij}\right) - \bar{C}_{ij}^T \Psi_1 \bar{C}_{ij}, \\
\Omega_{ij}^{12} &= \hat{P}\bar{D}_{1ij} - \bar{C}_{ij}^T \Psi_2 - \bar{C}_{ij}^T \Psi_1 \bar{D}_{2i}, \\
\Omega_{ij}^{13} &= -\mathbf{He}\left(\bar{D}_{2i}^T \Psi_2\right) - \Psi_3 - \bar{D}_{2i}^T \Psi_1 \bar{D}_{2i}.
\end{aligned}$$

In this case, the scalar δ involved in Definition 7.1 can be chosen as

$$\delta = -V(0). \tag{7.14}$$

Proof Consider the following Lyapunov function:

$$V(t) = \bar{x}^T(t)\hat{P}\bar{x}(t), \tag{7.15}$$

where $\hat{P} = \hat{P}^T > 0$, then the time derivative of $V(t)$ is expressed as

$$\dot{V}(t) = 2\bar{x}^T(t)\hat{P}\dot{\bar{x}}(t) = 2\sum_{i=1}^{r}\sum_{j=1}^{r} \tilde{w}_i \tilde{m}_j \bar{x}^T(t)\hat{P}\left[\bar{A}_{ij}\bar{x}(t) + \bar{D}_{1ij}w(t)\right].$$

Thus, we have

$$\begin{aligned}
\dot{V}(t) - J(t) \leq \sum_{i=1}^{r}\sum_{j=1}^{r} \tilde{w}_i \tilde{m}_j \{ & 2\bar{x}^T(t)\hat{P}\left[\bar{A}_{ij}\bar{x}(t) + \bar{D}_{1ij}w(t)\right] \\
& - \left[\bar{C}_{ij}\bar{x}(t) + \bar{D}_{2i}w(t)\right]^T \Psi_1 \left[\bar{C}_{ij}\bar{x}(t) + \bar{D}_{2i}w(t)\right]
\end{aligned}$$

$$-2\left[\bar{C}_{ij}\bar{x}(t)+\bar{D}_{2i}w(t)\right]^T\Psi_2 w(t)-w^T(t)\Psi_3 w(t)\}$$

$$=\xi^T(t)\left(\sum_{i=1}^{r}\sum_{j=1}^{r}\tilde{w}_i\tilde{m}_j\Xi_{ij}\right)\xi(t), \tag{7.16}$$

where

$$\xi(t)=\begin{bmatrix}\bar{x}(t)\\ w(t)\end{bmatrix},\quad \Xi_{ij}=\begin{bmatrix}\Omega_{ij}^{11} & \Omega_{ij}^{12}\\ * & \Omega_{ij}^{13}\end{bmatrix},$$
$$\Omega_{ij}^{11}=\mathbf{He}\left(\hat{P}\bar{A}_{ij}\right)-\bar{C}_{ij}^T\Psi_1\bar{C}_{ij},$$
$$\Omega_{ij}^{12}=\hat{P}\bar{D}_{1ij}-\bar{C}_{ij}^T\Psi_2-\bar{C}_{ij}^T\Psi_1\bar{D}_{2i},$$
$$\Omega_{ij}^{13}=-\mathbf{He}\left(\bar{D}_{2i}^T\Psi_2\right)-\Psi_3-\bar{D}_{2i}^T\Psi_1\bar{D}_{2i}.$$

It can be seen from the Eq. (7.16), if $\sum_{i=1}^{r}\sum_{j=1}^{r}\tilde{w}_i\tilde{m}_j\Xi_{ij}<0$ then $\dot{V}(t)-J(t)<0$. Consider $\sum_{i=1}^{r}\sum_{j=1}^{r}\tilde{w}_i\left(\tilde{w}_j-\tilde{m}_j\right)\Lambda_i=0$, where $\Lambda_i=\Lambda_i^T$ $(i=1,2,\ldots,r)$ is arbitrary matrix with appropriate dimensions. Then we have

$$\begin{aligned}\sum_{i=1}^{r}\sum_{j=1}^{r}\tilde{w}_i\tilde{m}_j\Xi_{ij}&=\sum_{i=1}^{r}\sum_{j=1}^{r}\tilde{w}_i\left(\tilde{w}_j-\tilde{m}_j+\rho_j\tilde{w}_j-\rho_j\tilde{w}_j\right)\Lambda_i+\sum_{i=1}^{r}\sum_{j=1}^{r}\tilde{w}_i\tilde{m}_j\Xi_{ij}\\&=\sum_{i=1}^{r}\sum_{j=i}^{r}\tilde{w}_i^2\left(\rho_i\Xi_{ii}-\rho_i\Lambda_i+\Lambda_i\right)\\&\quad+\sum_{i=1}^{r-1}\sum_{j=i+1}^{r}\tilde{w}_i\tilde{w}_j\left(\rho_j\Xi_{ij}-\rho_j\Lambda_i+\Lambda_i+\rho_i\Xi_{ji}-\rho_i\Lambda_j+\Lambda_j\right)\\&\quad+\sum_{i=1}^{r}\sum_{j=1}^{r}\tilde{w}_i\left(\tilde{m}_j-\rho_j\tilde{w}_j\right)\left(\Xi_{ij}-\Lambda_i\right).\end{aligned}$$

Under $\tilde{m}_j-\rho_j\tilde{w}_j\geq 0$ for all j, it can be seen from the inequalities (7.10)–(7.12) that

$$\dot{V}(t)-J(t)<\xi^T(t)\left(\sum_{i=1}^{r}\sum_{j=1}^{r}\tilde{w}_i\tilde{m}_j\Xi_{ij}\right)\xi(t)<0. \tag{7.17}$$

There always exists a sufficiently small scalar $c>0$ leading to $\Xi_{ij}\leq -cI$. This means that

$$\dot{V}(t)-J(t)\leq -c\left|\xi(t)\right|^2\leq -c\left|\bar{x}(t)\right|^2. \tag{7.18}$$

Thus $J(t) \geq \dot{V}(t)$ holds for any $t \geq 0$. For any $t \geq 0$, the following inequality holds

$$\int_0^t J(s)\,ds \geq V(t) - V(0). \tag{7.19}$$

From (7.8) and (7.14), it can be obtained

$$\int_0^t J(s)\,ds \geq \bar{x}^T(t)\,\hat{G}\bar{x}(t) + \delta, \quad \forall t \geq 0. \tag{7.20}$$

According to Definition 7.1, we need to prove that the following inequality holds for any matrices Φ, Ψ_1, Ψ_2 and Ψ_3 satisfying Assumption 3.1:

$$\int_0^t J(s)ds - e^T(t)\,\Phi e(t) \geq \delta. \tag{7.21}$$

To this end, we consider the two cases of $||\Phi|| = 0$ and $||\Phi|| \neq 0$, respectively. First, we consider the case when $||\Phi|| = 0$. It follows from (7.20), for any $t \geq 0$,

$$\int_0^t J(s)\,ds \geq \bar{x}^T(t)\,\hat{G}\bar{x}(t) + \delta \geq \delta, \tag{7.22}$$

which means (7.21) holds by noting that $e^T(t)\,\Phi e(t) \equiv 0$.

For the case of $||\Phi|| \neq 0$, under Assumption 3.1, it can be seen from $||\Psi_1|| + ||\Psi_2|| = 0$ and $\left\|\bar{D}_{2i}\right\| = 0$ that $\Psi_1 = 0$, $\Psi_2 = 0$ and $\Psi_3 > 0$. Thus, $J(t) = w^T(t)\,\Psi_3 w(t) \geq 0$. From (7.9), it is obtained that $\bar{C}_{ij}^T \Phi \bar{C}_{ij} < \hat{G}$. Then, for any $t \geq 0$, the following inequalities hold

$$\begin{aligned}
&\int_0^t J(s)\,ds - e^T(t)\,\Phi e(t) \\
&\quad \geq \int_0^t J(s)\,ds - \sum_{i=1}^{r}\sum_{j=1}^{r} \tilde{w}_i \tilde{m}_j \bar{x}^T(t)\,\hat{G}\bar{x}(t) \geq \delta.
\end{aligned}$$

Considering the two case of $||\Phi|| = 0$ and $||\Phi|| \neq 0$ as discussed above, we obtain that system (7.5) is extended dissipative in the sense of Definition 7.1. When $w(t) \equiv 0$, it follows from (7.18) that

$$\dot{V}(t) \leq e^T(t)\,\Psi_1 e(t) - c\,|\bar{x}(t)|^2. \tag{7.23}$$

Noticing that $\Psi_1 \leq 0$ under Assumption 3.1, we conclude that $\dot{V}(t) \leq -c\,|\bar{x}(t)|^2$, which means that the system (7.5) with $w(t) = 0$ is asymptotically stable. This completes the proof. □

In the following part, based on the condition in Theorem 7.3, we will solve the filtering problem for the IT2 fuzzy system (7.1). Recalling Assumption 3.1 and noting that $\Phi \geq 0$ and $\Psi_1 \leq 0$, there always exist matrices $\tilde{\Phi}$ and $\tilde{\Psi}_1$, such that

$$\Phi = \tilde{\Phi}^T \tilde{\Phi}, \quad \Psi_1 = -\tilde{\Psi}_1^T \tilde{\Psi}_1. \tag{7.24}$$

7.3.2 Filter Design

The existence condition of filter design for the IT2 fuzzy system (7.1) is presented in the following theorem.

Theorem 7.4 *Under the condition $\tilde{m}_j - \rho_j \tilde{w}_j \geq 0$ $(0 < \rho_j \leq 1)$ for all j, the filtering error system (7.5) is asymptotically stable and satisfies a new performance index in Definition 7.1 and the poles lie in the disc region $\mho(q, r)$, if there exist matrices $P > 0$, $F > 0$, $\tilde{G} = \tilde{G}^T = \begin{bmatrix} G_1 & G_2 \\ * & G_3 \end{bmatrix} > 0$, $\tilde{\Lambda}_i^T = \tilde{\Lambda}_i$, $\bar{A}_{fj}$, $\bar{B}_{fj}$, $\bar{C}_{fj}$, $\tilde{\Phi}$, $\tilde{\Psi}_1$, Ψ_2 and Ψ_3 with appropriate dimensions satisfying the following conditions:*

$$\tilde{G} - \tilde{P} < 0, \tag{7.25}$$

$$\tilde{\Theta} < 0, \tag{7.26}$$

$$\tilde{\Xi}_{ij} - \tilde{\Lambda}_i < 0, \tag{7.27}$$

$$\rho_i \tilde{\Xi}_{ii} - \rho_i \tilde{\Lambda}_i + \tilde{\Lambda}_i < 0, \tag{7.28}$$

$$\rho_j \tilde{\Xi}_{ij} + \rho_i \tilde{\Xi}_{ji} - \rho_j \tilde{\Lambda}_i - \rho_i \tilde{\Lambda}_j + \tilde{\Lambda}_i + \tilde{\Lambda}_j < 0, \quad i < j, \tag{7.29}$$

$$\begin{bmatrix} -\tilde{P} & \vartheta_{1ij} \\ * & -r^2 \tilde{P} \end{bmatrix} < 0, \tag{7.30}$$

where

$$\tilde{\Xi}_{ij} = \begin{bmatrix} \theta_{ij}^{11} & \theta_{ij}^{12} & \theta_{ij}^{13} \\ * & -\mathbf{He}\left(D_{2i}^T \Psi_2\right) - \Psi_3 & D_{2i}^T \tilde{\Psi}_1^T \\ * & * & -I \end{bmatrix},$$

$$\tilde{\Theta} = \begin{bmatrix} -G_1 & -G_2 & C_i^T \tilde{\Phi}^T \\ * & -G_3 & -\bar{C}_{fj}^T \tilde{\Phi}^T \\ * & * & -I \end{bmatrix},$$

$$\theta_{ij}^{11} = \begin{bmatrix} \mathbf{He}\left(PA_i + \bar{B}_{fj} E_i\right) & \bar{A}_{fj} + A_i^T F^T + E_i^T \bar{B}_{fj}^T \\ * & \mathbf{He}\left(\bar{A}_{fj}\right) \end{bmatrix},$$

$$\theta_{ij}^{12} = \begin{bmatrix} PD_{1i} + \bar{B}_{fj} D_{3i} - C_i^T \Psi_2 \\ FD_{1i} + \bar{B}_{fj} D_{3i} + \bar{C}_{fj}^T \Psi_2 \end{bmatrix},$$

$$\theta_{ij}^{13} = \begin{bmatrix} C_i^T \tilde{\Psi}_1^T \\ -\bar{C}_{fj}^T \tilde{\Psi}_1^T \end{bmatrix}, \quad \tilde{P} = \begin{bmatrix} P & F \\ F & F \end{bmatrix},$$

$$\vartheta_{1ij} = \begin{bmatrix} PA_i + \bar{B}_{fj}E_i - qP & \bar{A}_{fj} - qF \\ FA_i + \bar{B}_{fj}E_i - qF & \bar{A}_{fj} - qF \end{bmatrix}.$$

Moreover, the IT2 fuzzy filter parameters are given by

$$A_{fj} = F^{-1}\bar{A}_{fj}, \quad B_{fj} = F^{-1}\bar{B}_{fj}, \quad C_{fj} = \bar{C}_{fj}. \tag{7.31}$$

Proof From the inequality of (7.17), we know $\sum_{i=1}^{r}\sum_{j=1}^{r}\tilde{w}_i\tilde{m}_j\Xi_{ij} < 0$. Under the condition of $\Psi_1 = -\tilde{\Psi}_1^T\tilde{\Psi}_1$, then using Schur complement, one can obtain

$$\sum_{i=1}^{r}\sum_{j=1}^{r}\tilde{w}_i\tilde{m}_j\bar{\Xi}_{ij} = \sum_{i=1}^{r}\sum_{j=1}^{r}\tilde{w}_i\tilde{m}_j \begin{bmatrix} \mathbf{He}\left(\hat{P}\bar{A}_{ij}\right) & \hat{P}\bar{D}_{1ij} - \bar{C}_{ij}^T\Psi_2 & \bar{C}_{ij}^T\tilde{\Psi}_1^T \\ * & -\mathbf{He}\left(\bar{D}_{2i}^T\Psi_2\right) - \Psi_3 & \bar{D}_{2i}^T\tilde{\Psi}_1^T \\ * & * & -I \end{bmatrix}.$$

Partition as $\hat{P} = \begin{bmatrix} P & S \\ S^T & W \end{bmatrix}$, where $P > 0$, $W > 0$, and S is invertible. Let $H = \begin{bmatrix} I & 0 \\ 0 & SW^{-1} \end{bmatrix}$, $F = SW^{-1}S^T$, $\bar{A}_{fj} = SA_{fj}W^{-1}S^T$, $\bar{B}_{fj} = SB_{fj}$, $\bar{C}_{fj} = C_{fj}W^{-1}S^T$, $\tilde{\Lambda}_i = \text{diag}\{H, I, I\}$, $\bar{\Lambda}_i = \text{diag}\{H, I, I\}^T$, $\bar{\Lambda}_i$ is the matrix with appropriate dimensions. In (7.10)–(7.12), Λ_i is the matrix with appropriate dimensions. Replacing Ξ_{ij} and Λ_i with $\bar{\Xi}_{ij}$ and $\bar{\Lambda}_i$ in (7.10)–(7.12), these inequalities still hold. After replacing Ξ_{ij} and Λ_i with $\bar{\Xi}_{ij}$ and $\bar{\Lambda}_i$ in (7.10)–(7.12), then pre- and post-multiplying the inequalities by $\text{diag}\{H, I, I\}$ and its transpose, respectively, the inequalities (7.27)–(7.29) hold. Let $\tilde{G} = H\hat{G}H^T$, $\tilde{P} = H\hat{P}H^T$. Pre- and post-multiplying (7.8) by H and its transpose, one can have (7.25). Pre- and post-multiplying (7.13) by $\text{diag}\{H, H\}$ and its transpose, one can get (7.30). By Schur complement with the condition of $\Phi = \tilde{\Phi}^T\tilde{\Phi}$, it can be seen from (7.9) that

$$\begin{bmatrix} -\hat{G} & \bar{C}_{ij}^T\tilde{\Phi}^T \\ * & -I \end{bmatrix} < 0. \tag{7.32}$$

Pre- and post-multiplying (7.32) by $\text{diag}\{H, I\}$ and its transpose, respectively, we can have (7.26). If the inequalities (7.25)–(7.30) hold, the filter design problem is solvable, and the filter matrices are designed by

$$A_{fj} = S^{-1}\bar{A}_{fj}S^{-T}W, \quad B_{fj} = S^{-1}\bar{B}_{fj}, \quad C_{fj} = \bar{C}_{fj}S^{-T}W,$$

where matrices $W > 0$ and S are such that $F = SW^{-1}S^T$. Or equivalently under transformation $S^{-T}W\hat{x}(t)$, the filter parameters can be computed by

$$\begin{aligned} A_{fj} &= S^{-T}W\left(S^{-1}\bar{A}_{fj}S^{-T}W\right)W^{-1}S^T = F^{-1}\bar{A}_{fj}, \\ B_{fj} &= S^{-T}W\left(S^{-1}\bar{B}_{fj}\right) = F^{-1}\bar{B}_{fj}, \quad C_{fj} = \left(\bar{C}_{fj}S^{-T}W\right)W^{-1}S^T = \bar{C}_{fj}. \end{aligned}$$

This completes the proof. □

7.4 Simulation Results

In this section, an example is used to illustrate the effectiveness of the proposed results.

Example 7.5 Consider a 2-rule IT2 fuzzy system in the form of (7.2), the matrices are listed below:

$$
\begin{aligned}
A_1 &= \begin{bmatrix} -1 & 0.1 \\ 0.1 & -1.4 \end{bmatrix}, \quad A_2 = \begin{bmatrix} -2 & 0.2 \\ -1 & -1 \end{bmatrix}, \quad D_{11} = \begin{bmatrix} 0.03 \\ 0.01 \end{bmatrix}, \\
D_{12} &= \begin{bmatrix} 0.03 \\ 0.02 \end{bmatrix}, \quad C_1 = \begin{bmatrix} 0.01 & 0.01 \end{bmatrix}, \quad C_2 = \begin{bmatrix} 0.2 & -0.1 \end{bmatrix}, \\
E_1 &= \begin{bmatrix} 0.01 & 0.01 \end{bmatrix}, \quad E_2 = \begin{bmatrix} 0.02 & 0.01 \end{bmatrix}, \\
D_{21} &= 0.01, \quad D_{22} = 0.02, \quad D_{31} = 0.01, \quad D_{32} = 0.02.
\end{aligned}
$$

Membership functions for Rules 1 and 2 are given as follows:

$$
\begin{aligned}
\tilde{w}_1(x_1) &= 1 - \frac{1}{e^{(-x_1 - 0.1 \times \sin(x_1))}}, \\
\tilde{w}_2(x_1) &= 1 - \tilde{w}_1(x_1), \\
\underline{m}_1(x_1) &= 1 - \frac{1}{e^{\frac{-x_1 - 0.25}{4}}}, \quad \bar{m}_1(x_1) = 1 - \frac{1}{e^{\frac{-x_1 + 0.25}{4}}}, \\
\underline{m}_2(x_1) &= 1 - \bar{m}_1(x_1), \quad \bar{m}_2(x_1) = 1 - \underline{m}_1(x_1), \\
\tilde{m}_j(x_1) &= \frac{\underline{\beta}_j \underline{m}_j(x_1) + \bar{\beta}_j \bar{m}_j(x_1)}{\sum_{k=1}^{2} \left(\underline{\beta}_k \underline{m}_k(x_1) + \bar{\beta}_k \bar{m}_k(x_1) \right)}, \quad j = 1, 2.
\end{aligned}
$$

In this chapter, we consider the $\mathcal{D}$-stability constraints and design the filter such that the filtering error system (7.5) lies in a disk region with center $q = (-10, 0)$ and radius $r = 9.9$. Figure 7.1 plots the disk region. Under a unified frame, the fuzzy filter is designed to satisfy H_∞, L_2-L_∞, passive and dissipativity performances. Due to limited space, we only consider L_2-L_∞ performance in this example. In the LMIs conditions of Theorem 7.4, let $\rho_1 = 0.2$, $\rho_2 = 0.6$, $\Phi = I$, $\Psi_1 = 0$, $\Psi_2 = 0$, $\Psi_3 = \gamma^2 I$, $\delta = 0$, $\underline{\beta}_j = 0.6$ and $\bar{\beta}_j = 0.4$, it can be found that the minimized L_2-L_∞ performance index $\gamma = 0.0024$ and the L_2-L_∞ filter parameters in (7.4) are listed below:

$$
\begin{aligned}
A_{f1} &= \begin{bmatrix} -1.2119 & 0.1034 \\ -0.3199 & -1.6193 \end{bmatrix}, \quad B_{f1} = \begin{bmatrix} -2.0923 \\ -1.2293 \end{bmatrix}, \\
A_{f2} &= \begin{bmatrix} -1.1966 & 0.0804 \\ -0.3241 & -1.6611 \end{bmatrix}, \quad B_{f2} = \begin{bmatrix} -2.0726 \\ -1.2430 \end{bmatrix}, \\
C_{f1} &= \begin{bmatrix} -0.0990 & 0.0267 \end{bmatrix}, \quad C_{f2} = \begin{bmatrix} -0.0990 & 0.0267 \end{bmatrix}.
\end{aligned}
$$

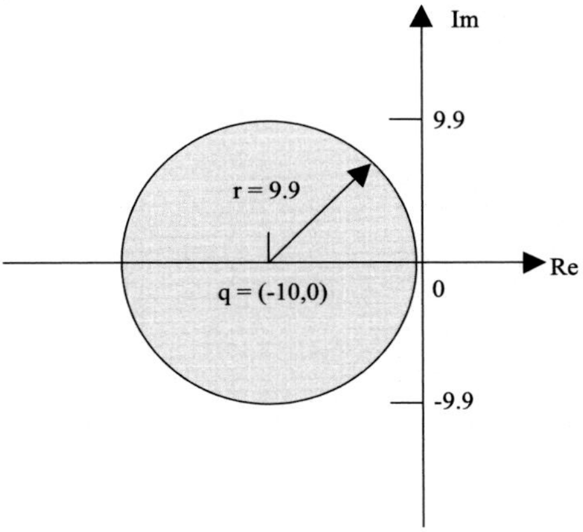

Fig. 7.1 Disk region. © [2014] IEEE. Reprinted, with permission, from ref. [6]

Using the parameters of the filter, it can be seen from Fig. 7.2 that all the poles of filtering error system lie in the region $\mho(q, r)$.

Suppose the disturbance $w(t)$ be

$$w(t) = \begin{cases} 0.1\sin(2t), & t \leq 5, \\ 0, & \text{else.} \end{cases}$$

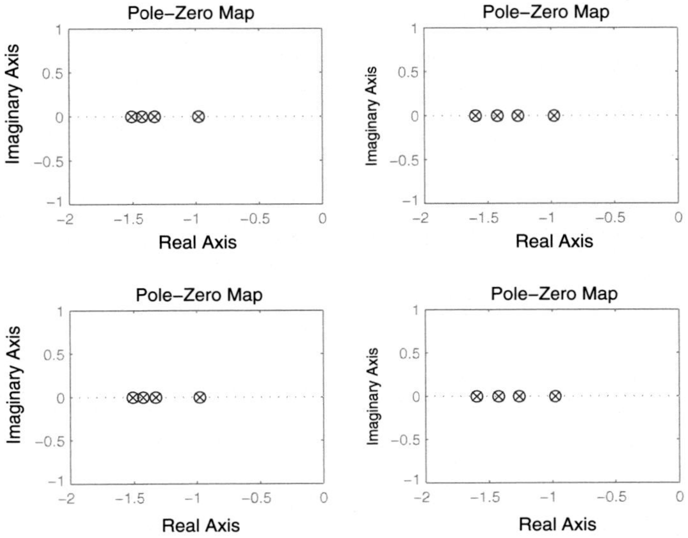

Fig. 7.2 Poles of the filtering error system

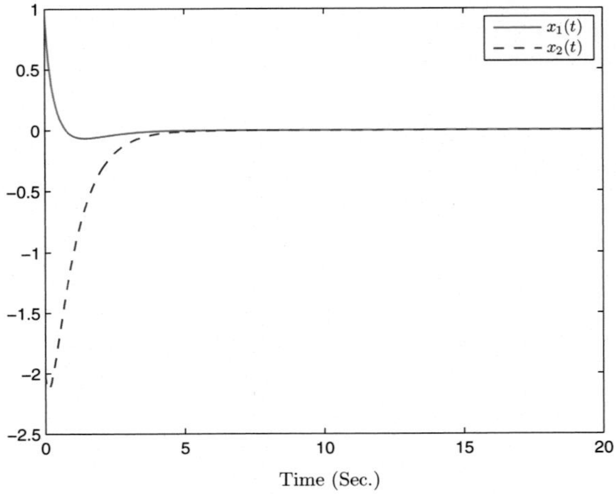

Fig. 7.3 Responses of state $x_1(t)$ and $x_2(t)$

Figures 7.3 and 7.4 plot the state responses of $x(t)$ and $x_f(t)$, under the initial condition of $x(t) = [1 \quad -2]^T$, $\hat{x}(t) = [1 \quad -2]^T$, respectively. Figure 7.5 depicts the responses of $z(t)$ and $z_f(t)$, and Fig. 7.6 shows the error response of $z(t) - z_f(t)$ under the initial condition of $z(t) = 0$, $z_f(t) = 0$.

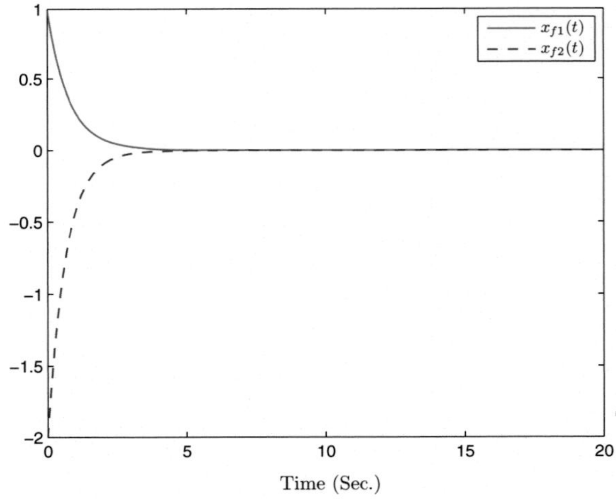

Fig. 7.4 Responses of state $x_{f1}(t)$ and $x_{f2}(t)$

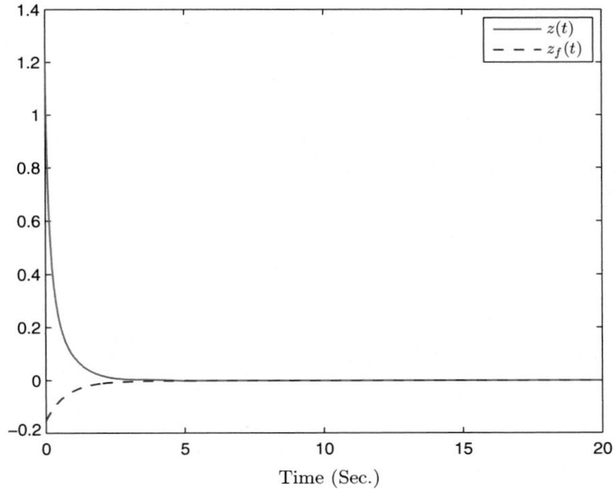

Fig. 7.5 Responses of $z(t)$ and $z_f(t)$

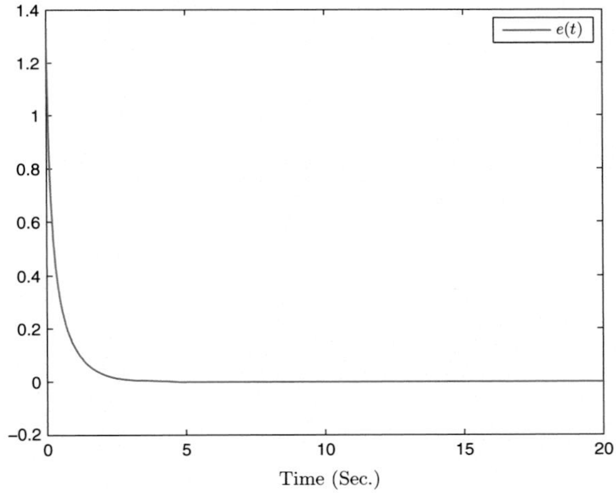

Fig. 7.6 Error response of $e(t)$

7.5 Conclusion

This chapter has addressed the filter design problem for IT2 fuzzy systems with $\mathcal{D}$-stability constraints. Under a unified frame, using a new performance index, the fuzzy filter has been designed for IT2 fuzzy systems with $\mathcal{D}$-stability constraints. This new performance index contains H_∞, L_2-L_∞, passive and dissipativity performances. The existence condition of the filter has been expressed by the convex optimization problem. Some simulation results have been provided to illustrate the effectiveness of the proposed results.

Chapter 8
Fault Detection of Interval Type-2 Fuzzy-Model-Based Systems

8.1 Introduction

In practical systems, various faults are likely to be encountered, especially faults from actuators and sensors [203]. In the past few years, the investigation of fault detection has been developed [27, 44, 54]. It is necessary and critical to detect the occurred faults immediately for the stability and the performance of the systems. This chapter investigates the fault detection problem for the IT2 fuzzy systems subject to sensor nonlinearities. The output considered in this chapter of IT2 fuzzy systems is a general sector-bounded nonlinearities. The IT2 fuzzy model and IT2 fuzzy fault detection filter do not require to share the same LMFs and UMFs. By using a general observer-based fault detection filter as a residual generator, the fault detection problem is described as a filter design problem. The fault detection filter is designed to guarantee the prescribed H_∞ performance level. A decomposition approach is employed to handle the characteristic of sensor saturation. Using Lyapunov stability theory, a novel type of IT2 fault detection filter is designed to guarantee that the fault detection system is asymptotically stable with an H_∞ performance. In the design procedure, the parameters of the IT2 filter can be solved by the standard software.

8.2 Problem Formulation

Consider the following IT2 fuzzy model that represents a continuous-time nonlinear system:

♦ **Plant Form:**

Rule i: IF $F_1(x(t))$ is $\tilde{M}_1^i$ and ... and $F_p(x(t))$ is $\tilde{M}_p^i$, THEN,

$$\begin{cases} \dot{x}(t) = A_i x(t) + B_i w(t) + B_{1i} f(t), \\ y(t) = \phi(C_i x(t)) + D_i w(t) + D_{1i} f(t), \end{cases} \tag{8.1}$$

© Springer Science+Business Media Singapore 2016
H. Li et al., *Analysis and Synthesis for Interval Type-2*
Fuzzy-Model-Based Systems, DOI 10.1007/978-981-10-0593-0_8

where $\tilde{M}_a^i$ is an IT2 fuzzy set of rule i corresponding to the function $F_a(x(t))$, $i = 1, 2, \ldots, r; a = 1, 2, \ldots, p, p$ is a positive integer, $x(t) \in \mathbf{R}^n$ is the system state vector, $w(t) \in \mathbf{R}^q$ is the disturbance input and $f(t) \in \mathbf{R}^m$ is the fault to be detected, $y(t) \in \mathbf{R}^l$ is the measure output, $A_i, B_i, B_{1i}, C_i, D_i$ and D_{1i} are the known matrices with appropriate dimensions. The firing strength of the ith rule is the following interval set:

$$W_i(x(t)) = \left[\underline{w}_i(x(t)), \overline{w}_i(x(t))\right], \quad i = 1, 2, \ldots, r, \tag{8.2}$$

where $\underline{w}_i(x(t)) = \prod_{a=1}^{p} \underline{\mu}_{\tilde{M}_a^i}(F_a(x(t))) \geq 0$ denotes the lower grade of membership, and $\overline{w}_i(x(t)) = \prod_{a=1}^{p} \bar{\mu}_{\tilde{M}_a^i}(F_a(x(t))) \geq 0$ denotes the upper grades of membership. $\underline{\mu}_{\tilde{M}_a^i}(F_a(x(t))) \geq 0$ and $\bar{\mu}_{\tilde{M}_a^i}(F_a(x(t))) \geq 0$ stand for the LMFs and UMFs, respectively. Therefore, it can be found that $\bar{\mu}_{\tilde{M}_a^i}(F_a(x(t))) \geq \underline{\mu}_{\tilde{M}_a^i}(F_a(x(t)))$ and $\overline{w}_i(x(t)) \geq \underline{w}_i(x(t))$ for all i. Then, the IT2 T–S fuzzy system is described as follows:

$$\begin{cases} \dot{x}(t) = \sum_{i=1}^{r} \tilde{w}_i(x(t)) \left[A_i x(t) + B_i w(t) + B_{1i} f(t)\right], \\ y(t) = \sum_{i=1}^{r} \tilde{w}_i(x(t)) \left[\phi(C_i x(t)) + D_i w(t) + D_{1i} f(t)\right], \end{cases} \tag{8.3}$$

where

$$\tilde{w}_i(x(t)) = \underline{a}_i(x(t))\, \underline{w}_i(x(t)) + \bar{a}_i(x(t))\, \overline{w}_i(x(t)) \geq, \quad \forall i,$$

$$\sum_{i=1}^{r} \tilde{w}_i(x(t)) = 1,$$

$0 \leq \underline{a}_i(x(t)) \leq 1$ and $0 \leq \bar{a}_i(x(t)) \leq 1$ are nonlinear functions and possess the trait of $\underline{a}_i(x(t)) + \bar{a}_i(x(t)) = 1$ for all i.

Many actual applications will inevitably result in the nonlinear characteristic of sensors. Here, the function $\phi(u)$ in system (8.3) is assumed to belong to $[K_1, K_2]$, for some given diagonal matrices $K_1 \geq 0$ and $K_2 \geq 0$ with $K_2 > K_1$, and satisfies the following sector condition:

$$(\phi(u) - K_1 u)^T (\phi(u) - K_2 u) \leq 0, \quad \forall u \in \mathbf{R}^l. \tag{8.4}$$

An IT2 fuzzy filter with r rules is constructed as follows:

♦ **Filter Form:**

Rule j: IF $g_1(x(t))$ is $\tilde{N}_1^j$ and ... and $g_l(x(t))$ is $\tilde{N}_l^j$, THEN,

$$\begin{cases} \dot{\hat{x}}(t) = A_{fj}\hat{x}(t) + B_{fj}y(t), \\ z_f(t) = C_{fj}\hat{x}(t), \end{cases} \tag{8.5}$$

where $\tilde{N}^j_\beta$ is an IT2 fuzzy set of rule j corresponding to the function $g_\beta(x(t))$, $j = 1, 2, \ldots, r$; $\beta = 1, 2, \ldots, l$; l is a positive integer; A_{fj}, B_{fj} and C_{fj} are the filter parameters to be designed. The firing strength of the jth rule is the following interval set:

$$M_j(x(t)) = \left[\underline{m}_j(x(t)), \overline{m}_j(x(t))\right], \quad j = 1, 2, \ldots, r,$$

where $\underline{m}_j(x(t)) = \prod_{\beta=1}^{l} \underline{\mu}_{\tilde{N}^j_\beta}(g_\beta(x(t)))$ and $\overline{m}_j(x(t)) = \prod_{\beta=1}^{l} \bar{\mu}_{\tilde{N}^j_\beta}(g_\beta(x(t)))$ stand for the lower and upper grade of membership, respectively. The LMF and UMF are represented by $\underline{\mu}_{\tilde{N}^j_\beta}(g_\beta(x(t))) \geq 0$ and $\bar{\mu}_{\tilde{N}^j_\beta}(g_\beta(x(t))) \geq 0$, respectively. Here, $\bar{\mu}_{\tilde{N}^j_\beta}(g_\beta(x(t))) \geq \underline{\mu}_{\tilde{N}^j_\beta}(g_\beta(x(t)))$ leading to $\overline{m}_j(x(t)) \geq \underline{m}_j(x(t))$ for all j. Assume that $x(t)$ is available for filter. The overall IT2 fuzzy filter is proposed as follows:

$$\begin{cases} \dot{\hat{x}}(t) = \sum_{j=1}^{r} \tilde{m}_j(x(t)) \left[A_{fj}\hat{x}(t) + B_{fj}y(t)\right], \\ z_f(t) = \sum_{j=1}^{r} \tilde{m}_j(x(t)) C_{fj}\hat{x}(t), \end{cases} \tag{8.6}$$

where

$$\tilde{m}_j(x(t)) = \frac{\underline{\beta}_j(x(t))\underline{m}_j(x(t)) + \bar{\beta}_j(x(t))\overline{m}_j(x(t))}{\sum_{k=1}^{r}\left(\underline{\beta}_k(x(t))\underline{m}_k(x(t)) + \bar{\beta}_k(x(t))\overline{m}_k(x(t))\right)} \geq 0, \quad \forall j,$$

$$\sum_{j=1}^{r} \tilde{m}_j(x(t)) = 1,$$

in which $0 \leq \underline{\beta}_j(x(t)) \leq 1$ and $0 \leq \bar{\beta}_j(x(t)) \leq 1$ are predefined functions and possess the trait of $\underline{\beta}_j(x(t)) + \bar{\beta}_j(x(t)) = 1$ for all j.

To improve the performance of the fault detection system, we add a weighting matrix function into the fault $f(s)$. Here, $f_w(s) = W(s)f(s)$. One state-space realization of $f_w(s) = W(s)f(s)$ can be described as:

$$\begin{cases} \dot{x}_w(t) = A_w x_w(t) + B_w f(t), \\ f_w(t) = C_w x_w(t), \\ x_w(0) = 0, \end{cases}$$

where $x_w(t) \in \mathbf{R}^h$ is the state vector, and A_w, B_w, C_w are constant matrices. Now, we decompose the nonlinear function $\phi(u)$ as follows [147, 148]:

$$\phi(u) = \phi_s(u) + K_1 u,$$

where the nonlinearity $\phi_s(u)$ belongs to the set Φ_s given by

$$\Phi_s = \{\phi_s : \phi_s^T(u)(\phi_s(u) - Ku) \leq 0\},$$

with $K = K_2 - K_1 > 0$.

In order to have a simple description, we denote $\tilde{w}_i(x(t)) = \tilde{w}_i$ and $\tilde{m}_j(x(t)) = \tilde{m}_j$, $i, j = 1, 2, \ldots, r$.

It can be seen from (8.3), (8.6) and (8.7) that the fault detection system can be given as follows:

$$\begin{cases} \dot{\bar{x}}(t) = \sum_{i=1}^{r}\sum_{j=1}^{r} \tilde{w}_i\tilde{m}_j\left[\bar{A}_{ij}\bar{x}(t) + \bar{B}_{ij}\zeta(t) + \bar{B}_{1ij}\phi_s(\bar{C}_{ij}\bar{x}(t))\right], \\ e(t) = \sum_{i=1}^{r}\sum_{j=1}^{r} \tilde{w}_i\tilde{m}_j\bar{C}_{1ij}\bar{x}(t), \end{cases} \tag{8.7}$$

where

$$\sum_{i=1}^{r}\tilde{w}_i = \sum_{j=1}^{r}\tilde{m}_j = \sum_{i=1}^{r}\sum_{j=1}^{r}\tilde{w}_i\tilde{m}_j = 1, \quad e(t) = z_f(t) - f_w(t),$$

$$\bar{x}(t) = \begin{bmatrix} x^T(t) & \hat{x}^T(t) & x_w^T(t) \end{bmatrix}^T, \quad \zeta(t) = \begin{bmatrix} w^T(t) & f^T(t) \end{bmatrix}^T,$$

$$\bar{A}_{ij} = \begin{bmatrix} A_i & 0 & 0 \\ B_{fj}K_1C_i & A_{fj} & 0 \\ 0 & 0 & A_w \end{bmatrix}, \quad \bar{B}_{ij} = \begin{bmatrix} B_i & B_{1i} \\ B_{fj}D_i & B_{fj}D_{1i} \\ 0 & B_w \end{bmatrix}, \quad \bar{B}_{1ij} = \begin{bmatrix} 0 \\ B_{fj} \\ 0 \end{bmatrix},$$

$$\bar{C}_{ij} = \begin{bmatrix} C_i & 0 & 0 \end{bmatrix}, \quad \bar{C}_{1ij} = \begin{bmatrix} 0 & C_{fj} & -C_w \end{bmatrix}.$$

Therefore, the fault detection problem to be addressed in this chapter can be summarized as follows: (1) The fault detection system (8.7) is asymptotically stable with an H_∞ performance level $\gamma > 0$. (2) Set up a fault detection measure. Select an evaluation function and a threshold. In this chapter, a residual evaluation function $\mathcal{J}(z_f)$ and a threshold $\mathcal{J}_{th}$ are given by

$$\mathcal{J}(z_f) = \left(\int_{t_0}^{t_0+t} z_f^T(t)\, z_f(t)\, dt\right)^{1/2}, \tag{8.8}$$

$$\mathcal{J}_{th} = \sup_{0 \neq w \in \mathcal{L}_2, f=0} \mathcal{J}(z_f), \tag{8.9}$$

where t_0 denotes the initial evaluation time instant and t stands for the evaluation time instant. Based on this, the occurrence of faults can be detected by comparing $\mathcal{J}(z_f)$ and $\mathcal{J}_{th}$ according to the following test:

$$\begin{aligned}\mathcal{J}(z_f) > \mathcal{J}_{th} &\Rightarrow \text{ with faults } \Rightarrow \text{ alarm,}\\ \mathcal{J}(z_f) \le \mathcal{J}_{th} &\Rightarrow \text{ no faults.}\end{aligned}$$

8.3 Main Results

8.3.1 Stability Analysis

In this section, the stability condition with H_∞ performance of the fault detection system (8.7) is first presented in the following theorem.

Theorem 8.1 *The membership functions satisfy* $\tilde{m}_j - \rho_j \tilde{w}_j \ge 0$ $(0 < \rho_j \le 1)$ *and the fault detection system (8.7) is asymptotically stable with an* H_∞ *performance level* γ*, if there exist matrices* $\bar{P} > 0$ *and* $\Theta_i > 0$ *with appropriate dimensions, such that the following LMIs hold for* $i, j = 1, 2, \ldots, r$*:*

$$\Omega_{ij} - \Theta_i < 0, \tag{8.10}$$

$$\rho_i \Omega_{ii} - \rho_i \Theta_i + \Theta_i < 0, \tag{8.11}$$

$$\rho_j \Omega_{ij} + \rho_i \Omega_{ji} - \rho_j \Theta_i - \rho_i \Theta_j + \Theta_i + \Theta_j \le 0, \quad i < j, \tag{8.12}$$

where

$$\Omega_{ij} = \begin{bmatrix} \mathbf{He}(\bar{P}\bar{A}_{ij}) + \bar{C}_{1ij}^T \bar{C}_{1ij} & \bar{P}\bar{B}_{ij} & \bar{P}\bar{B}_{1ij} + \bar{C}_{ij}^T K^T \\ * & -\gamma^2 I & 0 \\ * & * & -2I \end{bmatrix}.$$

Proof Consider the Lyapunov function as follows:

$$V(t) = \bar{x}^T(t)\bar{P}\bar{x}(t). \tag{8.13}$$

Then, the time derivative of $V(t)$ is expressed as:

$$\begin{aligned}\dot{V}(t) &= 2\bar{x}^T(t)\bar{P}\dot{\bar{x}}(t)\\ &= 2\sum_{i=1}^{r}\sum_{j=1}^{r}\tilde{w}_i\tilde{m}_j\bar{x}^T(t)\bar{P}\left[\bar{A}_{ij}\bar{x}(t) + \bar{B}_{ij}\zeta(t) + \bar{B}_{1ij}\phi_s\left(\bar{C}_{ij}\bar{x}(t)\right)\right].\end{aligned}$$

Then, we can have

$$\begin{aligned}
&\dot{V}(t) + e^T(t)e(t) - \gamma^2\zeta^T(t)\zeta(t) \\
&\leq \sum_{i=1}^{r}\sum_{j=1}^{r}\tilde{w}_i\tilde{m}_j\{2\bar{x}^T(t)\bar{P}\left[\bar{A}_{ij}\bar{x}(t) + \bar{B}_{ij}\zeta(t) + \bar{B}_{1ij}\phi_s\left(\bar{C}_{ij}\bar{x}(t)\right)\right] \\
&\quad + \bar{x}^T(t)\bar{C}_{1ij}^T\bar{C}_{1ij}\bar{x}(t) - \gamma^2\zeta^T(t)\zeta(t) \\
&\quad - 2\phi_s^T\left(\bar{C}_{ij}\bar{x}(t)\right)\left(\phi_s\left(\bar{C}_{ij}\bar{x}(t)\right) - K\bar{C}_{ij}\bar{x}(t)\right)\} \\
&= \xi^T(t)\left(\sum_{i=1}^{r}\sum_{j=1}^{r}\tilde{w}_i\tilde{m}_j\Omega_{ij}\right)\xi(t),
\end{aligned} \tag{8.14}$$

where

$$\xi(t) = \left[\bar{x}^T(t)\ \zeta^T(t)\ \phi_s^T\left(\bar{C}_{ij}\bar{x}(t)\right)\right]^T,$$

$$\Omega_{ij} = \begin{bmatrix} \mathbf{He}\left(\bar{P}\bar{A}_{ij}\right) + \bar{C}_{1ij}^T\bar{C}_{1ij} & \bar{P}\bar{B}_{ij} & \bar{P}\bar{B}_{1ij} + \bar{C}_{ij}^T K^T \\ * & -\gamma^2 I & 0 \\ * & * & -2I \end{bmatrix}.$$

It could be seen from the Eq. (8.14), if $\sum_{i=1}^{r}\sum_{j=1}^{r}\tilde{w}_i\tilde{m}_j\Omega_{ij} < 0$ then $\dot{V}(t) + e^T(t)e(t) - \gamma^2\zeta^T(t)\zeta(t) < 0$. Consider $\sum_{i=1}^{r}\sum_{j=1}^{r}\tilde{w}_i\left(\tilde{w}_j - \tilde{m}_j\right)\Theta_i = 0$, where $\Theta_i = \Theta_i^T$ $(i = 1, 2, \ldots, r)$ is arbitrary matrix with appropriate dimensions. Then, we have

$$\begin{aligned}
&\sum_{i=1}^{r}\sum_{j=1}^{r}\tilde{w}_i\tilde{m}_j\Omega_{ij} \\
&= \sum_{i=1}^{r}\sum_{j=1}^{r}\tilde{w}_i\left(\tilde{w}_j - \tilde{m}_j + \rho_j\tilde{w}_j - \rho_j\tilde{w}_j\right)\Theta_i + \sum_{i=1}^{r}\sum_{j=1}^{r}\tilde{w}_i\tilde{m}_j\Omega_{ij} \\
&= \sum_{i=1}^{r}\sum_{j=1}^{r}\tilde{w}_i\left(\tilde{m}_j + \rho_j\tilde{w}_j - \rho_j\tilde{w}_j\right)\Omega_{ij} + \sum_{i=1}^{r}\sum_{j=1}^{r}\tilde{w}_i\left(\tilde{w}_j - \rho_j\tilde{w}_j\right)\Theta_i \\
&\quad - \sum_{i=1}^{r}\sum_{j=1}^{r}\tilde{w}_i\left(\tilde{m}_j - \rho_j\tilde{w}_j\right)\Theta_i \\
&= \sum_{i=1}^{r}\sum_{j=1}^{r}\tilde{w}_i\tilde{w}_j\left(\rho_j\Omega_{ij} - \rho_j\Theta_i + \Theta_i\right) + \sum_{i=1}^{r}\sum_{j=1}^{r}\tilde{w}_i\left(\tilde{m}_j - \rho_j\tilde{w}_j\right)\left(\Omega_{ij} - \Theta_i\right)
\end{aligned}$$

$$= \sum_{i=1}^{r}\sum_{j=i}^{r} \tilde{w}_i^2 \left(\rho_i \Omega_{ii} - \rho_i \Theta_i + \Theta_i\right) + \sum_{i=1}^{r-1}\sum_{j=i+1}^{r} \tilde{w}_i \tilde{w}_j \big(\rho_j \Omega_{ij} - \rho_j \Theta_i + \Theta_i$$

$$+ \rho_i \Omega_{ji} - \rho_i \Theta_j + \Theta_j\big) + \sum_{i=1}^{r}\sum_{j=1}^{r} \tilde{w}_i \left(\tilde{m}_j - \rho_j \tilde{w}_j\right)\left(\Omega_{ij} - \Theta_i\right).$$

Under $\tilde{m}_j - \rho_j \tilde{w}_j \geq 0$ for all j, it can be seen from the inequalities (8.10)–(8.12) that

$$\dot{V}(t) + e^T(t)\, e(t) - \gamma^2 \zeta^T(t)\, \zeta(t) < 0. \tag{8.15}$$

Under zero initial conditions, integrating both sides of (8.15) yields $\|e(t)\|_2 < \gamma \|\zeta(t)\|_2$ for all nonzero $\zeta(t) \in [0, \infty)$. In addition, when $\zeta(t) \equiv 0$, it can be seen from the conditions in Theorem 8.1 that $\dot{V}(t) < 0$, which means that the system (8.7) with $\zeta(t) = 0$ is asymptotically stable. The proof is completed. □

8.3.2 Fault Detection Filter Design

The existence condition of the H_∞ fault detection filter for the IT2 fuzzy system (8.1) is presented in the following theorem.

Theorem 8.2 *Under the condition $\tilde{m}_j - \rho_j \tilde{w}_j \geq 0$ $(0 < \rho_j \leq 1)$ for all j, the fault detection system (8.7) is asymptotically stable with an H_∞ performance level γ, if there exist matrices $P > 0$, $F > 0$, $V > 0$, $\tilde{\Theta}_i^T = \tilde{\Theta}_i$, $\bar{A}_{fj}$, $\bar{B}_{fj}$ and $\bar{C}_{fj}$ with appropriate dimensions such that the following LMIs hold for $i, j = 1, 2, \ldots, r$:*

$$\tilde{\Omega}_{ij} - \tilde{\Theta}_i < 0, \tag{8.16}$$

$$\rho_i \tilde{\Omega}_{ii} - \rho_i \tilde{\Theta}_i + \tilde{\Theta}_i < 0, \tag{8.17}$$

$$\rho_j \tilde{\Omega}_{ij} + \rho_i \tilde{\Omega}_{ji} - \rho_j \tilde{\Theta}_i - \rho_i \tilde{\Theta}_j + \tilde{\Theta}_i + \tilde{\Theta}_j \leq 0, \quad i < j, \tag{8.18}$$

where

$$\tilde{\Omega}_{ij} = \begin{bmatrix} \theta_{ij}^{11} & \theta_{ij}^{12} & \theta_{ij}^{13} & \theta_{ij}^{14} \\ * & -\gamma^2 I & 0 & 0 \\ * & * & -2I & 0 \\ * & * & * & -I \end{bmatrix}, \quad \theta_{ij}^{14} = \begin{bmatrix} 0 \\ \bar{C}_{fj}^T \\ -C_w^T \end{bmatrix},$$

$$\theta_{ij}^{11} = \begin{bmatrix} \mathbf{He}\left(PA_i + \bar{B}_{fj} K_1 C_i\right) & \bar{A}_{fj} + A_i^T F^T + C_i^T K_1^T \bar{B}_{fj}^T & 0 \\ * & \mathbf{He}\left(\bar{A}_{fj}\right) & 0 \\ * & * & \mathbf{He}\left(V A_w\right) \end{bmatrix},$$

$$\theta_{ij}^{12} = \begin{bmatrix} PB_i + \bar{B}_{fj} D_i & PB_{1i} + \bar{B}_{fj} D_{1i} \\ FB_i + \bar{B}_{fj} D_i & FB_{1i} + \bar{B}_{fj} D_{1i} \\ 0 & V B_w \end{bmatrix}, \quad \theta_{ij}^{13} = \begin{bmatrix} \bar{B}_{fj} + C_i^T K^T \\ \bar{B}_{fj} \\ 0 \end{bmatrix}.$$

The IT2 fuzzy filter parameters are given by:

$$A_{fj} = F^{-1}\bar{A}_{fj}, \quad B_{fj} = F^{-1}\bar{B}_{fj}, \quad C_{fj} = \bar{C}_{fj}. \tag{8.19}$$

Proof From the inequality of (8.15), we know $\sum_{i=1}^{r}\sum_{j=1}^{r}\tilde{w}_i\tilde{m}_j\Omega_{ij} < 0$. Using Schur complement to $\sum_{i=1}^{r}\sum_{j=1}^{r}\tilde{w}_i\tilde{m}_j\Omega_{ij}$, one can obtain:

$$\begin{aligned}
&\sum_{i=1}^{r}\sum_{j=1}^{r}\tilde{w}_i\tilde{m}_j\bar{\Omega}_{ij}\\
&=\sum_{i=1}^{r}\sum_{j=1}^{r}\tilde{w}_i\tilde{m}_j\begin{bmatrix} He\left(\bar{P}\bar{A}_{ij}\right) & \bar{P}\bar{B}_{ij} & \bar{P}\bar{B}_{1ij}+\bar{C}_{ij}^TK^T & \bar{C}_{1ij}^T \\ * & -\gamma^2 I & 0 & 0 \\ * & * & -2I & 0 \\ * & * & * & -I \end{bmatrix}\\
&< 0.
\end{aligned} \tag{8.20}$$

Let $\bar{P} = \begin{bmatrix} \hat{P} & 0 \\ 0 & V \end{bmatrix}$, $\hat{P} = \begin{bmatrix} P & S \\ S^T & W \end{bmatrix}$, where $\hat{P} > 0$, $V > 0$, $P > 0$, $W > 0$, and S is invertible. The matrices $\hat{H}$ and H are partitioned as $\hat{H} = \begin{bmatrix} H & 0 \\ 0 & I \end{bmatrix}$, $H = \begin{bmatrix} I & 0 \\ 0 & SW^{-1} \end{bmatrix}$, respectively. After replacing Ω_{ij} and Θ_i with $\bar{\Omega}_{ij}$ and $\bar{\Theta}_i$ in inequalities (8.10)–(8.12), $\bar{\Theta}_i$ is the matrix with appropriate dimensions, then performing a congruence transformation to (8.20) by diagonal matrix diag$\{\hat{H}, I, I, I\}$, we obtain

$$\tilde{\Omega}_{ij} = \begin{bmatrix} \hat{\theta}_{ij}^{11} & \hat{\theta}_{ij}^{12} & \hat{\theta}_{ij}^{13} & \hat{\theta}_{ij}^{14} \\ * & -\gamma^2 I & 0 & 0 \\ * & * & -2I & 0 \\ * & * & * & -I \end{bmatrix} < 0, \tag{8.21}$$

where

$$\begin{aligned}
\hat{\theta}_{ij}^{11} &= \begin{bmatrix} \mathbf{He}\left(PA_i + SB_{fj}K_1C_i\right) & \check{\theta}_{ij} & 0 \\ * & \mathbf{He}\left(SA_{fj}W^{-1}S^T\right) & 0 \\ * & * & \mathbf{He}\left(VA_w\right) \end{bmatrix},\\
\hat{\theta}_{ij}^{12} &= \begin{bmatrix} PB_i + SB_{fj}D_i & PB_{1i} + SB_{fj}D_{1i} \\ FB_i + SB_{fj}D_i & FB_{1i} + SB_{fj}D_{1i} \\ 0 & VB_w \end{bmatrix}, \quad \hat{\theta}_{ij}^{13} = \begin{bmatrix} SB_{fj} + C_i^TK^T \\ SB_{fj} \\ 0 \end{bmatrix},\\
\hat{\theta}_{ij}^{14} &= \begin{bmatrix} 0 \\ SW^{-1}C_{fj}^T \\ -\bar{C}_w^T \end{bmatrix}, \quad \check{\theta}_{ij} = SA_{fj}W^{-1}S^T + A_i^TF^T + C_i^TK_1^TB_{fj}^TS^T.
\end{aligned}$$

The filter matrices are given as follows:

$$A_{fj} = S^{-1}\bar{A}_{fj}S^{-T}W, \quad B_{fj} = S^{-1}\bar{B}_{fj}, \quad C_{fj} = \bar{C}_{fj}S^{-T}W,$$

where the matrices $W > 0$ and S satisfy the condition $F = SW^{-1}S^T$. Or equivalently under transformation $S^{-T}W\hat{x}(t)$, the filter parameters can be yielded in the following form:

$$\begin{aligned} A_{fj} &= S^{-T}W\left(S^{-1}\bar{A}_{fj}S^{-T}W\right)W^{-1}S^T = F^{-1}\bar{A}_{fj}, \\ B_{fj} &= S^{-T}W(S^{-1}\bar{B}_{fj}) = F^{-1}\bar{B}_{fj}, \quad C_{fj} = \left(\bar{C}_{fj}S^{-T}W\right)W^{-1}S^T = \bar{C}_{fj}. \end{aligned}$$

Based on the above discussion, define $\tilde{\Theta}_i = \text{diag}\{\hat{H}, I, I, I\}$, $\bar{\Theta}_i = \text{diag}\{\hat{H}^T, I, I, I\}$, it can be seen from the conditions (8.10)–(8.12) that the conditions (8.16)–(8.18) hold. This completes the proof. □

In this section, we will consider the IT2 fuzzy system without sensor nonlinearities and give the following results. First, we present the IT2 fuzzy system which can be described by the following IT2 fuzzy model.

$$\begin{cases} \dot{x}(t) = \sum_{i=1}^{r} \tilde{w}_i \left[A_i x(t) + B_i w(t) + B_{1i} f(t)\right], \\ y(t) = \sum_{i=1}^{r} \tilde{w}_i \left[C_i x(t) + D_i w(t) + D_{1i} f(t)\right]. \end{cases} \tag{8.22}$$

It can be seen from (8.22), (8.6) and (8.7) that the fault detection system can be given as follows:

$$\begin{cases} \dot{\bar{x}}(t) = \sum_{i=1}^{r}\sum_{j=1}^{r} \tilde{w}_i \tilde{m}_j \left[\hat{A}_{ij}\bar{x}(t) + \hat{B}_{ij}\zeta(t)\right], \\ e(t) = \sum_{i=1}^{r}\sum_{j=1}^{r} \tilde{w}_i \tilde{m}_j \hat{C}_{1ij}\bar{x}(t), \end{cases} \tag{8.23}$$

where

$$\hat{A}_{ij} = \begin{bmatrix} A_i & 0 & 0 \\ B_{fj}C_i & A_{fj} & 0 \\ 0 & 0 & A_w \end{bmatrix}, \quad \hat{B}_{ij} = \begin{bmatrix} B_i & B_{1i} \\ B_{fj}D_i & B_{fj}D_{1i} \\ 0 & B_w \end{bmatrix},$$
$$\hat{C}_{1ij} = \begin{bmatrix} 0 & C_{fj} & -C_w \end{bmatrix}.$$

Then, the stability condition with H_∞ performance of the fault detection system (8.23) is proposed in the following corollary.

Corollary 8.3 *The membership functions satisfy* $\tilde{m}_j - \rho_j\tilde{w}_j \geq 0$ $(0 < \rho_j \leq 1)$ *and the fault detection system* (8.23) *is asymptotically stable with an* H_∞ *performance* *level* γ, *if there exist matrices* $\bar{P} > 0$ *and* $\Theta_i > 0$ *with appropriate dimensions, such that the following conditions hold for* $i, j = 1, 2, \ldots, r$:

$$\begin{aligned} \Delta_{ij} - \Theta_i &< 0, \\ \rho_i \Delta_{ii} - \rho_i \Theta_i + \Theta_i &< 0, \\ \rho_j \Delta_{ij} + \rho_i \Delta_{ji} - \rho_j \Theta_i - \rho_i \Theta_j + \Theta_i + \Theta_j &\le 0, \quad i < j, \end{aligned}$$

where

$$\Delta_{ij} = \begin{bmatrix} \mathbf{He}\left(\bar{P}\bar{A}_{ij}\right) + \bar{C}_{1ij}^T \bar{C}_{1ij} & \bar{P}\bar{B}_{ij} \\ * & -\gamma^2 I \end{bmatrix}.$$

Proof This proof is similar to the proof of Theorem 8.1. The detailed procedure is omitted here. □

The existence condition of the H_∞ fault detection filter for the IT2 fuzzy system (8.22) is presented in the following corollary.

Corollary 8.4 *Under the condition* $\tilde{m}_j - \rho_j \tilde{w}_j \ge 0$ $(0 < \rho_j \le 1)$ *for all* j, *the fault detection system (8.23) is asymptotically stable with an* H_∞ *performance level* γ, *if there exist matrices* $P > 0$, $F > 0$, $V > 0$, $\tilde{\Theta}_i^T = \tilde{\Theta}_i$, $\bar{A}_{fj}$, $\bar{B}_{fj}$ *and* $\bar{C}_{fj}$ *with appropriate dimensions such that the following LMIs hold for* $i, j = 1, 2, \ldots, r$:

$$\begin{aligned} \tilde{\Delta}_{ij} - \tilde{\Theta}_i &< 0, \\ \rho_i \tilde{\Delta}_{ii} - \rho_i \tilde{\Theta}_i + \tilde{\Theta}_i &< 0, \\ \rho_j \tilde{\Delta}_{ij} + \rho_i \tilde{\Delta}_{ji} - \rho_j \tilde{\Theta}_i - \rho_i \tilde{\Theta}_j + \tilde{\Theta}_i + \tilde{\Theta}_j &\le 0, \quad i < j, \end{aligned}$$

where

$$\begin{aligned} \tilde{\Delta}_{ij} &= \begin{bmatrix} \Xi_{ij}^{11} & \Xi_{ij}^{12} & \Xi_{ij}^{13} \\ * & -\gamma^2 I & 0 \\ * & * & -I \end{bmatrix}, \\ \Xi_{ij}^{11} &= \begin{bmatrix} \mathbf{He}\left(PA_i + \bar{B}_{fj} C_i\right) & \bar{A}_{fj} + A_i^T F^T + C_i^T \bar{B}_{fj}^T & 0 \\ * & \mathbf{He}\left(\bar{A}_{fj}\right) & 0 \\ * & * & \mathbf{He}\left(V A_w\right) \end{bmatrix}, \\ \Xi_{ij}^{12} &= \begin{bmatrix} PB_i + \bar{B}_{fj} D_i & PB_{1i} + \bar{B}_{fj} D_{1i} \\ FB_i + \bar{B}_{fj} D_i & FB_{1i} + \bar{B}_{fj} D_{1i} \\ 0 & VB_w \end{bmatrix}, \quad \Xi_{ij}^{13} = \begin{bmatrix} 0 \\ \bar{C}_{fj}^T \\ -C_w^T \end{bmatrix}. \end{aligned}$$

The IT2 fuzzy filter parameters are given by:

$$A_{fj} = F^{-1} \bar{A}_{fj}, \quad B_{fj} = F^{-1} \bar{B}_{fj}, \quad C_{fj} = \bar{C}_{fj}.$$

Proof The corollary can be proved by the similar line of the proof of Theorem 8.2. □

8.4 Simulation Results

In this section, an example is used to illustrate the effectiveness of the proposed method.

Example 8.5 Consider a 2-rule IT2 fuzzy system in the form of (8.3) with (8.7), the matrices are listed below:

$$
\begin{aligned}
A_1 &= \begin{bmatrix} -1 & 0.2 \\ -0.9 & 0.15 \end{bmatrix}, \quad A_2 = \begin{bmatrix} -0.4 & 0.2 \\ -0.8 & -1.1 \end{bmatrix}, \quad B_1 = \begin{bmatrix} 0.1 \\ 0.2 \end{bmatrix}, \quad B_2 = \begin{bmatrix} 0.4 \\ 0.9 \end{bmatrix}, \\
B_{11} &= \begin{bmatrix} -0.1 & 0.01 \end{bmatrix}, \quad B_{12} = \begin{bmatrix} -0.1 & 0.02 \end{bmatrix}, \quad C_1 = \begin{bmatrix} 0.2 & 0.1 \end{bmatrix}, \\
C_2 &= \begin{bmatrix} 0.1\ 0.2 \end{bmatrix}, \quad D_1 = 0.01, \quad D_2 = 0.02, \quad D_{11} = 0.01, \quad D_{12} = 0.02, \\
A_w &= -5, \quad B_w = 5, \quad C_w = 1, \quad K_1 = 0.6, \quad K_2 = 1, \\
\phi(u) &= \frac{K_1 + K_2}{2} u + \frac{K_2 - K_1}{2} \sin(u).
\end{aligned}
$$

Membership functions for Rules 1 and 2 are given as follows:

$$
\begin{aligned}
\tilde{w}_1(x_1) &= 1 - \frac{1}{1 + e^{(-x_1 + 4 + \delta(x_1))}}, \quad \tilde{w}_2(x_1) = 1 - \tilde{w}_1(x_1), \\
\underline{m}_1(x_1) &= 1 - \frac{1}{e^{\frac{-x_1 - 0.25}{2}}}, \quad \bar{m}_1(x_1) = 1 - \frac{1}{e^{\frac{-x_1 + 0.25}{2}}}, \\
\underline{m}_2(x_1) &= 1 - \bar{m}_1(x_1), \quad \bar{m}_2(x_1) = 1 - \underline{m}_1(x_1), \\
\tilde{m}_j(x_1) &= \frac{\underline{\beta}_j \underline{m}_j(x_1) + \bar{\beta}_j \bar{m}_j(x_1)}{\sum_{k=1}^{2} \left(\underline{\beta}_k \underline{m}_k(x_1) + \bar{\beta}_k \bar{m}_k(x_1) \right)}, \quad j = 1, 2.
\end{aligned}
$$

In the membership functions, $\delta(x_1) = 0.1 \sin(x_1) \in [-0.1, 0.1]$ represents the parameter uncertainty. Let $\underline{\beta}_j = 0.5$ and $\bar{\beta}_j = 0.5$. By solving LMIs (8.16)–(8.18) in Theorem 8.2 , it can be found that the minimized H_∞ performance index $\gamma = 1.0020$ and the H_∞ filter parameters in (8.6) are listed below:

$$
\begin{aligned}
A_{f1} &= \begin{bmatrix} -2.8809 & 0.4141 \\ -5.6598 & -3.9717 \end{bmatrix}, \quad A_{f2} = \begin{bmatrix} -1.2269 & -0.6565 \\ 11.9116 & -12.4249 \end{bmatrix}, \\
B_{f1} &= \begin{bmatrix} -0.9782 \\ -45.9352 \end{bmatrix}, \quad B_{f2} = \begin{bmatrix} -6.0443 \\ -9.5273 \end{bmatrix}, \\
C_{f1} &= \begin{bmatrix} 0.0011 & -0.0051 \end{bmatrix}, \quad C_{f2} = \begin{bmatrix} 0.0054 & 0.0016 \end{bmatrix}.
\end{aligned}
$$

It is assumed that the disturbance $w(t)$ is

$$
w(t) = \begin{cases} 0.1 \sin(2t), & 5 \le t \le 15, \\ 0, & \text{else.} \end{cases}
$$

The fault signal is set up as

$$f(t) = \begin{cases} 3, & 5 \leq t \leq 15, \\ 0, & \text{else.} \end{cases}$$

Figure 8.1 plots the response of weighting fault signal (reference signal) and Fig. 8.2 plots the responses of the residual signal without fault case and with fault case, respectively. The different error values $e(t) = z_f(t) - f_w(t)$ are depicted in Figs. 8.3 and 8.4 under the initial conditions of $x(t) = [-0.5 \quad 1]^T$, $\hat{x}(t) = [-0.5 \quad 1]^T$, and $x_w(t) = 0$, respectively. When the residual signal is generated, next step is to set up the fault detection measure. Form the threshold of (8.9), $\mathcal{J}_{th} = 0.1040$, the results show that $\left(\int_0^{12.52} z_f^T(t) z_f(t) dt\right)^{1/2} = 0.1043 > \mathcal{J}_{th}$. Thus, the appeared fault can be detected after 12.52 s. Figure 8.6 illustrates that the fault detection filter can detect the fault immediately and effectively when fault occurs under the disturbance input. Finally, Fig. 8.5 plots the states responses of the filter. These simulation results show the effectiveness of the proposed fault detection method.

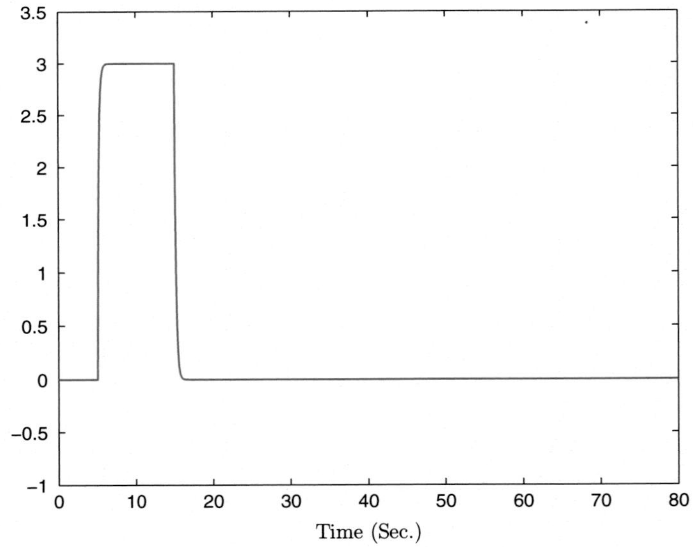

Fig. 8.1 Weighting fault signal (reference signal) $f_w(t)$

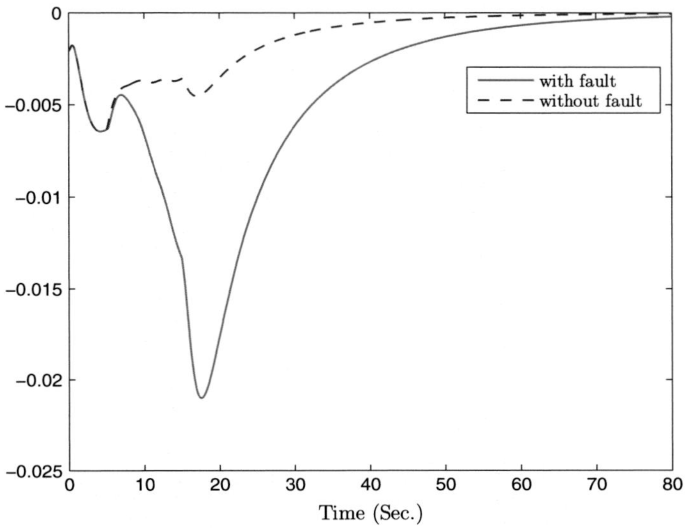

Fig. 8.2 Residual signal $z_f(t)$

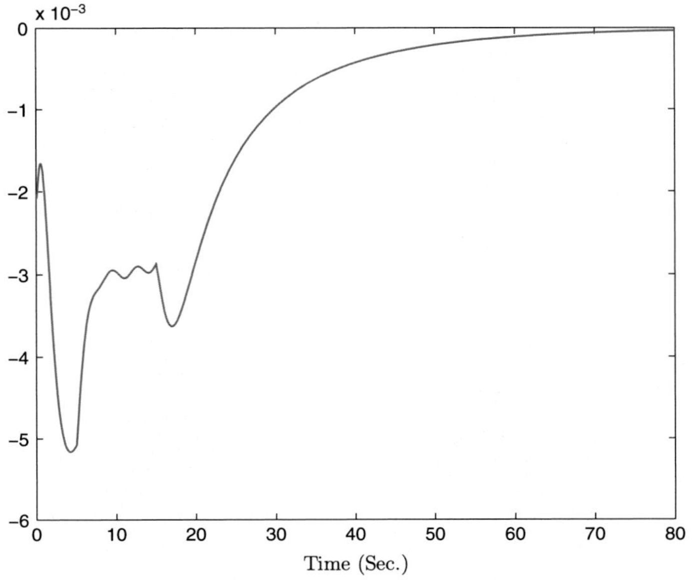

Fig. 8.3 The error value $e(t)$ without fault

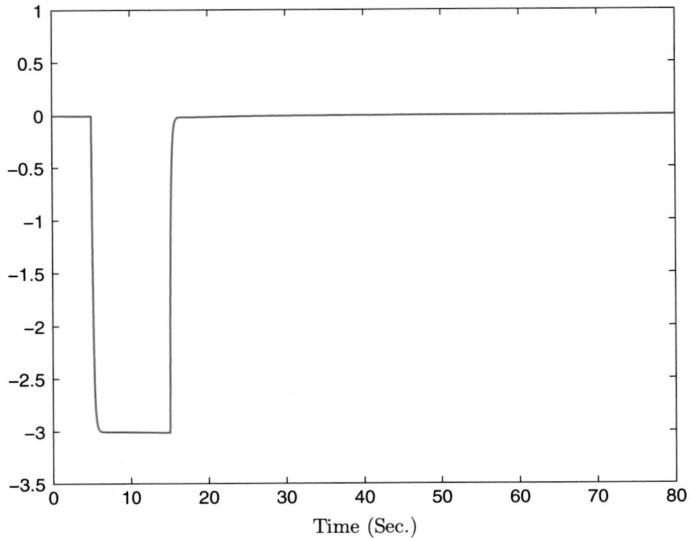

Fig. 8.4 The error value $e(t)$ with fault

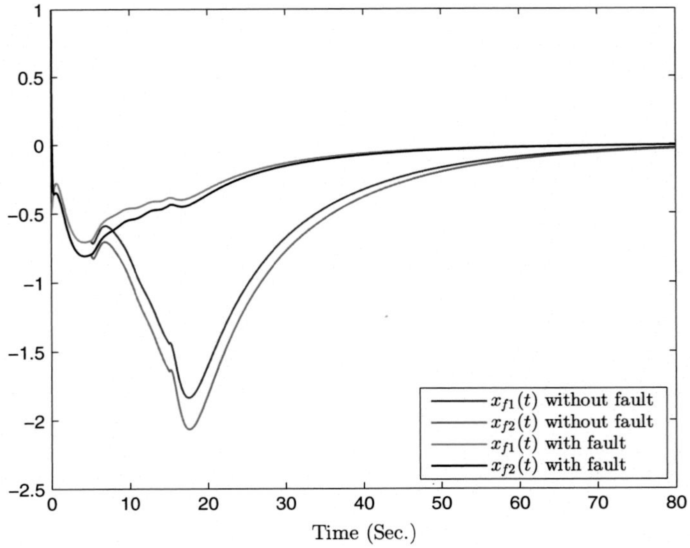

Fig. 8.5 State responses of the fault detection filter

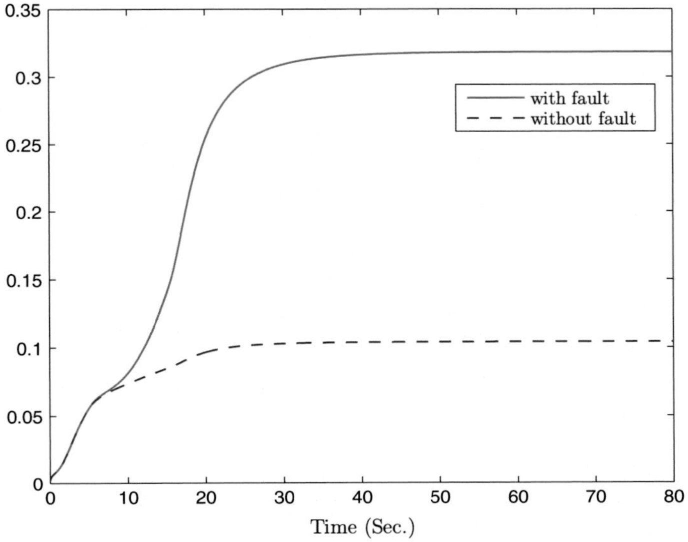

Fig. 8.6 Responses of evaluation function $\mathcal{J}(z_f)$

8.5 Conclusion

In this chapter, the H_∞ fault detection problem has been considered for a class of IT2 fuzzy systems with sensor nonlinearities. The IT2 fuzzy systems and the fault detection filter have been constructed. By using a general observer-based fault detection filter as a residual generator, the fault detection problem has been described as a filter design problem. The fault detection filter has been designed to guarantee the prescribed H_∞ performance level. In the design procedure, the parameters of the IT2 filter can be solved by the standard software. The IT2 fuzzy model and IT2 fuzzy filter do not require to share the same LMFs and UMFs. A numerical example has been given to demonstrate the merits of the proposed approach. In future work, we will attempt to solve the model reduction problem for IT2 fuzzy systems.

Chapter 9
Model Reduction of Interval Type-2 Fuzzy-Model-Based Systems

9.1 Introduction

This chapter is concerned with the problem of H_∞ model reduction for IT2 fuzzy systems with $\mathcal{D}$-stability constraints. In this chapter, the main advantages are as follows: (1) The problem of H_∞ model reduction based on IT2 fuzzy model is first proposed for nonlinear systems with parameter uncertainties. (2) By using LMFs and UMFs, the parameter uncertainties of the plants can be solved. (3) The membership functions and the number of fuzzy rules of the reduced-order system to be designed are independent of those of the original system, which can enhance the flexibility of model reduction and result in less conservativeness.

9.2 Problem Formulation

In this chapter, we consider the following IT2 T–S fuzzy model that represents a continuous-time nonlinear system

♦ **Plant Form:**

Rule i: IF $f_1(x(t))$ is $\tilde{M}_1^i$ and ... and $f_p(x(t))$ is $\tilde{M}_p^i$, THEN

$$\begin{cases} \dot{x}(t) = A_i x(t) + B_i w(t), \\ y(t) = C_i x(t), \end{cases} \tag{9.1}$$

where $f_a(x(t))$ is the premise variable and $\tilde{M}_a^i$ is an IT2 fuzzy set, $i = 1, 2, \ldots, r$, $a = 1, 2, \ldots, p$, p is a positive integer, $x(t) \in \mathbf{R}^n$ denotes the system state vector, $w(t) \in \mathbf{R}^{\bar{q}}$ stands for the disturbance input, $y(t) \in \mathbf{R}^{\bar{l}}$ is the output, A_i, B_i and C_i are known matrices with appropriate dimensions. The ith fuzzy rule can be described by the following interval set:

H. Li et al., *Analysis and Synthesis for Interval Type-2 Fuzzy-Model-Based Systems*, DOI 10.1007/978-981-10-0593-0_9

$$W_i(x(t)) = \left[\underline{w}_i(x(t)),\quad \overline{w}_i(x(t))\right],\quad i = 1, 2, \ldots, r,$$

where $\underline{w}_i(x(t)) = \prod_{a=1}^{p} \underline{\mu}_{\tilde{M}_a^i}(f_a(x(t)))$ and $\overline{w}_i(x(t)) = \prod_{a=1}^{p} \bar{\mu}_{\tilde{M}_a^i}(f_a(x(t)))$ stand for the lower and upper grades of membership, respectively. $\underline{\mu}_{\tilde{M}_a^i}(f_a(x(t))) \geq 0$ and $\bar{\mu}_{\tilde{M}_a^i}(f_a(x(t))) \geq 0$ stands for the LMF and UMF, respectively, respectively. Therefore, it can be found that $\bar{\mu}_{\tilde{M}_a^i}(f_a(x(t))) \geq \underline{\mu}_{\tilde{M}_a^i}(f_a(x(t)))$ and $\overline{w}_i(x(t)) \geq \underline{w}_i(x(t))$ for all i. Then, the IT2 T–S fuzzy system is described as follows:

$$\begin{cases} \dot{x}(t) = \sum_{i=1}^{r} \tilde{w}_i(x(t))\left[A_i x(t) + B_i w(t)\right], \\ y(t) = \sum_{i=1}^{r} \tilde{w}_i(x(t))\, C_i x(t), \end{cases} \tag{9.2}$$

where

$$\hat{w}_i(x(t)) = \underline{a}_i(x(t))\,\underline{w}_i(x(t)) + \bar{a}_i(x(t))\,\overline{w}_i(x(t)) \geq 0,\quad \forall i,$$

$$\tilde{w}_i(x(t)) = \frac{\hat{w}_i(x(t))}{\sum_{\bar{s}=1}^{r} \hat{w}_{\bar{s}}(x(t))},\quad \sum_{i=1}^{r} \tilde{w}_i(x(t)) = 1.$$

The nonlinear functions $\underline{a}_i(x(t))$ and $\bar{a}_i(x(t))$ satisfy: (1) $0 \leq \underline{a}_i(x(t)) \leq 1$ and $0 \leq \bar{a}_i(x(t)) \leq 1$. (2) $\underline{a}_i(x(t)) + \bar{a}_i(x(t)) = 1$.

Here, we will approximate the system (9.2) by an IT2 fuzzy reduced-order system with c rules constructed as follows:

Rule j: IF $g_1(x(t))$ is $\tilde{N}_1^j$ and ... and $g_{\hat{l}}(x(t))$ is $\tilde{N}_{\hat{l}}^j$, THEN,

$$\begin{cases} \dot{\hat{x}}(t) = \hat{A}_j \hat{x}(t) + \hat{B}_j w(t), \\ \hat{y}(t) = \hat{C}_j \hat{x}(t), \end{cases} \tag{9.3}$$

where $g_{\hat{\beta}}(x(t))$ is the premise variable and $\tilde{N}_{\hat{\beta}}^j$ is an IT2 fuzzy set, $j = 1, 2, \ldots, c$; $\hat{\beta} = 1, 2, \ldots, \hat{l}$; $\hat{l}$ is a positive integer; $\hat{x}(t) \in \mathbf{R}^{\hat{k}}$ is the state vector of the reduced-order system with $\hat{k} < n$; $\hat{y}(t) \in \mathbf{R}^{t}$ is the output of the reduced-order system; $\hat{A}_j$, $\hat{B}_j$ and $\hat{C}_j$ are appropriately dimensioned matrices to be designed. The jth fuzzy rule can be described by the following interval set:

$$M_j(x(t)) = \left[\underline{m}_j(x(t)), \overline{m}_j(x(t))\right],\quad j = 1, 2, \ldots, c,$$

where $\underline{m}_j(x(t)) = \prod_{\hat{\beta}=1}^{\hat{l}} \underline{\mu}_{\tilde{N}_{\hat{\beta}}^j}\left(g_{\hat{\beta}}(x(t))\right)$ and $\overline{m}_j(x(t)) = \prod_{\hat{\beta}=1}^{\hat{l}} \bar{\mu}_{\tilde{N}_{\hat{\beta}}^j}\left(g_{\hat{\beta}}(x(t))\right)$

are the lower and upper grades of membership, respectively. $\underline{\mu}_{\tilde{N}^j_{\hat{\beta}}}\left(g_{\hat{\beta}}\left(x\left(t\right)\right)\right) \geq 0$ and $\bar{\mu}_{\tilde{N}^j_{\hat{\beta}}}(g_{\hat{\beta}}\left(x\left(t\right)\right)) \geq 0$ stands for the LMF and UMF, respectively. Here, $\bar{\mu}_{\tilde{N}^j_{\hat{\beta}}}(g_{\hat{\beta}}\left(x\left(t\right)\right)) \geq \underline{\mu}_{\tilde{N}^j_{\hat{\beta}}}(g_{\hat{\beta}}\left(x\left(t\right)\right))$ leads to $\overline{m}_j\left(x\left(t\right)\right) \geq \underline{m}_j\left(x\left(t\right)\right)$. The overall IT2 fuzzy reduced-order model is proposed as follows:

$$\begin{cases} \dot{\hat{x}}\left(t\right) = \sum_{j=1}^{c} \tilde{m}_j\left(x\left(t\right)\right)\left[\hat{A}_j\hat{x}\left(t\right) + \hat{B}_j w\left(t\right)\right], \\ \hat{y}\left(t\right) = \sum_{j=1}^{c} \tilde{m}_j\left(x\left(t\right)\right)\hat{C}_j\hat{x}\left(t\right), \end{cases} \tag{9.4}$$

where

$$\tilde{m}_j\left(x\left(t\right)\right) = \frac{\underline{\beta}_j\left(x\left(t\right)\right)\underline{m}_j\left(x\left(t\right)\right) + \bar{\beta}_j\left(x\left(t\right)\right)\overline{m}_j\left(x\left(t\right)\right)}{\sum_{\bar{k}=1}^{c}\left(\underline{\beta}_{\bar{k}}\left(x\left(t\right)\right)\underline{m}_{\bar{k}}\left(x\left(t\right)\right) + \bar{\beta}_{\bar{k}}\left(x\left(t\right)\right)\overline{m}_{\bar{k}}\left(x\left(t\right)\right)\right)} \geq 0, \quad \forall j,$$

$$\sum_{j=1}^{c} \tilde{m}_j\left(x\left(t\right)\right) = 1.$$

The predefined functions $\underline{\beta}_j\left(x\left(t\right)\right)$ and $\bar{\beta}_j\left(x\left(t\right)\right)$ satisfy: (1) $0 \leq \underline{\beta}_j\left(x\left(t\right)\right) \leq 1$ and $0 \leq \bar{\beta}_j\left(x\left(t\right)\right) \leq 1$. (2) $\underline{\beta}_j\left(x\left(t\right)\right) + \bar{\beta}_j\left(x\left(t\right)\right) = 1$.

Remark 9.1 The existing model reduction results for T–S fuzzy system [169, 204] translated the original system into a linear system. Since it is difficult to confirm the membership functions for the ideal low dimensional T–S fuzzy system. In this chapter, for solving this problem, the grades of membership of the IT2 reduced-order model (9.4) to be designed are not the same as those of the IT2 fuzzy system (9.2), which can enhance the flexibility of model reduction and result in less conservativeness.

It can be seen from (9.2) and (9.4) that the IT2 error system can be given as follows:

$$\begin{cases} \dot{\bar{x}}\left(t\right) = \sum_{i=1}^{r}\sum_{j=1}^{c} \tilde{w}_i\left(x\left(t\right)\right)\tilde{m}_j\left(x\left(t\right)\right)\left[\bar{A}_{ij}\bar{x}\left(t\right) + \bar{B}_{ij}w\left(t\right)\right], \\ e\left(t\right) = \sum_{i=1}^{r}\sum_{j=1}^{c} \tilde{w}_i\left(x\left(t\right)\right)\tilde{m}_j\left(x\left(t\right)\right)\bar{C}_{ij}\bar{x}\left(t\right), \end{cases} \tag{9.5}$$

where $\sum_{i=1}^{r}\tilde{w}_i(x(t)) = \sum_{j=1}^{c}\tilde{m}_j(x(t)) = \sum_{i=1}^{r}\sum_{j=1}^{c}\tilde{w}_i(x(t))\tilde{m}_j(x(t)) = 1$, $\bar{x}(t) = \left[x^T(t)\ \hat{x}^T(t)\right]^T$, $e(t) = y(t) - \hat{y}(t)$,

$$\bar{A}_{ij} = \begin{bmatrix} A_i & 0 \\ 0 & \hat{A}_j \end{bmatrix}, \quad \bar{B}_{ij} = \begin{bmatrix} B_i \\ \hat{B}_j \end{bmatrix}, \quad \bar{C}_{ij} = \begin{bmatrix} C_i & -\hat{C}_j \end{bmatrix}.$$

To investigate the model reduction problem for the IT2 error system (9.5), we need to introduce the following results (see [94]).

First, the state space Γ can be partitioned into following q connected sub-state spaces Γ_k $(k = 1, 2, \ldots, q)$, such that $\Gamma = \cup_{k=1}^{q}\Gamma_k$. Second, the FOU can be divided into $\varsigma+1$ sub-FOUs. For $l = 1, 2, \ldots, \varsigma+1$, the LMFs and UMFs in the lth sub-FOU are described as follows for $\forall\, i, j, k, l$:

$$\begin{aligned}
\underline{h}_{ijl}(x(t)) &= \sum_{k=1}^{q}\sum_{i_1=1}^{2}\sum_{i_2=1}^{2}\cdots\sum_{i_n=1}^{2}\prod_{r=1}^{n}\upsilon_{ri_rkl}(x_r(t))\,\underline{\vartheta}_{iji_1i_2\ldots i_nkl},\\
\overline{h}_{ijl}(x(t)) &= \sum_{k=1}^{q}\sum_{i_1=1}^{2}\sum_{i_2=1}^{2}\cdots\sum_{i_n=1}^{2}\prod_{r=1}^{n}\upsilon_{ri_rkl}(x_r(t))\,\overline{\vartheta}_{iji_1i_2\ldots i_nkl},\\
0 &\le \underline{h}_{ijl}(x(t)) \le \overline{h}_{ijl}(x(t)) \le 1,\\
0 &\le \underline{\vartheta}_{iji_1i_2\ldots i_nkl} \le \overline{\vartheta}_{iji_1i_2\ldots i_nkl} \le 1,
\end{aligned}$$

where $\underline{\vartheta}_{iji_1i_2\ldots i_nkl}$ and $\overline{\vartheta}_{iji_1i_2\ldots i_nkl}$ are constant scalars to be determined; $0 \le \upsilon_{ri_skl}(x_r(t)) \le 1$ and $\upsilon_{r1kl}(x_r(t)) + \upsilon_{r2kl}(x_r(t)) = 1$ for $r, s = 1, 2, \ldots, n$; $l = 1, 2, \ldots, \varsigma+1$; $i_r = 1, 2$; $x(t) \in \Gamma_k$; otherwise, $\upsilon_{ri_sk}(x_r(t)) = 0$. Then, we have $\sum_{k=1}^{q}\sum_{i_1=1}^{2}\sum_{i_2=1}^{2}\cdots\sum_{i_n=1}^{2}\prod_{r=1}^{n}\upsilon_{ri_rkl}(x_r(t)) = 1$ for all l, which is used in this chapter. Then, according to system (9.5), we can obtain the following system:

$$\begin{cases}
\dot{\bar{x}}(t) = \sum_{i=1}^{r}\sum_{j=1}^{c}h_{ij}(x(t))\left[\bar{A}_{ij}\bar{x}(t) + \bar{B}_{ij}w(t)\right],\\
e(t) = \sum_{i=1}^{r}\sum_{j=1}^{c}h_{ij}(x(t))\bar{C}_{ij}\bar{x}(t),
\end{cases} \tag{9.6}$$

where

$$\begin{aligned}
h_{ij}(x(t)) &= \tilde{w}_i(x(t))\tilde{m}_j(x(t))\\
&= \sum_{l=1}^{\varsigma+1}\Theta_{ijl}(x(t))\left(\underline{\zeta}_{ijl}(x(t))\underline{h}_{ijl}(x(t))\right.\\
&\quad\left.+\overline{\zeta}_{ijl}(x(t))\overline{h}_{ijl}(x(t))\right), \quad \forall i, j,
\end{aligned} \tag{9.7}$$

with

$$\sum_{i=1}^{r}\sum_{j=1}^{c} h_{ij}\left(x\left(t\right)\right)=1,$$

in which $\underline{\zeta}_{ijl}(x(t))$ and $\overline{\zeta}_{ijl}(x(t))$ are two functions and they satisfy $0 \le \underline{\zeta}_{ijl}(x(t)) \le \overline{\zeta}_{ijl}(x(t)) \le 1$, which are unnecessary to be known and possess the trait of $\underline{\zeta}_{ijl}(x(t)) + \overline{\zeta}_{ijl}(x(t)) = 1$ for all i, j and l; $\Theta_{ijl}(x(t)) = 1$ if the membership function $h_{ij}(x(t))$ is within the sub-FOU l, otherwise, $\Theta_{ijl}(x(t)) = 0$. For brevity, in the following part, the variables $\tilde{w}_i(x(t))$, $\tilde{m}_j(x(t))$, $\Theta_{ijl}(x(t))$, $\underline{\zeta}_{ijl}(x(t))$, $\overline{\zeta}_{ijl}(x(t))$, $\underline{h}_{ijl}(x(t))$, $\overline{h}_{ijl}(x(t))$ and $h_{ij}(x(t))$ are denoted by $\tilde{w}_i$, $\tilde{m}_j$, Θ_{ijl}, $\underline{\zeta}_{ijl}$, $\overline{\zeta}_{ijl}$, $\underline{h}_{ijl}$, $\overline{h}_{ijl}$ and h_{ij}, respectively.

9.3 Main Results

9.3.1 Stability Analysis

The problem of H_∞ model reduction will be solved in the following section. Firstly, the stability condition for the IT2 error system (9.6) with an H_∞ performance is presented in the following theorem.

Theorem 9.2 *Given a scalar $\gamma > 0$, the error system (9.6) is asymptotically stable with an H_∞ performance and the poles lie in the region $\mho(\hat{q}, \hat{r})$ or $\Psi(\hat{v}, \hat{u})$, if there exist matrices $P > 0$, $W_{ijl} = W_{ijl}^T$, $M = M^T$, $i = 1, 2, \ldots, r$; $j = 1, 2, \ldots, c$; $l = 1, 2, \ldots, \varsigma+1$, such that the following LMIs (9.8), (9.9)–(9.11) (or, (1.17)–(1.18) and (9.9)–(9.11)) hold:*

$$\begin{bmatrix} -P & P(\bar{A}_{ij}-\hat{q}I) \\ * & -\hat{r}^2 P \end{bmatrix} < 0, \tag{9.8}$$

$$W_{ijl} \ge 0, \quad \forall i, j, l \tag{9.9}$$

$$\Omega_{ij} + W_{ijl} + M > 0, \quad \forall i, j, l \tag{9.10}$$

$$\sum_{i=1}^{r}\sum_{j=1}^{c} Z_{ijl} - M < 0, \quad \forall i_1, \ldots, i_n, k, l, \tag{9.11}$$

where

$$\Omega_{ij} = \begin{bmatrix} \mathrm{He}(P\bar{A}_{ij}) + \bar{C}_{ij}^T\bar{C}_{ij} & P\bar{B}_{ij} \\ * & -\gamma^2 I \end{bmatrix},$$

$$Z_{ijl} = \overline{\vartheta}_{iji_1i_2\ldots i_nkl}\Omega_{ij} - (\underline{\vartheta}_{iji_1i_2\ldots i_nkl} - \overline{\vartheta}_{iji_1i_2\ldots i_nkl})W_{ijl} + \overline{\vartheta}_{iji_1i_2\ldots i_nkl}M.$$

Proof Consider the following Lyapunov function:

$$V(t) = \bar{x}^T(t) P \bar{x}(t).$$

Then, the time derivative of $V(t)$ is expressed as:

$$\dot{V}(t) = 2\bar{x}^T(t) P \dot{\bar{x}}(t) = 2\sum_{i=1}^{r}\sum_{j=1}^{c} h_{ij}\bar{x}^T(t) P \left[\bar{A}_{ij}\bar{x}(t) + \bar{B}_{ij}w(t)\right].$$

Then, we can have

$$\begin{aligned}
&\dot{V}(t) + e^T(t) e(t) - \gamma^2 w^T(t) w(t) \\
&\le \sum_{i=1}^{r}\sum_{j=1}^{c} h_{ij}\{2\bar{x}^T(t) P \left[\bar{A}_{ij}\bar{x}(t) + \bar{B}_{ij}w(t)\right] \\
&\quad + \bar{x}^T(t)\bar{C}_{ij}^T\bar{C}_{ij}\bar{x}(t) - \gamma^2 w^T(t) w(t)\} \\
&= \xi^T(t)\left(\sum_{i=1}^{r}\sum_{j=1}^{c} h_{ij}\Omega_{ij}\right)\xi(t),
\end{aligned} \tag{9.12}$$

where

$$\xi(t) = \left[\bar{x}^T(t)\ w^T(t)\right]^T, \quad \Omega_{ij} = \begin{bmatrix} \mathrm{He}(P\bar{A}_{ij}) + \bar{C}_{ij}^T\bar{C}_{ij} & P\bar{B}_{ij} \\ * & -\gamma^2 I \end{bmatrix}.$$

Considering following slack matrices which are used in the following inequalities:

$$\left(\sum_{i=1}^{r}\sum_{j=1}^{c}\sum_{l=1}^{\varsigma+1}\Theta_{ijl}\left(\underline{\zeta}_{ijl}\underline{h}_{ijl} + \overline{\zeta}_{ijl}\overline{h}_{ijl}\right) - 1\right) M = 0, \tag{9.13}$$

$$-\sum_{i=1}^{r}\sum_{j=1}^{c}\left(1 - \underline{\zeta}_{ijl}\right)\left(\underline{h}_{ijl} - \overline{h}_{ijl}\right) W_{ijl} \ge 0, \tag{9.14}$$

where $M = M^T$ and $0 \le W_{ijl} = W_{ijl}^T$ are matrices with appropriate dimensions. From (9.7), (9.12)–(9.14), we have

$$\begin{aligned}
&\dot{V}(t) + e^T(t) e(t) - \gamma^2 w^T(t) w(t) \\
&\le \xi^T(t)\left(\sum_{i=1}^{r}\sum_{j=1}^{c} h_{ij}\Omega_{ij}\right)\xi(t)
\end{aligned}$$

$$\leq \xi^T(t)\left[\sum_{i=1}^{r}\sum_{j=1}^{c}\left(\overline{\vartheta}_{iji_1i_2\ldots i_nkl}\Omega_{ij} - \left(\underline{\vartheta}_{iji_1i_2\ldots i_nkl}\right.\right.\right.$$
$$\left.- \overline{\vartheta}_{iji_1i_2\ldots i_nkl}\right) W_{ijl} + \overline{\vartheta}_{iji_1i_2\ldots i_nkl}M\Big) - M$$
$$\left.- \overline{\vartheta}_{iji_1i_2\ldots i_nkl}\right) W_{ijl} + \overline{\vartheta}_{iji_1i_2\ldots i_nkl}M\Big) - M\Big]\xi(t)$$
$$+\sum_{i=1}^{r}\sum_{j=1}^{c}\sum_{l=1}^{\varsigma+1}\Theta_{ijl}\underline{\zeta}_{ijl}\left(\underline{h}_{ijl} - \overline{h}_{ijl}\right)\xi^T(t)\left[\Omega_{ij}\right.$$
$$\left.+ W_{ijl} + M\right]\xi(t).$$

It can be seen from the inequalities (9.9)–(9.11) that

$$\dot{V}(t) + e^T(t)e(t) - \gamma^2 w^T(t)w(t) < 0. \tag{9.15}$$

Under zero initial conditions, integrating both sides of (9.15) yields $\|e(t)\|_2 < \gamma\|w(t)\|_2$ for all nonzero $w \in \mathcal{L}_2[0,\infty)$. In addition, when $w(t) \equiv 0$, it can be seen from the conditions in Theorem 9.2 that $\dot{V}(t) < 0$, which means that the system (9.6) with $w(t) = 0$ is asymptotically stable. The proof is completed. □

9.3.2 Reduced-Order Control

Based on the above result, we will present an approach to solve the problem of H_∞ model reduction for IT2 fuzzy system (9.2) by using the linearization procedure in the following theorem.

Theorem 9.3 *Consider the IT2 fuzzy system (9.2), there is a low dimensional system (9.4) that deals with the problem of H_∞ model reduction and the poles lie in the region $\mho(\hat{q},\hat{r})$ or $\Psi(\hat{v},\hat{u})$, if there exist matrices $\mathcal{P} > 0$, $\mathcal{Q} > 0$, $\tilde{W}_{ijl} = \tilde{W}_{ijl}^T$, $\tilde{M} = \tilde{M}^T$, $\bar{A}_j$, $\bar{B}_j$ and $\bar{C}_j$ with appropriate dimensions such that the following LMIs (9.16)–(9.19) (or, (9.16)–(9.18), (9.20) and (9.21)) hold for $i = 1,2,\ldots,r$; $j = 1,2,\ldots,c$; $l = 1,2,\ldots,\varsigma+1$;*

$$\tilde{W}_{ijl} \geq 0, \quad \forall i, j, l \tag{9.16}$$

$$\tilde{\Omega}_{ij} + \tilde{W}_{ijl} + \tilde{M} > 0, \quad \forall i, j, l \tag{9.17}$$

$$\sum_{i=1}^{r} \sum_{j=1}^{c} \tilde{Z}_{ijl} - \tilde{M} < 0, \quad \forall i_1, i_2, \ldots, i_n, k, l \tag{9.18}$$

$$\begin{bmatrix} -\tilde{P} & \tilde{\Delta}_{ij} \\ * & -\hat{r}^2 \tilde{P} \end{bmatrix} < 0, \tag{9.19}$$

$$\mathrm{He}\left(\Delta_{ij}\right) - 2\hat{u}\tilde{P} < 0, \tag{9.20}$$

$$\mathrm{He}\left(-\Delta_{ij}\right) + 2\hat{v}\tilde{P} < 0, \tag{9.21}$$

where

$$\tilde{\Omega}_{ij} = \begin{bmatrix} \theta_{ij}^{11} & \theta_{ij}^{12} & \theta_{ij}^{13} \\ * & -\gamma^2 I & 0 \\ * & * & -I \end{bmatrix}, \quad \mathcal{H} = \begin{bmatrix} I_{\hat{k}\times\hat{k}} \\ 0_{\left(n-\hat{k}\right)\times\hat{k}} \end{bmatrix},$$

$$\theta_{ij}^{11} = \begin{bmatrix} \mathbf{He}\left(\mathcal{P}A_i\right) & \mathcal{H}\bar{A}_j + A_i^T\mathcal{H}\mathcal{Q} \\ * & \mathbf{He}(\bar{A}_j) \end{bmatrix}, \quad \tilde{\Delta}_{ij} = \Delta_{ij} - \hat{q}\tilde{P},$$

$$\tilde{P} = \begin{bmatrix} \mathcal{P} & \mathcal{H}\mathcal{Q} \\ * & \mathcal{Q} \end{bmatrix}, \quad \theta_{ij}^{12} = \begin{bmatrix} \mathcal{P}B_i + \mathcal{H}\bar{B}_j \\ \mathcal{Q}\mathcal{H}^T B_i + \bar{B}_j \end{bmatrix},$$

$$\theta_{ij}^{13} = \begin{bmatrix} C_i^T \\ -\bar{C}_j^T \end{bmatrix}, \quad \Delta_{ij} = \begin{bmatrix} \mathcal{P}A_i & \mathcal{H}\bar{A}_j \\ \mathcal{Q}\mathcal{H}^T A_i & \bar{A}_j \end{bmatrix},$$

$$\tilde{Z}_{ijl} = \overline{\vartheta}_{iji_1i_2\ldots i_nkl}\tilde{\Omega}_{ij} - (\underline{\vartheta}_{iji_1i_2\ldots i_nkl} - \overline{\vartheta}_{iji_1i_2\ldots i_nkl})\tilde{W}_{ijl} + \overline{\vartheta}_{iji_1i_2\ldots i_nkl}\tilde{M}.$$

The matrix parameters of the admissible reduced-order model (9.4) can be obtained:

$$\begin{bmatrix} \hat{A}_j & \hat{B}_j \\ \hat{C}_j & 0 \end{bmatrix} = \begin{bmatrix} \mathcal{Q}^{-1} & 0 \\ 0 & I \end{bmatrix} \begin{bmatrix} \bar{A}_j & \bar{B}_j \\ \bar{C}_j & 0 \end{bmatrix}. \tag{9.22}$$

Proof From the condition of Theorem 9.2 that $P > 0$, we can have P is nonsingular. To have this, define P as

$$P = \left[\begin{array}{c|c} P_1 & P_2 \\ \hline * & P_3 \end{array}\right], \quad P_2 = \left[\begin{array}{c} P_4 \\ \hline 0_{\left(n-\hat{k}\right)\times\hat{k}} \end{array}\right],$$

where $P_1 \in \mathbf{R}^{n\times n}$ and $P_3 \in \mathbf{R}^{\hat{k}\times\hat{k}}$ are symmetric positive definite matrices; $P_2 \in \mathbf{R}^{n\times\hat{k}}$ and $P_4 \in \mathbf{R}^{\hat{k}\times\hat{k}}$. In general, we can assume P_4 is nonsingular. Then, define matrix $S \triangleq P + \hat{\alpha}T$, where $\hat{\alpha}$ is a positive scalar and

$$T = \left[\begin{array}{c|c} 0_{n\times n} & \mathcal{H} \\ \hline * & 0_{\hat{k}\times\hat{k}} \end{array}\right], \quad S = \left[\begin{array}{c|c} S_1 & S_2 \\ \hline * & S_3 \end{array}\right], \quad S_2 = \left[\begin{array}{c} S_4 \\ \hline 0_{(n-\hat{k})\times\hat{k}} \end{array}\right].$$

It can be seen from $P > 0$ and $\hat{\alpha} > 0$ that $S > 0$. Thus, there is an arbitrarily small $\hat{\alpha} > 0$ such that S_4 is nonsingular and (9.10)–(9.11) are feasible with P displaced by S. Since S_4 is nonsingular, we can assume that the matrix P_4 is nonsingular. Define the following nonsingular matrices:

$$G = \left[\begin{array}{c|c} I & 0 \\ \hline 0 & P_3^{-1}P_4^T \end{array}\right], \quad \mathcal{P} = P_1, \quad \mathcal{Q} = P_4P_3^{-1}P_4^T,$$

From the inequalities (9.12) and (9.15), we know $\sum_{i=1}^{r}\sum_{j=1}^{c} h_{ij}\Omega_{ij} < 0$. Using Schur complement to $\sum_{i=1}^{r}\sum_{j=1}^{c} h_{ij}\Omega_{ij}$, one can yield:

$$\sum_{i=1}^{r}\sum_{j=1}^{c} h_{ij}\bar{\Omega}_{ij} = \sum_{i=1}^{r}\sum_{j=1}^{c} h_{ij}\begin{bmatrix} \mathbf{He}(P\bar{A}_{ij}) & P\bar{B}_{ij} & \bar{C}_{ij}^T \\ * & -\gamma^2 I & 0 \\ * & * & -I \end{bmatrix} < 0. \tag{9.23}$$

Define

$$\begin{bmatrix} \bar{A}_j & \bar{B}_j \\ \bar{C}_j & 0 \end{bmatrix} = \begin{bmatrix} P_4 & 0 \\ 0 & I \end{bmatrix}\begin{bmatrix} \hat{A}_j & \hat{B}_j \\ \hat{C}_j & 0 \end{bmatrix}\begin{bmatrix} P_3^{-1}P_4^T & 0 \\ 0 & I \end{bmatrix}, \tag{9.24}$$
$$\bar{M} = \mathrm{diag}\{G^T, I, I\}\bar{M}\mathrm{diag}\{G, I, I\},$$
$$\bar{W}_{ijl} = \mathrm{diag}\{G^T, I, I\}\bar{W}_{ijl}\mathrm{diag}\{G, I, I\},$$

and $\bar{M}$ and $\bar{W}_{ijl}$ are the matrices with appropriate dimensions. After replacing Ω_{ij}, M, W_{ijl} with $\bar{\Omega}_{ij}$, $\bar{M}$, $\bar{W}_{ijl}$ in inequalities (9.9)–(9.11), these inequalities still hold. Then performing a congruence transformation to (9.9)–(9.11) by diagonal matrix $\mathrm{diag}\{G, I, I\}$, we can obtain (9.16)–(9.18).

Then, performing a congruence transformation to (9.8) by diagonal matrix $\mathrm{diag}\{G, G\}$, we can obtain (9.19). Performing a congruence transformation to (1.17) and (1.18) by diagonal matrix G, we can obtain (9.20) and (9.21). Based on the above discussion, it can be seen that the conditions (9.16)–(9.21) hold. According to (9.24), we can obtain

$$\begin{aligned} \begin{bmatrix} \hat{A}_j & \hat{B}_j \\ \hat{C}_j & 0 \end{bmatrix} &= \begin{bmatrix} P_4^{-1} & 0 \\ 0 & I \end{bmatrix}\begin{bmatrix} \bar{A}_j & \bar{B}_j \\ \bar{C}_j & 0 \end{bmatrix}\begin{bmatrix} P_4^{-T}P_3 & 0 \\ 0 & I \end{bmatrix} \\ &= \begin{bmatrix} (P_4^{-T}P_3)^{-1}\mathcal{Q}^{-1} & 0 \\ 0 & I \end{bmatrix}\begin{bmatrix} \bar{A}_j & \bar{B}_j \\ \bar{C}_j & 0 \end{bmatrix}\begin{bmatrix} P_4^{-T}P_3 & 0 \\ 0 & I \end{bmatrix}. \end{aligned} \tag{9.25}$$

The matrices $\hat{A}_j$, $\hat{B}_j$, and $\hat{C}_j$ in (9.3) are given by (9.25). Then, we may set $P_4^{-T}P_3 = I$, thus obtaining (9.22) (see [169]). Therefore, the matrices in (9.3) are given by (9.22). The proof is completed. □

9.4 Simulation Results

In this section, we will use an example to demonstrate the effectiveness of the proposed method.

Example 9.4 Consider the IT2 fuzzy system (9.2) with the model parameters given as follows:

$$A_1 = \begin{bmatrix} -2.2 & 0.3 & 0.2 & 0.1 \\ 0.2 & -4.2 & 0.2 & 0.1 \\ 0.3 & 0 & -3.1 & 0.2 \\ 0.4 & 0.2 & 0.1 & -1.5 \end{bmatrix}, \quad B_1 = \begin{bmatrix} 2.4 \\ 1 \\ 1.3 \\ 2 \end{bmatrix},$$
$$C_1 = \begin{bmatrix} 1.1 & 1.3 & 0.4 & 0.2 \end{bmatrix},$$
$$A_2 = \begin{bmatrix} -2.5 & 0.1 & 0.3 & 0.2 \\ 0.2 & -3.3 & 0.1 & 0.2 \\ 0.1 & 0 & -2.9 & 0.2 \\ 0.4 & 0.2 & 0.1 & -1.7 \end{bmatrix}, \quad B_2 = \begin{bmatrix} 2.2 \\ 1.1 \\ 1.2 \\ 2.2 \end{bmatrix},$$
$$C_2 = \begin{bmatrix} 1.1 & 1 & 0.5 & 0.3 \end{bmatrix},$$
$$A_3 = \begin{bmatrix} -2.3 & 0.1 & 0.3 & 0.2 \\ 0.2 & -3.7 & 0.1 & 0.2 \\ 0.1 & 0 & -2.7 & 0.2 \\ 0.4 & 0.2 & 0.1 & -1.2 \end{bmatrix}, \quad B_3 = \begin{bmatrix} 2.1 \\ 1.1 \\ 1.2 \\ 2.1 \end{bmatrix},$$
$$C_3 = \begin{bmatrix} 1.1 & 1 & 0.6 & 0.4 \end{bmatrix}.$$

The LMFs and UMFs for system (9.2) are given in Tables 9.1 and 9.2, while the weighting functions $\underline{a}_i\left(x\left(t\right)\right)$ and $\bar{a}_i\left(x\left(t\right)\right)$ are defined as $\underline{a}_i\left(x_1\left(t\right)\right) = 0.6\sin\left(2x_1\left(t\right)\right)^2$ and $\bar{a}_i\left(x_1\left(t\right)\right) = 1 - \underline{a}_i\left(x_1\left(t\right)\right)$. Let $r = 3$, $c = 2$. Then, a 2-rule IT2 fuzzy membership functions for the system (9.4) with the LMFs and UMFs are

Table 9.1 LMFs for systems (9.2) and (9.4)

LMFs for system (9.2)
$\underline{u}_{\tilde{M}_1^1}\left(x_1\right) = 1 - e^{-\frac{x_1^2}{1.5}}$
$\underline{u}_{\tilde{M}_1^2}\left(x_1\right) = \underline{u}_{\tilde{M}_1^1}\left(x_1\right)$
$\underline{u}_{\tilde{M}_1^3}\left(x_1\right) = 1 - \overline{u}_{\tilde{M}_1^1}\left(x_1\right)$
$\underline{u}_{\tilde{M}_2^1}\left(x_1\right) = 0.5e^{-\frac{x_1^2}{0.2}}$
$\underline{u}_{\tilde{M}_2^2}\left(x_1\right) = 1 - \overline{u}_{\tilde{M}_2^1}\left(x_1\right)$
$\underline{u}_{\tilde{M}_2^3}\left(x_1\right) = \underline{u}_{\tilde{M}_2^1}\left(x_1\right)$
LMFs for system (9.4)
$\underline{u}_{\tilde{N}_1^1}\left(x_1\right) = e^{-\frac{x_1^2}{0.5}}$
$\underline{u}_{\tilde{N}_1^2}\left(x_1\right) = 1 - \overline{u}_{\tilde{N}_1^1}\left(x_1\right)$

Table 9.2 UMFs for systems (9.2) and (9.4)

UMFs for system (9.2)
$\overline{u}_{\tilde{M}_1^1}(x_1) = 1 - 0.23e^{-\frac{x_1^2}{0.3}}$
$\overline{u}_{\tilde{M}_1^2}(x_1) = \overline{u}_{\tilde{M}_1^1}(x_1)$
$\overline{u}_{\tilde{M}_1^3}(x_1) = 1 - \underline{u}_{\tilde{M}_1^1}(x_1)$
$\overline{u}_{\tilde{M}_2^1}(x_1) = e^{-\frac{x_1^2}{2.5}}$
$\overline{u}_{\tilde{M}_2^2}(x_1) = 1 - \underline{u}_{\tilde{M}_2^1}(x_1)$
$\overline{u}_{\tilde{M}_2^3}(x_1) = \overline{u}_{\tilde{M}_2^1}(x_1)$
UMFs for system (9.4)
$\overline{u}_{\tilde{N}_1^1}(x_1) = \underline{u}_{\tilde{N}_1^1}(x_1)$
$\overline{u}_{\tilde{N}_1^2}(x_1) = 1 - \underline{u}_{\tilde{N}_1^1}(x_1)$

chosen in Tables 9.1 and 9.2, while $\underline{\beta}_j(x(t))$ and $\bar{\beta}_j(x(t))$ are defined as $\underline{\beta}_j = \bar{\beta}_j = 0.5$. In this chapter, we consider the $\mathcal{D}$-stability constraints such that all the eigenvalues of the error system (9.6) lies in a disk region or vertical strip region. Because of limited space, we only consider vertical strip region with the parameters $\hat{v} = -10$, $\hat{u} = -1$. Then, we can obtain the parameters in (9.4) are listed below. In this example, we consider $\varsigma = 0$, which means that $l = 1$. Then, we consider $x_1 \in [-10, 10]$ and divide the state space of x_1 into 20 equal-size regions, Γ_k: $\underline{x}_{1,k} \le x_1 \le \bar{x}_{1,k}$, $k = 1, 2, \ldots, 20$, where $\underline{x}_{1,k} = (k - 11)$ and $\bar{x}_{1,k} = (k - 10)$. The membership functions $\underline{h}_{ij}(x_1)$ and $\overline{h}_{ij}(x_1)$ are defined by choosing $\upsilon_{11k}(x_1) = 1 - \frac{x_1 - \underline{x}_{1,k}}{\bar{x}_{1,k} - \underline{x}_{1,k}}$ and $\upsilon_{12k}(x_1) = 1 - \upsilon_{11k}(x_1)$; and the constant scalars are given as $\underline{\vartheta}_{ij1k} = \underline{w}_i(\underline{x}_{1,k})\underline{m}_j(\underline{x}_{1,k})$, $\underline{\vartheta}_{ij2k} = \underline{w}_i(\bar{x}_{1,k})\underline{m}_j(\bar{x}_{1,k})$, $\overline{\vartheta}_{ij1k} = \overline{w}_i(\underline{x}_{1,k})\overline{m}_j(\underline{x}_{1,k})$, and $\overline{\vartheta}_{ij2k} = \overline{w}_i(\bar{x}_{1,k})\overline{m}_j(\bar{x}_{1,k})$, for all k.

Remark 9.5 The problem of H_∞ model reduction is first time proposed for nonlinear systems with parameter uncertainties on the basis of IT2 fuzzy model. By using lower and upper grades of membership and relevant weighting functions, the parameter uncertainties of the plants can be solved according to the forms (9.2) and (9.4). The authors in [97] used an example to demonstrate the process. The other existing results [169, 204] cannot deal with the model reduction problem for type-1 T–S fuzzy systems with parameter uncertainties.

Remark 9.6 The number of sub-regions can be any integer greater than 0. In [94], the authors chose it as 20 and 500, respectively. The greater value, to some degree, may result in less conservativeness. However, considering the burden of calculation, we choose it as 20 in this chapter.

It is interesting to find reduced order systems (Case 1: $\hat{k} = 3$; Case 2: $\hat{k} = 2$; Case 3: $\hat{k} = 1$;) (9.4) to approximate the system (9.2) in an H_∞ sense. It can be found from Theorem 9.3 that the results for different cases of IT2 fuzzy system are given as follows:

Case 1: when $\hat{k} = 3$, the minimum $\gamma_{\min} = 0.1283$, and

$$\left[\begin{array}{c|c} \hat{A}_1 & \hat{B}_1 \\ \hline \hat{C}_1 & 0 \end{array}\right] = \left[\begin{array}{ccc|c} 3.6020 & -5.0280 & -6.4077 & -2.2592 \\ 5.4018 & -8.4507 & -5.5494 & -1.2364 \\ 5.6161 & -4.9195 & -8.8248 & -1.3466 \\ \hline -3.2420 & 4.0067 & -1.7596 & 0 \end{array}\right],$$

$$\left[\begin{array}{c|c} \hat{A}_2 & \hat{B}_2 \\ \hline \hat{C}_2 & 0 \end{array}\right] = \left[\begin{array}{ccc|c} 0.1274 & -2.5731 & -1.8446 & -2.2072 \\ 2.5437 & -5.7761 & -1.6615 & -1.1021 \\ 2.5969 & -2.5845 & -4.8669 & -1.2050 \\ \hline -3.7922 & 2.3836 & 0.8570 & 0 \end{array}\right].$$

Case 2: when $\hat{k} = 2$, the minimum $\gamma_{\min} = 0.1322$, and

$$\left[\begin{array}{c|c} \hat{A}_1 & \hat{B}_1 \\ \hline \hat{C}_1 & 0 \end{array}\right] = \left[\begin{array}{cc|c} -0.5866 & -2.8883 & -2.1309 \\ 1.7015 & -6.4060 & -1.1148 \\ \hline -4.1160 & 3.8843 & 0 \end{array}\right],$$

$$\left[\begin{array}{c|c} \hat{A}_2 & \hat{B}_2 \\ \hline \hat{C}_2 & 0 \end{array}\right] = \left[\begin{array}{cc|c} -1.1316 & -1.8710 & -2.1858 \\ 1.2706 & -4.8107 & -1.0741 \\ \hline -3.2353 & 2.1394 & 0 \end{array}\right].$$

Case 3: when $\hat{k} = 1$, the minimum $\gamma_{\min} = 0.2012$, and

$$\left[\begin{array}{cc} \hat{A}_1 & \hat{B}_1 \\ \hat{C}_1 & 0 \end{array}\right] = \left[\begin{array}{cc} -1.6896 & -1.9926 \\ -2.5118 & 0 \end{array}\right],$$

$$\left[\begin{array}{cc} \hat{A}_2 & \hat{B}_2 \\ \hat{C}_2 & 0 \end{array}\right] = \left[\begin{array}{cc} -1.9265 & -2.1372 \\ -2.2068 & 0 \end{array}\right].$$

In order to validate the effectiveness of the IT2 model reduction results, some simulation results will be given in the following part. It is assumed that the disturbance is chosen as follows:

$$w(t) = \begin{cases} \sin(t), & 2.6 \leq t \leq 2.8, \\ 0, & \text{else.} \end{cases}$$

Figure 9.1 depicts the output trajectories of the original system and the different reduced-order models under the initial condition of $\bar{x}(0) = 0 \left(x(0) = 0, \hat{x}(0) = 0\right)$. The output errors between the original system and the reduced models are shown in Fig. 9.2.

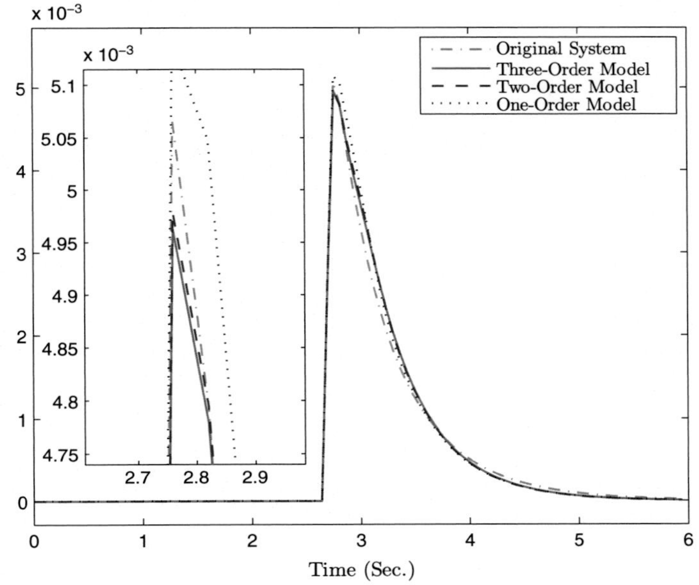

Fig. 9.1 Output trajectories

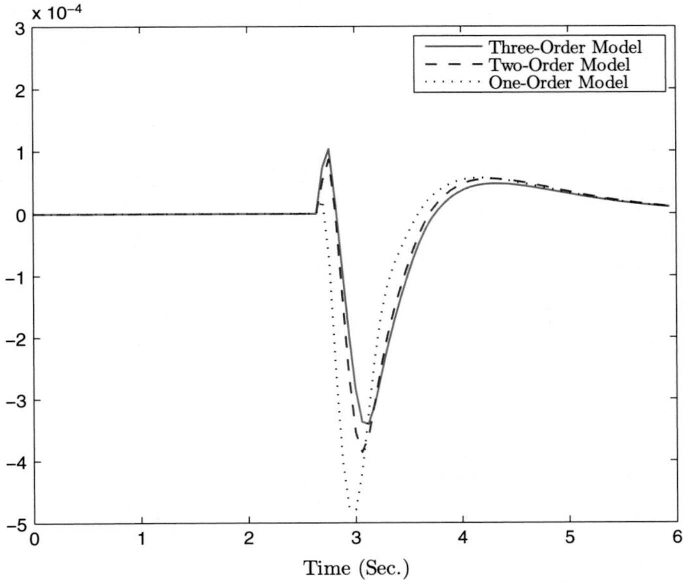

Fig. 9.2 Output errors

9.5 Conclusion

This chapter has considered the H_∞ model reduction problem for IT2 fuzzy systems with $\mathcal{D}$-stability constraints. The reduced-order model has been constructed to approximate the original system under an H_∞ performance. The membership functions and the number of rules are different between the original system and the reduced-order system. Sufficient conditions have been established to deal with the reduced-order models for IT2 fuzzy systems. The parameters of the reduced-order model can be solved by standard software. Finally, a numerical example has been provided to show the effectiveness of the proposed results.

Part II
Discrete-Time Systems

Chapter 10
Optimal Control of Interval Type-2 Fuzzy-Model-Based Systems

10.1 Introduction

This chapter focuses on designing a novel IT2 fuzzy H_∞ state-feedback controller for discrete-time IT2 FMB systems with $\mathcal{D}$-stability constraint. Firstly, the discrete-time IT2 FMB systems and a fuzzy controller are constructed for control design. Secondly, using Lyapunov stability theory, new stability conditions are derived for the closed-loop systems, and then, the desired IT2 fuzzy state-feedback controller is designed such that the closed-loop system is asymptotically stable with an H_∞ performance and all the poles of the fuzzy systems are rested in the disk region. Finally, simulation results for the inverted pendulum model are employed to validate the effectiveness and usefulness of the presented design methods.

10.2 Problem Formulation

Consider a p-rule IT2 T–S fuzzy system of discrete-time form for representing the dynamics of a nonlinear plant below:

♦ **Plant Form**:

Rule i: IF $f_1\left(\xi\left(k\right)\right)$ is $M_1^i, \ldots,$ and $f_\alpha\left(\xi\left(k\right)\right)$ is $M_\alpha^i, \ldots,$ and $f_\Theta\left(\xi\left(k\right)\right)$ is M_Θ^i, THEN,

$$\begin{cases} x\left(k+1\right) = A_i x\left(k\right) + B_i u\left(k\right) + B_{wi} w\left(k\right), \\ z\left(k\right) = C_i x\left(k\right) + D_i u\left(k\right) + D_{wi} w\left(k\right), \\ x\left(k\right) = \chi\left(k\right), \end{cases} \tag{10.1}$$

where M_α^i is an IT2 fuzzy set of ith rule ($i = 1, 2, \ldots, p$ and $\alpha = 1, 2, \ldots, \Theta$), $f_\alpha\left(\xi\left(k\right)\right)$ is the measurable premise variable, Θ is a positive integer, $x\left(k\right) \in \mathbf{R}^n$ stands for the system state vector, $u\left(k\right) \in \mathbf{R}^v$ is the control input, $w\left(k\right) \in \mathbf{R}^s$ is a disturbance input belonging to $\ell_2[0, \infty)$, $z\left(k\right) \in \mathbf{R}^m$ stands for the controlled output,

H. Li et al., *Analysis and Synthesis for Interval Type-2 Fuzzy-Model-Based Systems*, DOI 10.1007/978-981-10-0593-0_10

and $\chi(k)$ is a continuous vector-valued initial function. $A_i, B_i, B_{wi}, C_i, D_i$ and D_{wi} are system matrices with appropriate dimensions. The ith interval set of firing strength is:

$$\Phi_i(\xi(k)) = \left[\underline{\phi}_i(\xi(k)), \overline{\phi}_i(\xi(k))\right],$$

where

$$\underline{\phi}_i(\xi(k)) = \prod_{\alpha=1}^{\Theta} \underline{\mu}_{M_\alpha^i}(f_\alpha(\xi(k))) \geq 0,$$
$$\overline{\phi}_i(\xi(k)) = \prod_{\alpha=1}^{\Theta} \overline{\mu}_{M_\alpha^i}(f_\alpha(\xi(k))) \geq 0,$$

$\overline{\mu}_{M_\alpha^i}(f_\alpha(\xi(k))) \geq \underline{\mu}_{M_\alpha^i}(f_\alpha(\xi(k))) \geq 0$, and $\overline{\phi}_i(\xi(k)) \geq \underline{\phi}_i(\xi(k)) \geq 0$, in which $\underline{\mu}_{M_\alpha^i}(f_\alpha(\xi(k))) \in [0, 1]$ and $\overline{\mu}_{M_\alpha^i}(f_\alpha(\xi(k))) \in [0, 1]$ are the LMFs and UMFs, respectively. $\underline{\phi}_i(\xi(k))$ and $\overline{\phi}_i(\xi(k))$ stands for the lower and upper grades of membership, respectively. Then the discrete-time IT2 T–S fuzzy system in (10.1) can be defined as:

$$\begin{cases} x(k+1) = \sum_{i=1}^{p} \phi_i(\xi(k))(A_i x(k) + B_i u(k) + B_{wi} w(k)), \\ z(k) = \sum_{i=1}^{p} \phi_i(\xi(k))(C_i x(k) + D_i u(k) + D_{wi} w(k)), \\ x(k) = \chi(k), \end{cases} \tag{10.2}$$

where

$$\begin{aligned} \phi_i(\xi(k)) &= \underline{\alpha}_i(\xi(k))\underline{\phi}_i(\xi(k)) + \overline{\alpha}_i(\xi(k))\overline{\phi}_i(\xi(k)) \geq 0, \\ 0 &\leq \underline{\alpha}_i(\xi(k)) \leq 1, \quad 0 \leq \overline{\alpha}_i(\xi(k)) \leq 1, \\ 1 &= \underline{\alpha}_i(\xi(k)) + \overline{\alpha}_i(\xi(k)), \end{aligned} \tag{10.3}$$

in which $\underline{\alpha}_i(\xi(k))$ and $\overline{\alpha}_i(\xi(k))$ are weighting coefficient functions that are not necessary to be determined in real systems. $\phi_i(\xi(k))$ denotes the grade of membership of the real membership function. Similar to the form of fuzzy state-feedback controller proposed in [94], the fuzzy state-feedback controller with c rules is represented as:

♦ **Controller Form**:

Rule j: IF $g_1(\xi(k))$ is $N_1^j, \ldots,$ and $g_\beta(\xi(k))$ is $N_\beta^j, \ldots,$ and $g_\Omega(\xi(k))$ is N_Ω^j, THEN,

$$u(k) = G_j x(k), \tag{10.4}$$

where N_β^j denotes an IT2 fuzzy set of jth rule ($j = 1, 2, \ldots, c$ and $\beta = 1, 2, \ldots, \Omega$), Ω is a positive integer, $g_\beta(\xi(k))$ is the measurable premise variable, and G_j stands for the control gain to be designed. The interval set of firing strength of the jth rule is:

$$\Psi_j(\xi(k)) = \left[\underline{\psi}_j(\xi(k)), \overline{\psi}_j(\xi(k))\right],$$

where

$$\underline{\psi}_j(\xi(k)) = \prod_{\beta=1}^{\Omega} \underline{\mu}_{N_\beta^j}\left(g_\beta(\xi(k))\right) \geq 0,$$
$$\overline{\psi}_j(\xi(k)) = \prod_{\beta=1}^{\Omega} \overline{\mu}_{N_\beta^j}\left(g_\beta(\xi(k))\right) \geq 0,$$

$\overline{\mu}_{N_\beta^j}\left(g_\beta(\xi(k))\right) \geq \underline{\mu}_{N_\beta^j}\left(g_\beta(\xi(k))\right) \geq 0$, and $\overline{\psi}_j(\xi(k)) \geq \underline{\psi}_j(\xi(k)) \geq 0$, in which $\underline{\mu}_{N_\beta^j}\left(g_\beta(\xi(k))\right) \in [0, 1]$ and $\overline{\mu}_{N_\beta^j}\left(g_\beta(\xi(k))\right) \in [0, 1]$ is the LMF and UMF, respectively. $\underline{\psi}_j(\xi(k))$ and $\overline{\psi}_j(\xi(k))$ are the lower and upper grades of membership, respectively. Then the fuzzy controller in (10.4) is expressed as:

$$u(k) = \sum_{j=1}^{c} \psi_j(\xi(k))\, G_j x(k), \tag{10.5}$$

where for $j = 1, 2, \ldots, c$,

$$\psi_j(\xi(k)) = \frac{\underline{\beta}_j(\xi(k))\,\underline{\psi}_j(\xi(k)) + \overline{\beta}_j(\xi(k))\,\overline{\psi}_j(\xi(k))}{\sum_{\kappa=1}^{c}\left(\underline{\beta}_\kappa(\xi(k))\,\underline{\psi}_\kappa(\xi(k)) + \overline{\beta}_\kappa(\xi(k))\,\overline{\psi}_\kappa(\xi(k))\right)}, \tag{10.6}$$
$$0 \leq \underline{\beta}_j(\xi(k)) \leq 1, \quad 0 \leq \overline{\beta}_j(\xi(k)) \leq 1,$$
$$1 = \underline{\beta}_j(\xi(k)) + \overline{\beta}_j(\xi(k)),$$

in which $\underline{\beta}_j(\xi(k))$ and $\overline{\beta}_j(\xi(k))$ are predefined functions; $\psi_j(\xi(k))$ denotes the grade of membership of the real membership function. From (10.2) and (10.5), we have

$$\sum_{i=1}^{p} \phi_i(\xi(k)) = \sum_{j=1}^{c} \psi_j(\xi(k)) = \sum_{i=1}^{p}\sum_{j=1}^{c} \phi_i(\xi(k))\,\psi_j(\xi(k)) = 1.$$

Thus, the closed-loop IT2 fuzzy control system is rewritten as:

$$\begin{cases} x(k+1) = \sum_{i=1}^{p}\sum_{j=1}^{c}\phi_i(\xi(k))\,\psi_j(\xi(k))\left[\left(A_i + B_iG_j\right)x(k) + B_{wi}w(k)\right], \\ z(k) = \sum_{i=1}^{p}\sum_{j=1}^{c}\phi_i(\xi(k))\,\psi_j(\xi(k))\left[\left(C_i + D_iG_j\right)x(k) + D_{wi}w(k)\right]. \end{cases} \tag{10.7}$$

Thus, any membership function within the FOU [93] can be reconstructed by the LMFs and UMFs. According to [94], we divide the state space and the FOU for developing the main results. The descriptions of them are as follows:

- The state space Δ is divided into q sub-state spaces $\Delta_{\tilde{k}}$ $\left(\tilde{k} = 1, 2, \ldots, q\right)$, such that $\Delta = \bigcup_{\tilde{k}=1}^{q} \Delta_{\tilde{k}}$.
- Furthermore, to consider more information in membership functions, local LMFs and UMFs within the FOU are introduced. The FOU is divided into $\vartheta + 1$ sub-FOUs. For $l = 1, 2, \ldots, \vartheta + 1$, the LMFs and UMFs in the lth sub-FOU are defined, respectively, as follows for $\forall i, j, \tilde{k}, l$:

$$\underline{h}_{ijl}(\xi(k)) = \sum_{\tilde{k}=1}^{q}\sum_{i_1=1}^{2}\sum_{i_2=1}^{2}\cdots\sum_{i_n=1}^{2}\prod_{r=1}^{n} v_{ri_r\tilde{k}l}(x_r(k))\,\underline{\epsilon}_{iji_1i_2\ldots i_n\tilde{k}l}, \tag{10.8}$$

$$\overline{h}_{ijl}(\xi(k)) = \sum_{\tilde{k}=1}^{q}\sum_{i_1=1}^{2}\sum_{i_2=1}^{2}\cdots\sum_{i_n=1}^{2}\prod_{r=1}^{n} v_{ri_r\tilde{k}l}(x_r(k))\,\overline{\epsilon}_{iji_1i_2\ldots i_n\tilde{k}l}, \tag{10.9}$$

where $\underline{\epsilon}_{iji_1i_2\ldots i_n\tilde{k}l}$ and $\overline{\epsilon}_{iji_1i_2\ldots i_n\tilde{k}l}$ are constants to be determined, and $0 \le \underline{\epsilon}_{iji_1i_2\ldots i_n\tilde{k}l} \le \overline{\epsilon}_{iji_1i_2\ldots i_n\tilde{k}l} \le 1$; for $r, s = 1, 2, \ldots, n$, $i_r, i_s = 1, 2$, $\tilde{k} = 1, 2, \ldots, q$, and $x(k) \in \Delta_{\tilde{k}}$, we have $0 \le v_{ri_r\tilde{k}l}(x_r(k)) \le 1$ and $v_{r1\tilde{k}l}(x_r(k)) + v_{r2\tilde{k}l}(x_r(k)) = 1$; and $v_{ri_r\tilde{k}l}(x_r(k)) = 0$ if otherwise. Thus, for all l,

$$\sum_{k=1}^{q}\sum_{i_1=1}^{2}\sum_{i_2=1}^{2}\cdots\sum_{i_n=1}^{2}\prod_{r=1}^{n} v_{ri_r\tilde{k}l}(x_r(k)) = 1. \tag{10.10}$$

Then, we rewrite the IT2 FMB control system (10.7) as follows:

$$\begin{cases} x(k+1) = \sum_{i=1}^{p}\sum_{j=1}^{c} h_{ij}(\xi(k))\left[\left(A_i + B_iG_j\right)x(k) + B_{wi}w(k)\right], \\ z(k) = \sum_{i=1}^{p}\sum_{j=1}^{c} h_{ij}(\xi(k))\left[\left(C_i + D_iG_j\right)x(k) + D_{wi}w(k)\right], \end{cases} \tag{10.11}$$

where

$$h_{ij}(\xi(k)) = \phi_i(\xi(k))\,\psi_j(\xi(k))$$
$$= \sum_{l=1}^{\vartheta+1} \sigma_{ijl}(\xi(k)) \left(\underline{\delta}_{ijl}(\xi(k))\, \underline{h}_{ijl}(\xi(k)) + \overline{\delta}_{ijl}(\xi(k))\, \overline{h}_{ijl}(\xi(k)) \right),$$

with $\sum_{i=1}^{p}\sum_{j=1}^{c} h_{ij}(\xi(k)) = 1$, the two functions $0 \le \underline{\delta}_{ijl}(\xi(k)) \le \overline{\delta}_{ijl}(\xi(k)) \le 1$ satisfy the property $\underline{\delta}_{ijl}(\xi(k)) + \overline{\delta}_{ijl}(\xi(k)) = 1$, and are not necessarily known, and $\sigma_{ijl}(\xi(k)) = 1$ if the membership function $h_{ijl}(\xi(k))$ is within the lth sub-FOU; otherwise, $\sigma_{ijl}(\xi(k)) = 0$.

Remark 10.1 For the discrete-time IT2 fuzzy systems in (10.2), our purpose is to confirm the system stability with an H_∞ performance by determining the feedback gains G_j. Based on closed-loop system (10.11), sufficient conditions of the stability for the discrete-time IT2 fuzzy control system (10.7) with H_∞ performance can be derived in next section.

The control objective of this chapter is to design an IT2 fuzzy state-feedback controller such that system (10.7) is asymptotically stable and satisfies a prescribed H_∞ performance and $\mathcal{D}$-stability, simultaneously, under input constraint. The detailed requirements are listed as follows:

1. The closed-loop system in (10.11) is asymptotically stable.
2. The disturbance input $w(k)$ to system output $z(k)$ is attenuated below a desired level in the H_∞ sense that is, for a given scalar $\gamma > 0$, under zero initial condition, it holds that
$$\|z\|_2 < \gamma \|w\|_2, \quad \forall 0 \ne w \in \ell_2[0, \infty), \tag{10.12}$$
where $\|z\|_2 = \sqrt{\sum_{k=0}^{\infty} z^T(k)\, z(k)}$.
3. The control input is subject to the constraint of $|u(k)| \le u_{\max}$ ($u_{\max}$ is a given positive scalar).
4. $\mathcal{D}$-stability, specifically, the eigenvalues of matrices $\mathcal{A}_{ij} = A_i + B_i G_j \in \mathbf{R}^{n\times n}$ all belong to the closed disk region $\mathcal{D}(\varrho, \tau)$ (ϱ is a given negative scalar denoting the center of the disk; τ is a given positive scalar denoting the radius of the disk).

10.3 Main Results

10.3.1 Stability Analysis

In this section, by using the Lyapunov functional approach, a sufficient criterion is first given to satisfy four requirements mentioned above.

Theorem 10.2 *Considering the discrete-time IT2 FMB system in (10.11) with FOU being divided into $\vartheta + 1$ sub-FOUs and the state space being divided into q connected sub-state spaces, for given input constraint $u_{\max} > 0$, scalars $\rho > 0$, $\gamma > 0$ and disk region $\mathcal{D}(\varrho, \tau)$, system (10.11) with the input constraint is asymptotically stable with an H_∞ performance index γ with all the poles resting in the disk region $\mathcal{D}(\varrho, \tau)$, if there exist symmetric matrices $P > 0$, $X_{ijl} > 0$, $Y_{ijl} > 0$, $U_{ijl} > 0$, $V_{ijl} > 0$, $R_{ijl} > 0$ $(i = 1, 2, \ldots, p, j = 1, 2, \ldots, c, l = 1, 2, \ldots, \vartheta + 1)$ and matrix Q such that:*

$$\sum_{i=1}^{p}\sum_{j=1}^{c} Z_{ij} - Q < 0, \quad \forall i_1, i_2, \ldots, i_n, k, l, \tag{10.13}$$

$$\Xi_{2ijl} + R_{ijl} + Q > 0, \quad \forall i, j, l, \tag{10.14}$$

$$\begin{bmatrix} -\rho^{-1}I & G_j \\ * & -u_{\max}^2 P \end{bmatrix} < 0, \quad \forall j, \tag{10.15}$$

$$\begin{bmatrix} -P & P\left(A_{ij} - \varrho I\right) \\ * & -\tau^2 P \end{bmatrix} < 0, \quad \forall i, j, \tag{10.16}$$

where

$$\Xi_{1ijl} = \begin{bmatrix} A_{ij}^T \Upsilon_{11} A_{ij} + C_{ij}^T \Upsilon_{12} C_{ij} - P & A_{ij}^T \Upsilon_{11} B_{wi} + C_{ij}^T \Upsilon_{12} D_{wi} \\ * & B_{wi}^T \Upsilon_{11} B_{wi} + D_{wi}^T \Upsilon_{12} D_{wi} - \gamma^2 I \end{bmatrix},$$

$$\Xi_{2ijl} = \begin{bmatrix} A_{ij}^T \Upsilon_{21} A_{ij} + C_{ij}^T \Upsilon_{22} C_{ij} - P & A_{ij}^T \Upsilon_{21} B_{wi} + C_{ij}^T \Upsilon_{22} D_{wi} \\ * & B_{wi}^T \Upsilon_{21} B_{wi} + D_{wi}^T \Upsilon_{22} D_{wi} - \gamma^2 I \end{bmatrix},$$

$$Z_{ij} = \overline{\epsilon}_{iji_1i_2\ldots i_n\tilde{k}l}\Xi_{1ijl} - \left(\underline{\epsilon}_{iji_1i_2\ldots i_n\tilde{k}l} - \overline{\epsilon}_{iji_1i_2\ldots i_n\tilde{k}l}\right) R_{ijl} + \overline{\epsilon}_{iji_1i_2\ldots i_n\tilde{k}l} Q,$$

$$A_{ij} = A_i + B_i G_j, \quad C_{ij} = C_i + D_i G_j, \quad \Upsilon_{11} = P + U_{ijl},$$

$$\Upsilon_{12} = I + V_{ijl}, \quad \Upsilon_{21} = P - X_{ijl}, \quad \Upsilon_{22} = I - Y_{ijl}.$$

Proof Consider some slack matrices in the following inequalities under the *S-procedure* [17]:

$$\left[\sum_{i=1}^{p}\sum_{j=1}^{c}\sum_{l=1}^{\vartheta+1} \sigma_{ijl}(\xi(k)) \left(\underline{\delta}_{ijl}(\xi(k))\,\underline{h}_{ijl}(\xi(k)) + \overline{\delta}_{ijl}(\xi(k))\,\overline{h}_{ijl}(\xi(k))\right) - 1\right] Q = 0, \tag{10.17}$$

$$-\sum_{i=1}^{p}\sum_{j=1}^{c}\sum_{l=1}^{\vartheta+1} \sigma_{ijl}(\xi(k)) \left(1 - \underline{\delta}_{ijl}(\xi(k))\right)\left(\underline{h}_{ijl}(\xi(k)) - \overline{h}_{ijl}(\xi(k))\right) R_{ijl} \geq 0, \tag{10.18}$$

$$\sum_{i=1}^{p}\sum_{j=1}^{c}\sum_{l=1}^{\vartheta+1} \sigma_{ijl}(\xi(k))\,\overline{h}_{ijl}(\xi(k))\, U_{ijl} \geq 0, \tag{10.19}$$

$$-\sum_{i=1}^{p}\sum_{j=1}^{c}\sum_{l=1}^{\vartheta+1} \sigma_{ijl}(\xi(k))\,\underline{\delta}_{ijl}(\xi(k))\left(\underline{h}_{ijl}(\xi(k)) - \overline{h}_{ijl}(\xi(k))\right) X_{ijl} \geq 0, \tag{10.20}$$

$$\sum_{i=1}^{p}\sum_{j=1}^{c}\sum_{l=1}^{\vartheta+1}\sigma_{ijl}(\xi(k))\overline{h}_{ijl}(\xi(k))V_{ijl}\geq 0, \tag{10.21}$$

$$-\sum_{i=1}^{p}\sum_{j=1}^{c}\sum_{l=1}^{\vartheta+1}\sigma_{ijl}(\xi(k))\underline{\delta}_{ijl}(\xi(k))\left(\underline{h}_{ijl}(\xi(k))-\overline{h}_{ijl}(\xi(k))\right)Y_{ijl}\geq 0. \tag{10.22}$$

Define the Lyapunov function $V(k)$ for the system in (10.11) as follows:

$$V(k)=x^T(k)Px(k), \tag{10.23}$$

where symmetric matrix $P>0$. From the system (10.11) and the function (10.23), based on Lemma 1.7 and the inequalities (10.17)–(10.20), we have

$$\begin{aligned}
\Delta V(k) &= x^T(k+1)Px(k+1)-x^T(k)Px(k)\\
&=\sum_{i=1}^{p}\sum_{j=1}^{c}\sum_{\kappa=1}^{p}\sum_{\iota=1}^{c}h_{ij}(\xi(k))h_{\kappa\iota}(\xi(k))\left[A_{ij}x(k)+B_{wi}w(k)\right]^T\\
&\quad\times P\left[A_{\kappa\iota}x(k)+B_{w\kappa}w(k)\right]-x^T(k)Px(k)\\
&\leq\frac{1}{2}\sum_{i=1}^{p}\sum_{j=1}^{c}\sum_{\kappa=1}^{p}\sum_{\iota=1}^{c}h_{ij}(\xi(k))h_{\kappa\iota}(\xi(k))\Big\{\left[A_{ij}x(k)+B_{wi}w(k)\right]^TP\\
&\quad\times\left[A_{ij}x(k)+B_{wi}w(k)\right]+\left[A_{\kappa\iota}x(k)+B_{w\kappa}w(k)\right]^TP\\
&\quad\times\left[A_{\kappa\iota}x(k)+B_{w\kappa}w(k)\right]\Big\}-x^T(k)Px(k)\\
&=\sum_{i=1}^{p}\sum_{j=1}^{c}h_{ij}(\xi(k))\eta^T(k)\begin{bmatrix}A_{ij}^TPA_{ij}-P & A_{ij}^TPB_{wi}\\ * & B_{wi}^TPB_{wi}\end{bmatrix}\eta(k)\\
&=\sum_{i=1}^{p}\sum_{j=1}^{c}\sum_{l=1}^{\vartheta+1}\sigma_{ijl}(\xi(k))\left(\underline{\delta}_{ijl}(\xi(k))\underline{h}_{ijl}(\xi(k))+\overline{\delta}_{ijl}(\xi(k))\overline{h}_{ijl}(\xi(k))\right)\\
&\quad\times\eta^T(k)\begin{bmatrix}A_{ij}^TPA_{ij}-P & A_{ij}^TPB_{wi}\\ * & B_{wi}^TPB_{wi}\end{bmatrix}\eta(k)\\
&\leq\sum_{i=1}^{p}\sum_{j=1}^{c}\sum_{l=1}^{\vartheta+1}\sigma_{ijl}(\xi(k))\left(\underline{\delta}_{ijl}(\xi(k))\underline{h}_{ijl}(\xi(k))+\overline{\delta}_{ijl}(\xi(k))\overline{h}_{ijl}(\xi(k))\right)\\
&\quad\times\eta^T(k)\begin{bmatrix}A_{ij}^TPA_{ij}-P & A_{ij}^TPB_{wi}\\ * & B_{wi}^TPB_{wi}\end{bmatrix}\eta(k)+\Bigg[\sum_{i=1}^{p}\sum_{j=1}^{c}\sum_{l=1}^{\vartheta+1}\sigma_{ijl}(\xi(k))\\
&\quad\times\left(\underline{\delta}_{ijl}(\xi(k))\underline{h}_{ijl}(\xi(k))+\overline{\delta}_{ijl}(\xi(k))\overline{h}_{ijl}(\xi(k))\right)-1\Bigg]\eta^T(k)Q\eta(k)
\end{aligned}$$

$$
\begin{aligned}
&- \sum_{i=1}^{p} \sum_{j=1}^{c} \sum_{l=1}^{\vartheta+1} \sigma_{ijl}(\xi(k)) \left(1 - \underline{\delta}_{ijl}(\xi(k))\right) \left(\underline{h}_{ijl}(\xi(k)) - \overline{h}_{ijl}(\xi(k))\right) \\
&\times \eta^T(k) R_{ijl} \eta(k) \\
&+ \sum_{i=1}^{p} \sum_{j=1}^{c} \sum_{l=1}^{\vartheta+1} \sigma_{ijl}(\xi(k)) \overline{h}_{ijl}(\xi(k)) x^T(k+1) U_{ijl} x(k+1) \\
&- \sum_{i=1}^{p} \sum_{j=1}^{c} \sum_{l=1}^{\vartheta+1} \sigma_{ijl}(\xi(k)) \underline{\delta}_{ijl}(\xi(k)) \left(\underline{h}_{ijl}(\xi(k)) - \overline{h}_{ijl}(\xi(k))\right) \\
&\times x^T(k+1) X_{ijl} x(k+1) \\
&= \eta^T(k) \Bigg\{ \sum_{i=1}^{p} \sum_{j=1}^{c} \sum_{l=1}^{\vartheta+1} \sigma_{ijl}(\xi(k)) \\
&\times \left[\overline{h}_{ijl} \tilde{\Xi}_{1ijl} - \left(\underline{h}_{ijl} - \overline{h}_{ijl}\right) R_{ijl} + \overline{h}_{ijl} Q\right] - Q \Bigg\} \eta(k) \\
&+ \eta^T(k) \sum_{i=1}^{p} \sum_{j=1}^{c} \sum_{l=1}^{\vartheta+1} \sigma_{ijl}(\xi(k)) \underline{\delta}_{ijl}(\xi(k)) \left(\underline{h}_{ijl} - \overline{h}_{ijl}\right) \\
&\times \left(\tilde{\Xi}_{2ijl} + R_{ijl} + Q\right) \eta(k),
\end{aligned} \tag{10.24}
$$

where $\eta(k) = \begin{bmatrix} x^T(k) & w^T(k) \end{bmatrix}^T$, and

$$
\tilde{\Xi}_{1ijl} = \begin{bmatrix} A_{ij}^T \left(P + U_{ijl}\right) A_{ij} - P & A_{ij}^T \left(P + U_{ijl}\right) B_{wi} \\ * & B_{wi}^T \left(P + U_{ijl}\right) B_{wi} \end{bmatrix},
$$

$$
\tilde{\Xi}_{2ijl} = \begin{bmatrix} A_{ij}^T \left(P - X_{ijl}\right) A_{ij} - P & A_{ij}^T \left(P - X_{ijl}\right) B_{wi} \\ * & B_{wi}^T \left(P - X_{ijl}\right) B_{wi} \end{bmatrix}.
$$

When $w(k) = 0$, we know that $\Delta V(k) < 0$ from (10.13) to (10.14).

Then, by considering H_∞ performance defined in (10.12) with the inequality in (10.20), under the zero initial condition, it follows

$$
\begin{aligned}
J &= \sum_{k=0}^{\infty} \left[z^T(k) z(k) - \gamma^2 w^T(k) w(k)\right] \le J + V(\infty) - V(0) \\
&= \sum_{k=0}^{\infty} \left[z^T(k) z(k) - \gamma^2 w^T(k) w(k) + \Delta V(k)\right]
\end{aligned} \tag{10.25}
$$

$$
\begin{aligned}
&\leq \sum_{k=0}^{\infty}\Bigg\{\sum_{i=1}^{p}\sum_{j=1}^{c}\sum_{l=1}^{\vartheta+1}\sigma_{ijl}\left(\xi\left(k\right)\right) \\
&\times\left(\underline{\delta}_{ijl}\left(\xi\left(k\right)\right)\underline{h}_{ijl}\left(\xi\left(k\right)\right)+\overline{\delta}_{ijl}\left(\xi\left(k\right)\right)\overline{h}_{ijl}\left(\xi\left(k\right)\right)\right) \\
&\times\eta^{T}\left(k\right)\begin{bmatrix} C_{ij}^{T}C_{ij} & C_{ij}^{T}D_{wi} \\ * & D_{wi}^{T}D_{wi}-\gamma^{2}I\end{bmatrix}\eta\left(k\right) \\
&+\Delta V\left(k\right)+\sum_{i=1}^{p}\sum_{j=1}^{c}\sum_{l=1}^{\vartheta+1}\sigma_{ijl}\left(\xi\left(k\right)\right)\overline{h}_{ijl}\left(\xi\left(k\right)\right)z^{T}\left(k\right)V_{ijl}z\left(k\right) \\
&-\sum_{i=1}^{p}\sum_{j=1}^{c}\sum_{l=1}^{\vartheta+1}\sigma_{ijl}\left(\xi\left(k\right)\right)\underline{\delta}_{ijl}\left(\xi\left(k\right)\right)\left(\underline{h}_{ijl}\left(\xi\left(k\right)\right)-\overline{h}_{ijl}\left(\xi\left(k\right)\right)\right) \\
&\times z^{T}\left(k\right)Y_{ijl}z\left(k\right)\Bigg\} \\
&=\sum_{k=0}^{\infty}\eta^{T}\left(k\right)\Bigg\{\sum_{i=1}^{p}\sum_{j=1}^{c}\sum_{l=1}^{\vartheta+1}\sigma_{ijl}\left(\xi\left(k\right)\right) \\
&\times\left[\overline{h}_{ijl}\Xi_{1ijl}-\left(\underline{h}_{ijl}-\overline{h}_{ijl}\right)R_{ijl}+\overline{h}_{ijl}Q\right]-Q\Bigg\}\eta\left(k\right) \\
&+\sum_{k=0}^{\infty}\eta^{T}\left(k\right)\sum_{i=1}^{p}\sum_{j=1}^{c}\sum_{l=1}^{\vartheta+1}\sigma_{ijl}\left(\xi\left(k\right)\right) \\
&\times\underline{\delta}_{ijl}\left(\xi\left(k\right)\right)\left(\underline{h}_{ijl}-\overline{h}_{ijl}\right)\left(\Xi_{2ijl}+R_{ijl}+Q\right)\eta\left(k\right). \qquad (10.26)
\end{aligned}
$$

Thus, $J < 0$ in (10.26) can be obtained from the two sets of inequalities: $\Xi_{2ijl} + R_{ijl} + Q > 0$ (which is guaranteed by (10.14)), and

$$
\sum_{i=1}^{p}\sum_{j=1}^{c}\sum_{l=1}^{\vartheta+1}\sigma_{ijl}\left(\xi\left(k\right)\right)\left[\overline{h}_{ijl}\Xi_{1ijl}-\left(\underline{h}_{ijl}-\overline{h}_{ijl}\right)R_{ijl}+\overline{h}_{ijl}Q\right]-Q<0. \qquad (10.27)
$$

Notice that only one $\sigma_{ijl}\left(\xi\left(k\right)\right) = 1$ for each fixed combination of i and j at any time, and $\sum_{l=1}^{\vartheta+1}\sigma_{ijl}\left(\xi\left(k\right)\right) = 1$, the inequality in (10.27) holds under the following inequality

$$
\sum_{i=1}^{p}\sum_{j=1}^{c}\left[\overline{h}_{ijl}\Xi_{1ijl}-\left(\underline{h}_{ijl}-\overline{h}_{ijl}\right)R_{ijl}+\overline{h}_{ijl}Q\right]-Q<0. \qquad (10.28)
$$

Considering $\underline{h}_{ijl}$ in (10.8), $\overline{h}_{ijl}$ in (10.9), and the equalities in (10.10), the following inequality is equivalent to the inequalities in (10.28),

$$\sum_{k=1}^{q}\sum_{i_1=1}^{2}\sum_{i_2=1}^{2}\cdots\sum_{i_n=1}^{2}\prod_{r=1}^{n} v_{ri_r\tilde{k}l}\left(x_r\left(k\right)\right)\left(\sum_{i=1}^{p}\sum_{j=1}^{c} Z_{ij}-Q\right)<0. \quad (10.29)$$

Obviously, the inequalities in (10.29) is satisfied by (10.13). Hence, $\|z\|_2 < \gamma \|w\|_2$ as $J < 0$. Thus, for all nonzero $w = w(k) \in \ell_2[0, \infty)$, the conditions (10.13)–(10.16) can guarantee the asymptotic stability of system (10.11) with an H_∞ performance index γ.

In the following, the condition of the input constraint $|u(k)| \le u_{\max}$ is derived. From the inequality in (10.25), we know

$$\Delta V(k) - \gamma^2 w^T(k)\, w(k) + z^T(k)\, z(k) < 0.$$

Thus, $\Delta V(k) - \gamma^2 w^T(k)\, w(k) < 0$, which implies

$$V(k) - V(0) < \gamma^2 \sum_{i=0}^{k} w^T(i)\, w(i) < \gamma^2 \|w\|_2^2. \quad (10.30)$$

Considering $V(k) = x^T(k)\, Px(k) > 0$, it can be concluded that

$$x^T(k)\, Px(k) = V(k) < \gamma^2 \|w\|_2^2 + V(0) < \rho = \gamma^2 w_{\max} + V(0),$$

where $w_{\max} = \frac{\rho - V(0)}{\gamma^2}$ denotes the disturbance energy bound. Then, it follows that

$$\begin{aligned}\max_{k>0} |u(k)|^2 &= \max_{k>0} \left\| \left[\sum_{j=1}^{c} \psi_j(\xi(k))\, G_j x(k)\right]^T \left[\sum_{j=1}^{c} \psi_j(\xi(k))\, G_j x(k)\right] \right\|_2 \\ &= \max_{k>0} \left\| x^T(k)\, G_j^T G_j x(k) \right\|_2 \\ &= \max_{k>0} \left\| x^T(k)\, P^{\frac{1}{2}} P^{-\frac{1}{2}} G_j^T G_j P^{-\frac{1}{2}} P^{\frac{1}{2}} x(k) \right\|_2 \\ &< \rho \cdot \lambda_{\max}\left(P^{-\frac{1}{2}} G_j^T G_j P^{-\frac{1}{2}}\right), \quad j = 1, 2, \ldots, c,\end{aligned}$$

where $\lambda_{\max}(\cdot)$ represents maximal eigenvalue. From the above inequalities, we know that the input constraint is satisfied, if

$$\rho \cdot P^{-\frac{1}{2}} G_j^T G_j P^{-\frac{1}{2}} < u_{\max}^2 I. \quad (10.31)$$

By Schur complement, (10.31) is equivalent to (10.15).

In addition, considering the pole assignment for the closed-loop system in (10.11), by applying Lemma 1.5, we know the condition in (10.16) is satisfied such that the system in (10.11) is $\mathcal{D}$-stability. The proof is completed. □

Remark 10.3 Theorem 10.2 provides a sufficient criterion of asymptotic stability for the discrete-time IT2 fuzzy system in (10.11) with input constrain, H_∞ performance and $\mathcal{D}$-stability. If the disturbance input, the input constraint and $\mathcal{D}$-stability constraint are not considered in this chapter, the stability condition can be also presented directly from Theorem 10.2.

10.3.2 State-Feedback Control

The following theorem presents the existence condition of the IT2 fuzzy controller in the form of (10.4).

Theorem 10.4 *Considering the discrete-time IT2 FMB system in (10.11) with FOU being divided into $\vartheta + 1$ sub-FOUs and the state space being divided into q connected substate spaces, for given input constraint $u_{\max} > 0$, scalar $\rho > 0$, and disk region $\mathcal{D}(\varrho, \tau)$, and the scalar $\gamma > 0$, system (10.11) with the input constraint is asymptotically stable with an H_∞ performance index γ with all the poles resting in the disk region $\mathcal{D}(\varrho, \tau)$, if there exist symmetric matrices $\bar{P} > 0$, $\bar{R}_{1ijl} > 0$, $\bar{R}_{2ijl} > 0$, $\bar{R}_{3ijl} > 0$ ($i = 1, 2, \ldots, p$, $j = 1, 2, \ldots, c$, $l = 1, 2, \ldots, \vartheta + 1$), $\bar{X}_{ijl}$, $\bar{Y}_{ijl}$, $\bar{U}_{ijl}$, $\bar{V}_{ijl}$, $\bar{Q}_1$, $\bar{Q}_2$ and $\bar{Q}_3$ satisfy*

$$\forall\, i, j, i_1, i_2, \ldots, i_n, \tilde{k}, l,$$

$$\begin{bmatrix} \bar{\Gamma}_{1ijl} & \bar{\Gamma}_{2ijl} & \varpi\left(\bar{P}A_i^T + \bar{K}_j^T B_i^T\right) & \varpi\left(\bar{P}C_i^T + \bar{K}_j^T D_i^T\right) \\ * & \bar{\Gamma}_{3ijl} & \varpi B_{wi}^T & \varpi D_{wi}^T \\ * & * & -\bar{U}_{ijl} & 0 \\ * & * & * & -\bar{V}_{ijl} \end{bmatrix} < 0, \tag{10.32}$$

$$\forall\, i, j, l,$$

$$\begin{bmatrix} -\bar{P} + \bar{R}_{1ijl} + \bar{Q}_1 & \bar{R}_{2ijl} + \bar{Q}_2 & \bar{P}A_i^T + \bar{K}_j^T B_i^T & \bar{P}C_i^T + \bar{K}_j^T D_i^T \\ * & -\gamma^2 I + \bar{R}_{3ijl} + \bar{Q}_3 & B_{wi}^T & D_{wi}^T \\ * & * & \bar{X}_{ijl} & 0 \\ * & * & * & \bar{Y}_{ijl} \end{bmatrix} > 0, \tag{10.33}$$

$$\forall\, i, j, l,\ P - X_{ijl} < 0, \tag{10.34}$$

$$\forall\, i, j, l,\ I - Y_{ijl} < 0, \tag{10.35}$$

$$\forall\, j,\ \begin{bmatrix} -\rho^{-1}I & \bar{K}_j \\ * & -u_{\max}^2\bar{P} \end{bmatrix} < 0, \tag{10.36}$$

$$\forall\, i, j,\ \begin{bmatrix} -\bar{P} & A_i\bar{P} + B_i\bar{K}_j - \varrho\bar{P} \\ * & -\tau^2\bar{P} \end{bmatrix} < 0, \tag{10.37}$$

where

$$\begin{aligned}
\bar{\Gamma}_{1ijl} &= -\bar{\epsilon}_{iji_1 i_2 \ldots i_n \tilde{k} l}\bar{P} + \left(\bar{\epsilon}_{iji_1 i_2 \ldots i_n \tilde{k} l} - \underline{\epsilon}_{iji_1 i_2 \ldots i_n \tilde{k} l}\right)\bar{R}_{1ijl} + \left(\varpi^2 - \frac{1}{pc}\right)\bar{Q}_1, \\
\bar{\Gamma}_{2ijl} &= \left(\bar{\epsilon}_{iji_1 i_2 \ldots i_n \tilde{k} l} - \underline{\epsilon}_{iji_1 i_2 \ldots i_n \tilde{k} l}\right)\bar{R}_{2ijl} + \left(\varpi^2 - \frac{1}{pc}\right)\bar{Q}_2, \\
\bar{\Gamma}_{3ijl} &= -\bar{\epsilon}_{iji_1 i_2 \ldots i_n \tilde{k} l}\gamma^2 I + \left(\bar{\epsilon}_{iji_1 i_2 \ldots i_n \tilde{k} l} - \underline{\epsilon}_{iji_1 i_2 \ldots i_n \tilde{k} l}\right)\bar{R}_{3ijl} + \left(\varpi^2 - \frac{1}{pc}\right)\bar{Q}_3, \\
\varpi &= \sqrt{\bar{\epsilon}_{iji_1 i_2 \ldots i_n \tilde{k} l}}.
\end{aligned}$$

Moreover, if the above LMIs are feasible, then the control gains in (10.4) are

$$G_j = \bar{K}_j\bar{P}^{-1}. \tag{10.38}$$

Proof Letting the following matrices

$$\begin{aligned}
\bar{K}_j &= G_j P^{-1}, \quad \bar{R}_{1ijl} = P^{-T}R_{1ijl}P^{-1}, \quad \bar{R}_{2ijl} = P^{-T}R_{2ijl}, \\
\bar{R}_{3ijl} &= R_{3ijl}, \quad \bar{U}_{ijl} = P^{-T}\left(P + U_{ijl}\right)^{-1}P^{-1}, \quad \bar{V}_{ijl} = \left(I + V_{ijl}\right)^{-1}, \\
\bar{Q}_1 &= P^{-T}Q_1P^{-1}, \quad \bar{Q}_2 = P^{-T}Q_2, \quad \bar{Q}_3 = Q_3, \quad \bar{P} = P^{-1},
\end{aligned}$$

and then, performing a congruence transformation to (10.32) by diag $\{P, I, P, I\}$, we have

$$\begin{bmatrix} \Gamma_{1ijl} & \Gamma_{2ijl} & \varpi A_{ij}^T P & \varpi C_{ij}^T \\ * & \Gamma_{3ijl} & \varpi B_{wi}^T P & \varpi D_{wi}^T \\ * & * & -\bar{U}_{ijl} & 0 \\ * & * & * & -\bar{V}_{ijl} \end{bmatrix} < 0, \quad \forall\, i_1, i_2, \ldots, i_n, \tilde{k}, l, i, j, \tag{10.39}$$

where

$$\begin{aligned}
\Gamma_{1ijl} &= -\bar{\epsilon}_{iji_1 i_2 \ldots i_n \tilde{k} l}P + \left(\bar{\epsilon}_{iji_1 i_2 \ldots i_n \tilde{k} l} - \underline{\epsilon}_{iji_1 i_2 \ldots i_n \tilde{k} l}\right)R_{1ijl} + \left(\varpi^2 - \frac{1}{pc}\right)Q_1, \\
\Gamma_{2ijl} &= \left(\bar{\epsilon}_{iji_1 i_2 \ldots i_n \tilde{k} l} - \underline{\epsilon}_{iji_1 i_2 \ldots i_n \tilde{k} l}\right)R_{2ijl} + \left(\varpi^2 - \frac{1}{pc}\right)Q_2, \\
\Gamma_{3ijl} &= -\bar{\epsilon}_{iji_1 i_2 \ldots i_n \tilde{k} l}\gamma^2 I + \left(\bar{\epsilon}_{iji_1 i_2 \ldots i_n \tilde{k} l} - \underline{\epsilon}_{iji_1 i_2 \ldots i_n \tilde{k} l}\right)R_{3ijl} + \left(\varpi^2 - \frac{1}{pc}\right)Q_3.
\end{aligned}$$

Moreover, let the following matrices be partioned as:

$$Q = \begin{bmatrix} Q_1 & Q_2 \\ * & Q_3 \end{bmatrix}, \quad R_{ijl} = \begin{bmatrix} R_{1ijl} & R_{2ijl} \\ * & R_{3ijl} \end{bmatrix}. \tag{10.40}$$

Then based on inequalities in (10.39), and by Schur complement, it yields

$$Z_{ij} - \frac{1}{pc} Q < 0, \quad \forall\, i_1, i_2, \ldots, i_n, \tilde{k}, l, i, j.$$

Thus,

$$\sum_{i=1}^{p} \sum_{j=1}^{c} Z_{ij} - Q < 0, \quad \forall\, i_1, i_2, \ldots, i_n, \tilde{k}, l,$$

which meets the condition (10.13) in Theorem 10.2.

Similarly, for the set of inequalities in (10.33), letting $\bar{Y}_{ijl} = \left(Y_{ijl} - I\right)^{-1}$ and $\bar{X}_{ijl} = P^{-T}\left(X_{ijl} - P\right)^{-1} P^{-1}$ with matrices in (10.40), then performing a congruence transformation to (10.33) by diag $\{P, I, P, I\}$, one can obtain the condition (10.14) in Theorem 10.2. Thus, the condition in (10.33) is satisfied.

Meanwhile, by performing congruence transformations to (10.36) and (10.37) by diag $\{I, P\}$ and diag $\{P, P\}$, respectively, (10.15) and (10.16) in Theorem 10.2 can be obtained, respectively. Hence, all the conditions in Theorem 10.2 satisfy the conditions of Theorem 10.2. The proof is completed. □

Remark 10.5 From Theorem 10.4, the existence condition of the desired IT2 fuzzy controller in (10.4) is presented for discrete-time IT2 fuzzy system in (10.11). The advantage of the presented controller subject to the LMIs constraint can guarantee the poles of the real systems rest in a given disk, which is according to the need of engineering in practical applications.

10.4 Simulation Results

Example 10.6 In this section, we will used the inverted pendulum example to verify the advantages over the existing type-1 one [22], and further validate the effectiveness of the optimal control design method. Figure 2.1 shows the sketch of the inverted pendulum on a cart. The dynamical equation of the inverted pendulum is given below:

$$\ddot{\xi}(t) = \frac{g \sin(\xi(t)) - am_p L \left(\dot{\xi}(t)\right)^2 \sin(2\theta(t))/2 - a\cos(\xi(t))\, u(t)}{4L/3 - am_p L \cos^2(\xi(t))}, \tag{10.41}$$

where $\xi(t)$ denotes the angular displacement of the pendulum, $2L = 1\,\text{m}$ is the length of the pendulum, the gravity acceleration is $g = 9.8\,\text{m/s}^2$, m_p denotes the mass of the pendulum, m_c denotes the mass of the cart, $a = 1/\left(m_p + m_c\right)$, and $u(t)$ denotes the force applied to the cart. We mark $x(t) = \begin{bmatrix} x_1(t) & x_2(t) \end{bmatrix}^T = \begin{bmatrix} \xi(t) & \dot{\xi}(t) \end{bmatrix}^T$. Firstly, we use a type-1 T–S fuzzy system to model the dynamical equations (10.41) for the comparison with the presented control approach. In such case, $m_p = 2\,\text{kg}$ and $m_c = 4\,\text{kg}$. A two-rule T–S fuzzy system is established below.

Plant Rule 1 : IF $\xi(t)$ is about $\pm\frac{3}{8}\pi$, THEN

$$\dot{x}(t) = A_1 x(t) + B_1 u(t),$$

Plant Rule 2 : IF $\xi(t)$ is about 0, THEN

$$\dot{x}(t) = A_2 x(t) + B_2 u(t),$$

where

$$A_1 = \begin{bmatrix} 0 & 1 \\ f_1\left(\frac{3}{8}\pi\right) & 0 \end{bmatrix}, \quad A_2 = \begin{bmatrix} 0 & 1 \\ f_1(0) & 0 \end{bmatrix},$$
$$B_1 = \begin{bmatrix} 0 \; f_2\left(\frac{3}{8}\pi\right) \end{bmatrix}^T, \quad B_2 = \begin{bmatrix} 0 \; f_2(0) \end{bmatrix}^T,$$

with

$$f_1(x_1(t)) = \frac{g - a m_p L x_2^2(t)\cos(x_1(t))}{4L/3 - a m_p L \cos^2(x_1(t))}\left(\frac{\sin(x_1(t))}{x_1(t)}\right),$$
$$f_2(x_1(t)) = \frac{-a\cos(x_1(t))}{4L/3 - a m_p L \cos^2(x_1(t))}\left(\frac{\sin(x_1(t))}{x_1(t)}\right).$$

In order to obtain the discrete-time form of the fuzzy system above, we let time period $T = 0.01$ s. Then, considering the disturbance input, the fuzzy system of discrete-time form (see [22], system (8–9)) is established below:

$$\begin{cases} x(k+1) = \sum_{i=1}^{r} h_i(\xi(k))\left[A_i x(k) + B_i u(k) + B_{wi} w(k)\right], \\ z(k) = \sum_{i=1}^{r} h_i(\xi(k))\left[C_i x(k) + D_i u(k) + D_{wi} w(k)\right], \end{cases} \tag{10.42}$$

where the corresponding system parameters (annotated in [22] , (1–3) and (8–9)), are given as follows:

$$A_1 = \begin{bmatrix} 1.0006 & 0.0100 \\ 0.1186 & 1.0006 \end{bmatrix}, \quad A_2 = \begin{bmatrix} 1.0000 & 0.0100 \\ -0.0010 & 1.0000 \end{bmatrix},$$

$$B_1 = \left[9.0470 \times 10^{-4} \ 1.8096 \times 10^{-1} \right]^T,$$
$$B_2 = \left[-1.5385 \times 10^{-5} \ -3.0769 \times 10^{-3} \right]^T,$$

with the considered matrices

$$C_1 = C_2 = \left[0 \ 1 \right], \quad B_{w1} = B_{w2} = \left[0 \ 0.02 \right]^T,$$
$$D_1 = D_2 = 1, \quad D_{w1} = D_{w2} = 0.01,$$

and the grade of membership functions are given by

$$h_1 (\xi (k)) = \left(1 - \frac{1}{1 + \mathrm{e}^{(3(\xi(k) - \pi/2))}} \right) \frac{1}{1 + \mathrm{e}^{(3(\xi(k) + \pi/2))}},$$
$$h_2 (\xi (k)) = 1 - h_1 (\xi (k)) .$$

Next, for further comparing with the presented control approach, we apply the Corollary 3 in [22] to obtain the control gains. Moreover, we constrain an input condition $u (k) \leq u_{\max} = 1.2$ with parameter $\rho = 1$ (the condition can be derived directly as the proof of (10.36)). Simultaneously, we choose the same disturbance attenuation index $\gamma = 0.6439$, which is obtained from the Theorem 10.4 of the this chapter in the following context. Then the control gains can be obtained as follows.

$$K_1 = \left[-0.3244 \ -6.4246 \right], \quad K_2 = \left[-2.3126 \ -0.0443 \right].$$

Assume the initial state $x (0) = \left[\frac{\pi}{12} \ 0 \right]^T$, and a random disturbance input $w (k) = 0.1$ randn. The controller (see [22], (11)) is employed to control the system in (10.42). The system state responses are depicted in Fig. 10.1, and the control force is shown in Fig. 10.2.

On the other hand, we consider the parametric uncertainties m_p and m_c existing in the pendulum system satisfying $m_{p\min} = 1\,\text{kg} \leq m_p \leq m_{p\max} = 3\,\text{kg}$, $m_{c\min} = 3\,\text{kg} \leq m_c \leq m_{c\max} = 5\,\text{kg}$. We use a four-rule IT2 fuzzy model below to describe the inverted pendulum subject to parametric uncertainties.

Plant Rule i : IF $\xi(t)$ is M_1^i and $\xi(t)$ is M_2^i, THEN

$$x (t) = A_i x (t) + B_i u (t) , \tag{10.43}$$

where

$$A_1 = A_2 = \begin{bmatrix} 0 & 1 \\ f_{1\min} & 0 \end{bmatrix}, \quad A_3 = A_4 = \begin{bmatrix} 0 & 1 \\ f_{1\max} & 0 \end{bmatrix},$$
$$B_1 = B_3 = \left[0 \ f_{2\min} \right]^T, \quad B_2 = B_4 = \left[0 \ f_{2\max} \right]^T .$$

Assume that the inverted pendulum operates in the workplace described by $x_1 = \xi (t) \in [-3\pi/8, 3\pi/8]$ and $x_2 = \xi (t) \in [-3, 3]$. Thus, $f_{1\min} = 11.1261$,

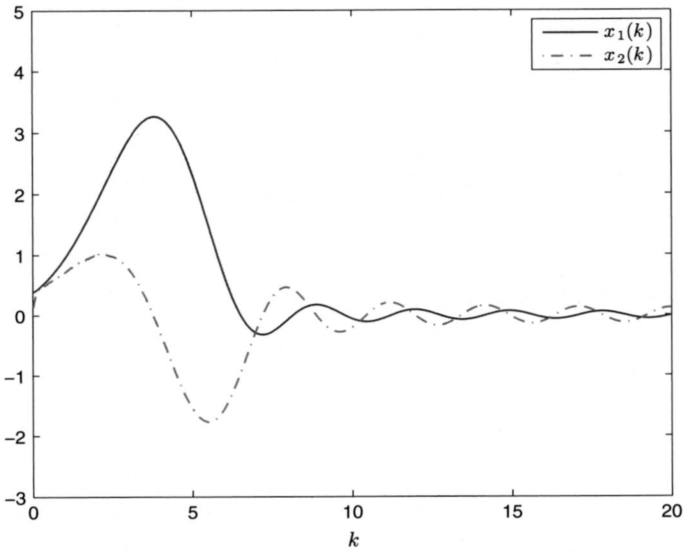

Fig. 10.1 States under the type-1 case in [22]

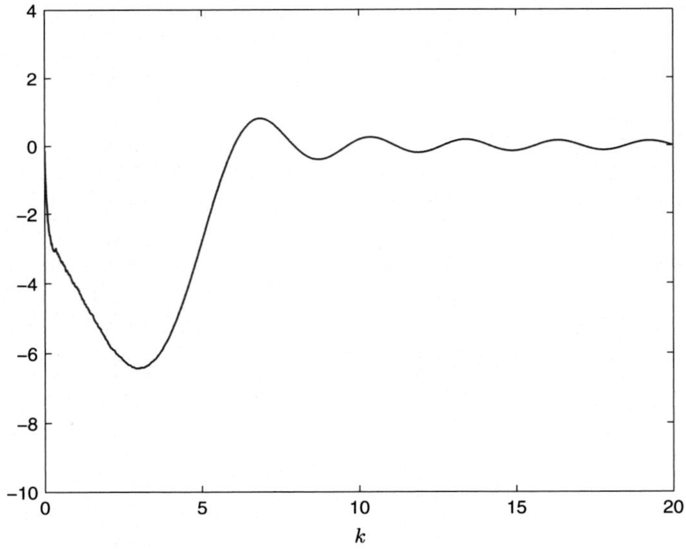

Fig. 10.2 Control input under the type-1 case in [22]

$f_{1\max} = 21.3333$, $f_{2\min} = -0.4615$, and $f_{2\max} = -0.0748$. In order to obtain the discrete-time form of IT2 fuzzy model and compare with the type-1 in (10.42), we use the same sampling time period to obtain the discrete-time IT2 fuzzy system with disturbance input in (10.2), which is expressed below.

Plant Rule i : IF $\xi(k)$ is M_1^i and $\xi(k)$ is M_2^i, THEN

$$\begin{cases} x(k+1) = A_i x(k) + B_i u(k) + B_{wi} w(k), \\ z(k) = C_i x(k) + D_i u(k) + D_{wi} w(k), \end{cases} \tag{10.44}$$

where

$$A_1 = A_2 = \begin{bmatrix} 1.0006 & 0.0100 \\ 0.1113 & 1.0006 \end{bmatrix}, \quad A_3 = A_4 = \begin{bmatrix} 1.0012 & 0.0100 \\ 0.2353 & 1.0012 \end{bmatrix},$$
$$B_1 = B_3 = \begin{bmatrix} -0.23 \times 10^{-4} & -0.46 \times 10^{-2} \end{bmatrix}^T,$$
$$B_2 = B_4 = \begin{bmatrix} -0.37 \times 10^{-5} & -0.75 \times 10^{-3} \end{bmatrix}^T,$$

with the comparative matrices

$$C_1 = C_2 = C_3 = C_4 = \begin{bmatrix} 0 & 1 \end{bmatrix}^T, \quad D_1 = D_2 = D_3 = D_4 = 1,$$
$$B_{w1} = B_{w2} = B_{w3} = B_{w4} = \begin{bmatrix} 0 & 0.020 \end{bmatrix}^T,$$
$$D_{w1} = D_{w2} = D_{w3} = D_{w4} = 0.010.$$

Besides, the LMFs and UMFs are given in Table 10.1, and for generality, we set $\underline{\alpha}_i(\xi(k)) \in [0, 1]$ ($\overline{\alpha}_i(\xi(k)) = 1 - \underline{\alpha}_i(\xi(k))$) for $i = 1, 2, 3, 4$, which obey the Gaussian distribution and satisfy $\sum_{i=1}^{4} \phi_i(\xi(k)) = 1$ to describe the parametric uncertainty. We use a two-rule IT2 fuzzy controller to stabilize the unstable system (10.44) via the LMFs and UMFs chosen in Table 10.1 and we choose $\underline{\beta}_j(x_1) = \overline{\beta}_j(x_1) = 0.5$ for simplicity.

We use one sub-FOU (i.e., $\tau = 0$, $l = 1$) and divide the state x_1 into 2000 equal-size sub-states (i.e., $k = 1, 2, \ldots, 2000$). Thus, the upper and lower bounds of kth state $x_1^{k,l}$ in the FOU l are defined as $\underline{x}_1^{k,l} = (3\pi/4)/2000(k - 101)$, $\overline{x}_1^{k,l} = (3\pi/4)/2000(k - 100)$. Then the constant scalars are determined by $\underline{\epsilon}_{ij1k1} = \underline{\phi}_i(\underline{x}_1^{k,l})\underline{\psi}_j(\underline{x}_1^{k,l})$, $\underline{\epsilon}_{ij2k1} = \underline{\phi}_i(\overline{x}_1^{k,l})\underline{\psi}_j(\overline{x}_1^{k,l})$, $\overline{\epsilon}_{ij1k1} = \overline{\phi}_i(\underline{x}_1^{k,l})\overline{\psi}_j(\underline{x}_1^{k,l})$, $\overline{\epsilon}_{ij2k1} = \overline{\phi}_i(\overline{x}_1^{k,l})\overline{\psi}_j(\overline{x}_1^{k,l})$. Moreover, the LMFs and UMFs $\underline{h}_{ij1}$ and $\overline{h}_{ij1}$ in the form of (10.8) and (10.9) are defined by choosing $v_{11k1}(x_1) = 1 - \left(x_1 - \underline{x}_1^{k,l}\right) / \left(\overline{x}_1^{k,l} - \underline{x}_1^{k,l}\right)$ and $v_{12k1}(x_1) = 1 - v_{11k1}(x_1)$, respectively. We remove the constraint (10.37) and use the same input constraint (10.36) for comparing with the type-1 one. Hence, the parameters for Theorem 10.4 are ready to derive the controller gains in the form of (10.38). By using the Robust Control Toolbox in MATLAB, a feasible solution for controller gains via Theorem 10.4 is listed as follows:

Table 10.1 LMFs and UMFs

LMFs of the pant	UMFs of the pant
$\underline{\mu}_{M_1^1}(x_1) = 1 - e^{\left(-\frac{x_1^2}{1.5}\right)}$	$\overline{\mu}_{M_1^1}(x_1) = 0.25e^{\left(-\frac{x_1^2}{0.3}\right)}$
$\underline{\mu}_{M_1^2}(x_1) = \underline{\mu}_{M_1^1}(x_1)$	$\overline{\mu}_{M_1^2}(x_1) = \overline{\mu}_{M_1^1}(x_1)$
$\underline{\mu}_{M_1^3}(x_1) = 1 - \overline{\mu}_{M_1^1}(x_1)$	$\overline{\mu}_{M_1^3}(x_1) = 1 - \underline{\mu}_{M_1^1}(x_1)$
$\underline{\mu}_{M_1^4}(x_1) = \underline{\mu}_{M_1^3}(x_1)$	$\overline{\mu}_{M_1^4}(x_1) = \overline{\mu}_{M_1^3}(x_1)$
$\underline{\mu}_{M_2^1}(x_1) = 0.4e^{\left(-\frac{x_1^2}{0.2}\right)}$	$\overline{\mu}_{M_2^1}(x_1) = e^{\left(-\frac{x_1^2}{2.5}\right)}$
$\underline{\mu}_{M_2^2}(x_1) = 1 - \overline{\mu}_{M_2^1}(x_1)$	$\overline{\mu}_{M_2^2}(x_1) = 1 - \underline{\mu}_{M_2^1}(x_1)$
$\underline{\mu}_{M_2^3}(x_1) = \underline{\mu}_{M_2^1}(x_1)$	$\overline{\mu}_{M_2^3}(x_1) = \overline{\mu}_{M_2^1}(x_1)$
$\underline{\mu}_{M_2^4}(x_1) = \underline{\mu}_{M_2^2}(x_1)$	$\overline{\mu}_{M_2^4}(x_1) = \overline{\mu}_{M_2^2}(x_1)$
LMFs of the controller	**UMFs of the controller**
$\underline{\mu}_{N_1^1}(x_1) = e^{\left(-\frac{x_1^2}{0.5}\right)}$	$\overline{\mu}_{N_1^1}(x_1) = \underline{\mu}_{N_1^1}(x_1)$
$\underline{\mu}_{N_1^2}(x_1) = \underline{\mu}_{N_1^1}(x_1)$	$\overline{\mu}_{N_1^2}(x_1) = \underline{\mu}_{N_1^2}(x_1)$

$$G_1 = \begin{bmatrix} 340.6544 & 77.4515 \end{bmatrix}, \quad G_2 = \begin{bmatrix} 340.6561 & 77.4520 \end{bmatrix}, \tag{10.45}$$

and the H_∞ performance index is $\gamma = 0.6439$.

Next, based on the gains in (10.45), we depict the responses of the closed-loop system under the same initial state and the same random disturbance input. Figures 10.3 and 10.4 show the simulation results. The system state responses are depicted in Fig. 10.3, and the control force is shown in Fig. 10.4. Apparently, the presented control method shows the superiority than the type-1 one.

In addition, we will append the constraint (10.37) to testify the presented optimal control scheme with the $\mathcal{D}$-stability constraint in the paper, and make a comparison with the removed one. For simplicity, we divide the state x_1 into 200 equal-size substates (i.e., $k = 1, 2, \ldots, 200$) in this case. We reset poles being in the disk region $\mathcal{D}\,(0.3, 0.6)$. Under the same conditions, via Theorem 10.4, the controller gains can be obtained as

$$G_1 = \begin{bmatrix} 2817.4000 & 281.4683 \end{bmatrix}, \quad G_2 = \begin{bmatrix} 2817.7519 & 281.5026 \end{bmatrix}. \tag{10.46}$$

Thus, based on the gains in (10.46), under the same conditions given above, the compared simulation results are provided in Figs. 10.5 and 10.6, which are depicted the different state trajectories of the two different cases. Obviously, the system responses with constraint case perform preferable than the one without the $\mathcal{D}$-stability constraint, which has also demonstrated the effectiveness of the presented optimal control scheme.

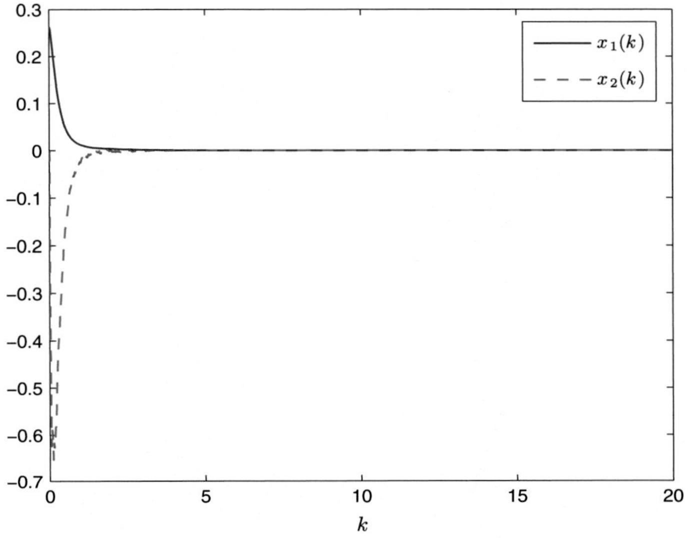

Fig. 10.3 States under the presented IT2 case

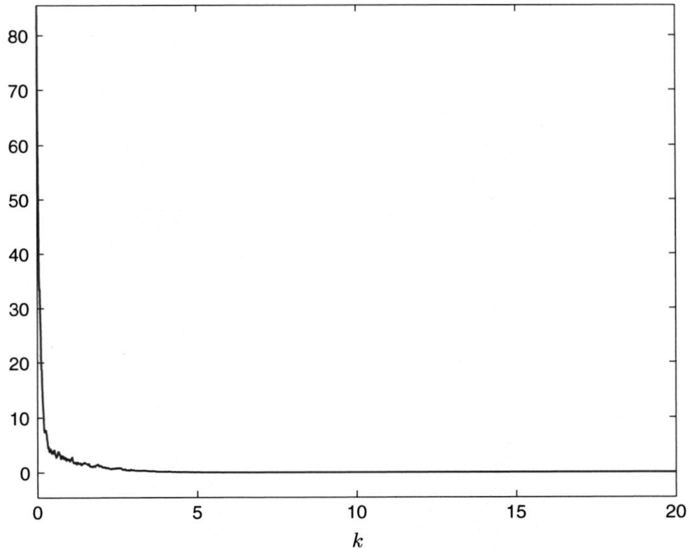

Fig. 10.4 Control input under the presented IT2 case

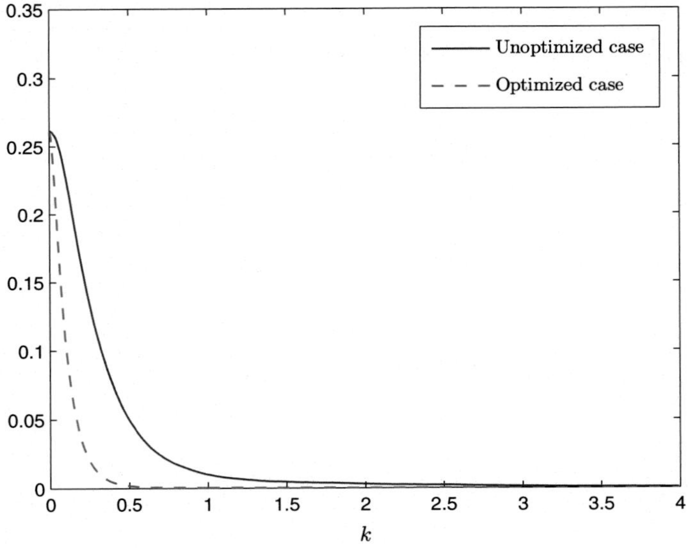

Fig. 10.5 State $x_1(k)$

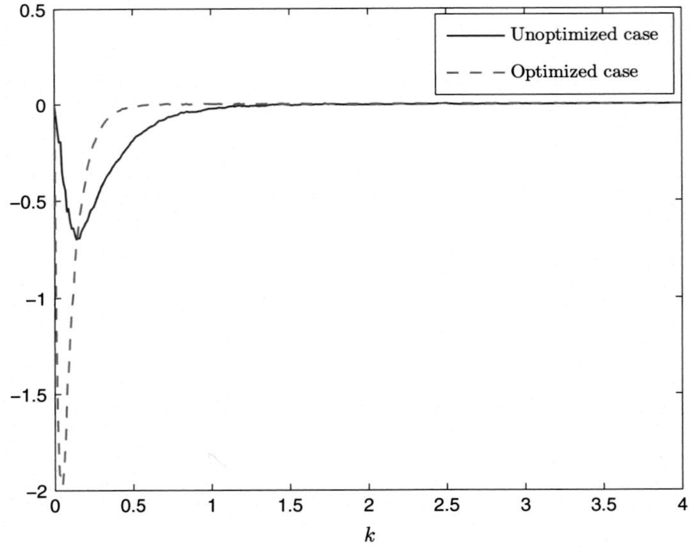

Fig. 10.6 State $x_2(k)$

10.5 Conclusion

This chapter has solved the problem of IT2 state-feedback control design for discrete-time IT2 fuzzy systems. The IT2 fuzzy systems have been considered with input constraint and $\mathcal{D}$-stability constraint. Sufficient criterions of asymptotic stability with H_∞ performance have been given for the discrete-time IT2 fuzzy systems with input and $\mathcal{D}$-stability constraints. The fuzzy state-feedback controller has been designed such that the resulted closed-loop system with input constraint is asymptotically stable with H_∞ performance, and the poles of the closed-loop system are all rested in a desired disk region. The inverted pendulum system has demonstrated the effectiveness and superiority of the presented optimal control design scheme.

Chapter 11
Fault-Tolerant Control of of Interval Type-2 Fuzzy-Model-Based Systems

11.1 Introduction

This chapter designs a fault-tolerant controller for discrete-time IT2 T–S fuzzy systems with time-varying delay and actuator faults. The nonlinear systems subject to parameter uncertainties are modeled by the IT2 T–S fuzzy model approach, in which the LMFs and UMFs are introduced to represent and capture the uncertainties. The IT2 fuzzy systems and the IT2 controller do not need to share the same membership functions and the number of rules, which makes the controller design more flexible. The time-varying delay and actuator fault are first taken into account for the IT2 fuzzy discrete-time systems. By developing some new techniques, a new type fault-tolerant controller is designed to guarantee that the closed-loop system is asymptotically stable under the actuator failures. The existence condition of the fault-tolerant controller can be expressed by a convex optimization problem. Finally, a numerical example is provided to demonstrate the effectiveness of the proposed results.

11.2 System Description and Preliminaries

We consider the following discrete-time IT2 T–S FMB systems with time-varying delay and faulty actuator to describe a nonlinear system.

♦ **Plant Form**:

Rule i: IF $f_1(k)$ is M_1^i and ... and $f_s(k)$ is M_s^i, THEN,

$$\begin{cases} x(k+1) = A_i x(k) + B_i u^F(k) + A_{di} x(k-d(k)) + B_{wi} w(k), \\ z(k) = E_i x(k) + G_i u^F(k) + E_{di} x(k-d(k)) + D_{wi} w(k), \end{cases} \tag{11.1}$$

where M_α^i is an IT2 fuzzy term of ith rule corresponding to the known function $f_\alpha(k)$, $i = 1, 2, \ldots, p$, $\alpha = 1, 2, \ldots, s$ with s is a positive integer, $x(k) \in \mathbf{R}^n$ is

H. Li et al., *Analysis and Synthesis for Interval Type-2 Fuzzy-Model-Based Systems*, DOI 10.1007/978-981-10-0593-0_11

the system state variable, $u^F(k) \in \mathbf{R}^q$ is the faulty control input defined below in (11.9), $w(k) \in \mathbf{R}^l$ represents the disturbance input vector that is assumed to belong to $\ell_2[0,\infty)$, $z(k) \in \mathbf{R}^m$ is the control output vector and $d(k)$ is the time-varying delay of the system, which is assumed to satisfy $d_m \le d(k) \le d_M$, where both $d_m > 0$ and $d_M > 0$ are known integers. Without loss of generality, we assume $d_M > d_m > 0$ throughout the paper. A_i, B_i, A_{di}, B_{wi}, E_i, G_i, E_{di} and D_{wi} are known constant matrices with appropriate dimensions. The firing strength of the ith rule is of the following interval sets:

$$\begin{aligned}\Theta_i(x(k)) &= \left[\prod_{\alpha=1}^{s} \underline{\mu}_{F_\alpha^i}(f_\alpha(k)), \prod_{\alpha=1}^{s} \overline{\mu}_{F_\alpha^i}(f_\alpha(k))\right] \\ &= \left[\underline{\theta}_i(x(k)), \overline{\theta}_i(x(k))\right], \end{aligned} \tag{11.2}$$

where

$$\begin{aligned}\overline{\mu}_{F_\alpha^i}(f_\alpha(k)) \ge \underline{\mu}_{F_\alpha^i}(f_\alpha(k)) \ge 0, \quad \overline{\theta}_i(x(k)) \ge \underline{\theta}_i(x(k)) \ge 0, \\ 1 \ge \overline{\mu}_{F_\alpha^i}(f_\alpha(k)) \ge 0, \quad 1 \ge \underline{\mu}_{F_\alpha^i}(f_\alpha(k)) \ge 0, \end{aligned} \tag{11.3}$$

with $\underline{\mu}_{F_\alpha^i}(f_\alpha(k))$ represents the LMF and $\overline{\mu}_{F_\alpha^i}(f_\alpha(k))$ represents the UMF. $\underline{\theta}_i(x(k))$ denotes the lower grade of membership and $\overline{\theta}_i(x(k))$ denotes the upper grade of membership. Then, the IT2 T–S fuzzy model from (11.1) can be defined as:

$$\begin{cases} x(k+1) = \sum_{i=1}^{p} \theta_i(x(k)) \big(A_i x(k) + B_i u^F(k) \\ \qquad + A_{di} x(k-d(k)) + B_{wi} w(k)\big), \\ z(k) = \sum_{i=1}^{p} \theta_i(x(k)) \big(E_i x(k) + G_i u^F(k) \\ \qquad + E_{di} x(k-d(k)) + D_{wi} w(k)\big), \end{cases} \tag{11.4}$$

where

$$\theta_i(x(k)) = \underline{\alpha}_i(x(k))\underline{\theta}_i(x(k)) + \overline{\alpha}_i(x(k))\overline{\theta}_i(x(k)) \ge 0, \tag{11.5}$$

$$\begin{aligned}&\underline{\alpha}_i(x(k)) \in [0,1], \quad \overline{\alpha}_i(x(k)) \in [0,1], \\ &1 = \underline{\alpha}_i(x(k)) + \overline{\alpha}_i(x(k)), \end{aligned} \tag{11.6}$$

with $\sum_{i=1}^{p} \theta_i(x(k)) = 1$, $\underline{\alpha}_i(x(k))$ and $\overline{\alpha}_i(x(k))$ might depend on parameter uncertainties and not imperative to give numerical values in the paper. $\theta_i(x(k))$ represents the grade of membership of the embedded membership function.

To discuss the fuzzy control problem in the presence of actuator faults, the fault model must be established at first. Let $u^F(k) = M_a u(k)$ be the faulted actuator. The actuator fault matrix M_a satisfies

$$M_a = \text{diag}\{m_{a1}, m_{a2}, \ldots, m_{aq}\},$$

where

$$0 \leq \underline{m}_{ai} \leq m_{ai} \leq \overline{m}_{ai}, \quad i = 1, 2, \ldots, q,$$

in which $\overline{m}_{ai}$ and $\underline{m}_{ai}$ represent the upper and lower bounds of the actuator failure with $0 \leq \underline{m}_{ai} \leq \overline{m}_{ai} \leq 1$, respectively. m_{ai} stands for the probable failure of the following actuator. The following conditions represent the following three different cases of the actuator.

1. When $\underline{m}_{ai} = \overline{m}_{ai} = 0$, so $m_{ai} = 0$, it covers the outage case.
2. When $\underline{m}_{ai} = \overline{m}_{ai} = 1$, so $m_{ai} = 1$, then it corresponds to the normal case $u^F(k) = u(k)$.
3. When $0 < \underline{m}_{ai} < \overline{m}_{ai} < 1$, there is a portion of failure in the corresponding actuator $u^F(k)$.

Then, let c be the fuzzy rules of the controller. The IT2 controller of jth rule is the following model:

♦ **Controller Form**:

Rule j: IF $g_1(k)$ is $N_1^j, \ldots$, and $g_\beta(k)$ is $N_\beta^j, \ldots$, and $g_z(k)$ is N_z^j, THEN,

$$u(k) = K_j x(k), \tag{11.7}$$

where N_β^j represents an IT2 fuzzy set of jth rule for $j = 1, 2, \ldots, c$ and $\beta = 1, 2, \ldots, z$ with z is a positive integer, $g_\beta(k)$ is the measurable premise variable, and K_j stands for the constant feedback gain to be determined. The emission intensity of the jth rule is of the following interval sets:

$$\begin{aligned} M_j(x(k)) &= \left[\prod_{\beta=1}^{z} \underline{\mu}_{N_\beta^j}\left(g_\beta(k)\right), \prod_{\beta=1}^{z} \overline{\mu}_{N_\beta^j}\left(g_\beta(k)\right)\right] \\ &= \left[\underline{m}_j(x(k)), \overline{m}_j(x(k))\right], \end{aligned} \tag{11.8}$$

where

$$\overline{\mu}_{N_\beta^j}\left(g_\beta(k)\right) \geq \underline{\mu}_{N_\beta^j}\left(g_\beta(k)\right) \geq 0, \quad \overline{m}_j(x(k)) \geq \underline{m}_j(x(k)) \geq 0,$$
$$1 \geq \underline{\mu}_{N_\beta^j}\left(g_\beta(k)\right) \geq 0, \quad 1 \geq \overline{\mu}_{N_\beta^j}\left(g_\beta(k)\right) \geq 0,$$

$\underline{\mu}_{N_\beta^j}\left(g_\beta(k)\right)$ represents the LMF and $\overline{\mu}_{N_\beta^j}\left(g_\beta(k)\right)$ represents the UMF. $\underline{m}_j(x(k))$ represents the lower grade of membership and $\overline{m}_j(x(k))$ represents the upper grade of membership. Then, the IT2 T–S fuzzy fault tolerant controller of (11.7) can be defined as:

$$u^F(k) = M_a u(k) = \sum_{j=1}^{c} m_j(x(k)) M_a K_j x(k), \quad (11.9)$$

where M_a is an unknown constant diagonal matrix [237], which can be expressed in (11.12) and (11.13) below, and

$$m_j(x(k)) = \frac{\underline{\beta}_j(x(k))\underline{m}_j(x(k)) + \overline{\beta}_j(x(k))\overline{m}_j(x(k))}{\sum_{\hat{k}=1}^{c}(\underline{\beta}_{\hat{k}}(x(k))\underline{m}_{\hat{k}}(x(k)) + \overline{\beta}_{\hat{k}}(x(k))\overline{m}_{\hat{k}}(x(k)))}, \quad (11.10)$$

$$\underline{\beta}_j(x(k)), \quad \overline{\beta}_j(x(k)) \in [0, 1], \quad 1 = \underline{\beta}_j(x(k)) + \overline{\beta}_j(x(k)), \quad (11.11)$$

where $\underline{\beta}_j(x(k))$ and $\overline{\beta}_j(x(k))$ may be dependent on the parameter uncertainties and not necessary to be known in this chapter. $m_j(x(k))$ represents the grade of membership of the embedded membership function.

Denote

$$M_a = M_{a0}(I + L_a), \quad M_{a0} = \text{diag}\{m_{a01}, m_{a02}, \ldots, m_{a0q}\}, \quad (11.12)$$

$$L_a = \text{diag}\{l_{a1}, l_{a2}, \ldots, l_{aq}\}, \quad \beta_1 = \text{diag}\{\beta_{10}, \beta_{20}, \ldots, \beta_{q0}\}, \quad (11.13)$$

where $m_{a0i} = (\underline{m}_{ai} + \overline{m}_{ai})/2$, $l_{ai} = (m_{ai} - m_{a0i})/m_{a0i}$, $\beta_{i0} = (\overline{m}_{ai} - \underline{m}_{ai})/(\underline{m}_{ai} + \overline{m}_{ai})$, $i = 1, 2, \ldots, q$. Then, it can be found that $|L_a| \leq |\beta_1| \leq I$.

Remark 11.1 For the uncertain nonlinear systems represented by IT2 T–S fuzzy model, LMFs and UMFs are introduced to express the real membership functions and further describe the parameter uncertainties. In detail, the weighting coefficients are used to determine the variety of the real membership functions, i.e., the variety of the uncertain parameters in the uncertain nonlinear systems. The weighting coefficients can be chosen from some time-varying functions to reflect the varying of the uncertain parameters, and the weighting coefficients $\underline{\alpha}_i(x(k))$ and $\overline{\alpha}_i(x(k))$ in (11.6), and $\underline{\beta}_j(x(k))$ and $\overline{\beta}_j(x(k))$ in (11.11), should satisfy the conditions in (11.5–11.6) and (11.8–11.11) respectively considering the whole fuzzy sets for the IT2 T–S fuzzy model.

For the sake of convenience, $\theta_i(x(k))$ and $m_j(x(k))$ are represented as θ_i and m_j respectivelyin the following analysis. From (11.3) and (11.10), we have

$$\sum_{i=1}^{p} \theta_i = \sum_{j=1}^{c} m_j = \sum_{i=1}^{p}\sum_{j=1}^{c} \theta_i m_j = 1.$$

Considering (11.4) and (11.9), the closed-loop system can be described as:

$$\begin{cases} x(k+1) = \sum_{i=1}^{p}\sum_{j=1}^{c}\theta_i m_j \left(A_i + B_i M_a K_j x(k)\right. \\ \qquad \left. + A_{di}x(k-d(k)) + B_{wi}w(k)\right), \\ z(k) = \sum_{i=1}^{p}\sum_{j=1}^{c}\theta_i m_j \left(E_{ij}x(k)\right. \\ \qquad \left. + E_{di}x(k-d(k)) + D_{wi}w(k)\right), \end{cases} \tag{11.14}$$

where $A_{ij} = A_i + B_i M_a K_j$ and $E_{ij} = E_i + G_i M_a K_j$. In order to consider more information of the IT2 membership functions, LMFs and UMFs within the FOU are recommended. The state space Ψ is separated into λ connected sub-state spaces defined as Ψ_τ, $\tau = 1, 2, \ldots, \lambda$ such that $\Psi = \cup_{\tau=1}^{\lambda}\Psi_\tau$. The LMFs and UMFs are defined as follows:

$$\underline{h}_{ij}(x(k)) = \sum_{\tau=1}^{\lambda}\sum_{i_1=1}^{2}\sum_{i_2=1}^{2}\cdots\sum_{i_n=1}^{2}\prod_{r=1}^{n} v_{ri_r\tau}(x_r(k))\,\underline{\varepsilon}_{iji_1i_2\ldots i_n\tau}, \tag{11.15}$$

$$\overline{h}_{ij}(x(k)) = \sum_{\tau=1}^{\lambda}\sum_{i_1=1}^{2}\sum_{i_2=1}^{2}\cdots\sum_{i_n=1}^{2}\prod_{r=1}^{n} v_{ri_r\tau}(x_r(k))\,\overline{\varepsilon}_{iji_1i_2\ldots i_n\tau}, \tag{11.16}$$

$$1 = v_{r1\tau}(x_r(k)) + v_{r2\tau}(x_r(k)),$$
$$\forall r = 1, 2, \ldots, n, \quad \tau = 1, 2, \ldots, q,$$
$$1 = \sum_{\tau=1}^{\lambda}\sum_{i_1=1}^{2}\sum_{i_2=1}^{2}\cdots\sum_{i_n=1}^{2}\prod_{r=1}^{n} v_{ri_r\tau}(x_r(k)), \tag{11.17}$$
$$0 \le v_{ri_r\tau}(x_r(k)) \le 1,$$

where $\underline{\varepsilon}_{iji_1i_2\ldots i_n\tau}$ and $\overline{\varepsilon}_{iji_1i_2\ldots i_n\tau}$ are scalars to be determined, and $0 \le \underline{\varepsilon}_{iji_1i_2\ldots i_n\tau} \le \overline{\varepsilon}_{iji_1i_2\ldots i_n\tau} \le 1$, for $r, s = 1, 2, \ldots, n$, $i_r, i_s = 1, 2$, $\tau = 1, 2, \ldots, \lambda$, and $x(k) \in \Psi_\tau$. In order to facilitate the analysis in next section, we reexpress the IT2 FMB control system (11.14) as follows:

$$\begin{cases} x(k+1) = \sum_{i=1}^{p}\sum_{j=1}^{c} h_{ij}(x(k))\left[A_{ij}x(k) + A_{di}x(k-d(k)) + B_{wi}w(k)\right], \\ z(k) = \sum_{i=1}^{p}\sum_{j=1}^{c} h_{ij}(x(k))\left[E_{ij}x(k) + E_{di}x(k-d(k)) + D_{wi}w(k)\right], \end{cases} \tag{11.18}$$

where

$$\begin{aligned} h_{ij}(x(k)) &\equiv \theta_i m_j \\ &= \underline{\gamma}_{ij}(x(k))\,\underline{h}_{ij}(x(k)) + \overline{\gamma}_{ij}(x(k))\,\overline{h}_{ij}(x(k)), \quad \forall i, j, \end{aligned}$$

$$1 = \sum_{i=1}^{p}\sum_{j=1}^{c} h_{ij}(x(k)), 1 = \underline{\gamma}_{ij}(x(k)) + \overline{\gamma}_{ij}(x(k)),$$
$$0 \le \underline{\gamma}_{ij}(x(k)) \le \overline{\gamma}_{ij}(x(k)) \le 1, \quad \forall i, j.$$

It should be mentioned that the $\underline{\gamma}_{ij}(x(k))$ and $\overline{\gamma}_{ij}(x(k))$ are two functions and not necessarily known. For convenience, the variables $\underline{h}_{ij}(x(k)), \overline{h}_{ij}(x(k)), h_{ij}(x(k))$, $\underline{\gamma}_{ij}(x(k)), \overline{\gamma}_{ij}(x(k))$, $\underline{\varepsilon}_{iji_1i_2\ldots i_{nT}}$ and $\overline{\varepsilon}_{iji_1i_2\ldots i_{nT}}$ are represented by $\underline{h}_{ij}$, $\overline{h}_{ij}$, h_{ij}, $\underline{\gamma}_{ij}$, $\overline{\gamma}_{ij}$, $\underline{\varepsilon}$ and $\overline{\varepsilon}$, respectively.

Remark 11.2 In this chapter, we divide the state space into λ sub-states for further reducing the conservativeness. However, the computation will be more complex when too many sub-states space are divided. In practical systems, it is better to choose a reasonable number of the sub-states considering the precision and the computational complexity.

We introduce a definition for developing the main results in the following part.

Definition 11.3 (*[215]*) For a given scalar $\gamma > 0$, the closed-loop system (11.18) is said to be asymptotically stable and satisfies H_∞ performance if the following inequality holds

$$\|z\|_2 < \gamma \|w\|_2, \quad \forall 0 \neq w \in \ell_2[0, \infty),$$

where $\|z\|_2 = \sqrt{\sum_{k=0}^{\infty} z^T(k) z(k)}$.

11.3 Main Results

In this section, under imperfect premise matching, a new IT2 fuzzy state-feedback controller is designed to ensure that the closed-loop system (11.18) is asymptotically stable and satisfies H_∞ performance. In this section, a novel approach handling the time delay is proposed. Firstly, for given controller gain K_j, we have the following theorem, in which the H_∞ performance analysis condition is given.

11.3.1 Stability Analysis

Theorem 11.4 *Considering the discrete-time IT2 fuzzy time-varying delay system in (11.18) under imperfect premise matching, system (11.18) is asymptotically stable with H_∞ performance γ if there exist matrices $P > 0$, $Q_1 > 0$, $Q_2 > 0$, $Q_3 > 0$, $R_1 > 0$, $R_2 > 0$, $X_{ij} > 0$, $Y_{ij} > 0$, $U_{ij} > 0$, $W_{ij} > 0$ and M with appropriate dimensions, for $i = 1, 2, \ldots, p$ and $j = 1, 2, \ldots, c$; satisfy the following conditions:*

$$\sum_{i=1}^{p}\sum_{j=1}^{c}\left(\bar{\varepsilon}\bar{\Phi}+\left(\bar{\varepsilon}-\underline{\varepsilon}\right)W_{ij}+\bar{\varepsilon}M\right)-M<0,\quad \forall i_1,i_2,\ldots,i_n,\tau,i,j, \tag{11.19}$$

$$\bar{\Phi}+W_{ij}+M-\Lambda_{1ij}-\Lambda_{2ij}-\Lambda_{3ij}>0,\quad \forall i,j, \tag{11.20}$$

where

$$\bar{\Phi}=\begin{bmatrix}\bar{\Phi}_{1ij} & R_1 & \bar{\Phi}_{2ij} & 0 & \bar{\Phi}_{3ij}\\ * & \Omega & 0 & R_2 & 0\\ * & * & \bar{\Phi}_{4ij} & 0 & \bar{\Phi}_{5ij}\\ * & * & * & -Q_2-R_2 & 0\\ * & * & * & * & \bar{\Phi}_{6ij}\end{bmatrix},$$

$$\bar{\Phi}_{1ij}=A_{ij}^TPA_{ij}-P+Q_1-R_1+Q_3+(A_{ij}-I)^T\tilde{R}(A_{ij}-I)+E_{ij}^TE_{ij},$$

$$\bar{\Phi}_{2ij}=A_{ij}^TPA_{di}+\left(A_{ij}-I\right)^T\tilde{R}A_{di}+E_{ij}^TE_{di},\quad \Omega=Q_2-Q_1-R_1-R_2,$$

$$\bar{\Phi}_{3ij}=A_{ij}^TPB_{wi}+\left(A_{ij}-I\right)^T\tilde{R}B_{wi}+E_{ij}^TD_{wi},\quad \tilde{R}=d_m^2R_1+d_M^2R_2,$$

$$\bar{\Phi}_{4ij}=A_{di}^TPA_{di}-Q_3+A_{di}^T\tilde{R}A_{di}+E_{di}^TE_{di},$$

$$\bar{\Phi}_{5ij}=A_{di}^TPB_{wi}+A_{di}^T\tilde{R}B_{wi}+E_{di}^TD_{wi},$$

$$\bar{\Phi}_{6ij}=B_{wi}^TPB_{wi}+B_{wi}^T\tilde{R}B_{wi}+D_{wi}^TD_{wi}-\gamma^2I,$$

$$\Lambda_{1ij}=\phi^T\begin{bmatrix}A_{ij}^TX_{ij}A_{ij} & A_{ij}^TX_{ij}A_{di} & A_{ij}^TX_{ij}B_{wi}\\ * & A_{di}^TX_{ij}A_{di} & A_{di}^TX_{ij}B_{wi}\\ * & * & B_{wi}^TX_{ij}B_{wi}\end{bmatrix}\phi,$$

$$\Lambda_{2ij}=\phi^T\begin{bmatrix}E_{ij}^TY_{ij}E_{ij} & E_{ij}^TY_{ij}E_{di} & E_{ij}^TY_{ij}D_{wi}\\ * & E_{di}^TY_{ij}E_{di} & E_{di}^TY_{ij}D_{wi}\\ * & * & D_{wi}^TY_{ij}D_{wi}\end{bmatrix}\phi,$$

$$\Lambda_{3ij}=\phi^T\begin{bmatrix}\left(A_{ij}-I\right)^TU_{ij}\left(A_{ij}-I\right) & \left(A_{ij}-I\right)^TU_{ij}A_{di} & \left(A_{ij}-I\right)^TU_{ij}B_{wi}\\ * & A_{di}^TU_{ij}A_{di} & A_{di}^TU_{ij}B_{wi}\\ * & * & B_{wi}^TU_{ij}B_{wi}\end{bmatrix}\phi,$$

$$e_1=[I_n\ \ 0\ \ 0\ \ 0\ \ 0]^T,\quad e_2=[0\ \ I_n\ \ 0\ \ 0\ \ 0]^T,$$

$$e_3=[0\ \ 0\ \ I_n\ \ 0\ \ 0]^T,\quad e_4=[0\ \ 0\ \ 0\ \ I_n\ \ 0]^T,$$

$$e_5=[0\ \ 0\ \ 0\ \ 0\ \ I_n]^T,\quad \phi=\left[e_1\ \ e_3\ \ e_5\right]^T.$$

Proof Consider the following Lyapunov–Krasovskii functional,

$$V(k)=V_1(k)+V_2(k)+V_3(k), \tag{11.21}$$

with

$$V_1(k) = x^T(k)Px(k),$$
$$V_2(k) = \sum_{\varrho=k-d_m}^{k-1} x^T(\varrho)Q_1x(\varrho) + \sum_{\varrho=k-d_M}^{k-d_m-1} x^T(\varrho)Q_2x(\varrho) + \sum_{\varrho=k-d(k)}^{k-1} x^T(\varrho)Q_3x(\varrho),$$
$$V_3(k) = d_m \sum_{\varrho=-d_m}^{-1}\sum_{\mu=k+\varrho}^{k-1} \Delta x^T(\mu)R_1\Delta x(\mu) + d\sum_{\varrho=-d_M}^{-d_m-1}\sum_{\mu=k+\varrho}^{k-1} \Delta x^T(\mu)R_2\Delta x(\mu),$$

where $d = d_M - d_m$, $\Delta x(k) = x(k+1) - x(k)$ and $\Delta V(k) = V(k+1) - V(k)$. Define the following new variables:

$$\zeta(k) = \left[x^T(k) \quad x^T(k-d_m) \quad x^T(k-d(k)) \quad x^T(k-d_M) \quad w^T(k)\right]^T.$$

By using Lemma 1.7, we can have

$$\begin{aligned}
\Delta V_1(k) &= x^T(k+1)Px(k+1) - x^T(k)Px(k) \\
&= \sum_{i=1}^{p}\sum_{j=1}^{c}\sum_{\imath=1}^{p}\sum_{\jmath=1}^{c} h_{ij}h_{\imath\jmath}\left[A_{ij}x(k) + A_{di}(k-d(k)) + B_{wi}w(k)\right]^T \\
&\quad \times P\left[A_{\imath\jmath}x(k) + A_{d\imath}(k-d(k)) + B_{w\imath}w(k)\right] - x^T(k)Px(k) \\
&\leq \sum_{i=1}^{p}\sum_{j=1}^{c} h_{ij}\left[A_{ij}x(k) + A_{di}(k-d(k)) + B_{wi}w(k)\right]^T P \\
&\quad \times \left[A_{ij}x(k) + A_{di}(k-d(k)) + B_{wi}w(k)\right] - x^T(k)Px(k) \\
&= \zeta^T(k)\Gamma_1\zeta(k), \qquad (11.22) \\
\Delta V_2(k) &= \zeta^T(k)\left[e_1(Q_1+Q_3)e_1^T - e_2(Q_1-Q_2)e_2^T \right.\\
&\quad \left. - e_3Q_3e_3^T - e_4Q_2e_4^T\right]\zeta(k) \\
&= \zeta^T(k)\Gamma_2\zeta(k), \qquad (11.23) \\
\Delta V_3(k) &= \Delta x^T(k)\left(d_m^2R_1 + d^2R_2\right)\Delta x(k) - d_m\sum_{\varrho=k-d_m}^{k-1} \\
&\quad \times \Delta x^T(\varrho)R_1\Delta x(\varrho) - d\sum_{\varrho=k-d_M}^{k-d_m-1}\Delta x^T(\varrho)R_2\Delta x(\varrho). \qquad (11.24)
\end{aligned}$$

The terms $-d_m\sum_{\varrho=k-d_m}^{k-1}\Delta x^T(\varrho)R_1\Delta x(\varrho)$ and $-d\sum_{\varrho=k-d_M}^{k-d_m-1}\Delta x^T(\varrho)R_2\Delta x(\varrho)$ in (11.24) are bounded as

$$-d_m\sum_{\varrho=k-d_m}^{k-1}\Delta x^T(\varrho)R_1\Delta x(\varrho) \leq -\left[\sum_{\varrho=k-d_m}^{k-1}\Delta x(\varrho)\right]^T R_1\left[\sum_{\varrho=k-d_m}^{k-1}\Delta x(\varrho)\right]$$

$$\begin{aligned}
&= -\zeta^T(k)(e_1 - e_2) R_1 (e_1 - e_2)^T \zeta(k) \\
&= \zeta^T(k) \Gamma_4 \zeta(k),
\end{aligned}$$

$$\begin{aligned}
-d \sum_{\varrho=k-d_M}^{k-d_m-1} \Delta x^T(\varrho) R_2 \Delta x(\varrho) &\leq -\left[\sum_{\varrho\tau=k-d_M}^{k-d_m-1} \Delta x(\varrho)\right]^T R_2 \left[\sum_{\varrho\tau=k-d_M}^{k-d_m-1} \Delta x(\varrho)\right] \\
&= -\zeta^T(k)(e_2 - e_4) R_2 (e_2 - e_4)^T \zeta(k) \\
&= \zeta^T(k) \Gamma_5 \zeta(k).
\end{aligned}$$

Note that (11.24) can be rewritten as

$$\begin{aligned}
\Delta V_3(k) &= \left[x^T(k+1) - x^T(k)\right]\left(d_m^2 R_1 + d^2 R_2\right)[x(k+1) - x(k)] \\
&\quad + \zeta^T(k)(\Gamma_4 + \Gamma_5)\zeta(k) \\
&= \zeta^T(k)(\Gamma_3 + \Gamma_4 + \Gamma_5)\zeta(k),
\end{aligned}$$

where

$$\begin{aligned}
\Gamma_1 &= \phi^T \begin{bmatrix} A_{ij}^T P A_{ij} - P & A_{ij}^T P A_{di} & A_{ij}^T P B_{wi} \\ * & A_{di}^T P A_{di} & A_{di}^T P B_{wi} \\ * & * & B_{wi}^T P B_{wi} \end{bmatrix} \phi, \\
\Gamma_2 &= e_1 (Q_1 + Q_3) e_1^T - e_2 (Q_1 - Q_2) e_2^T - e_3 Q_3 e_3^T - e_4 Q_2 e_4^T, \\
\Gamma_3 &= \phi^T \begin{bmatrix} \left(A_{ij} - I\right)^T \tilde{R} \left(A_{ij} - I\right) & \left(A_{ij} - I\right)^T \tilde{R} A_{di} & \left(A_{ij} - I\right)^T \tilde{R} B_{wi} \\ * & A_{di}^T \tilde{R} A_{di} & A_{di}^T \tilde{R} B_{wi} \\ * & * & B_{wi}^T \tilde{R} B_{wi} \end{bmatrix} \phi, \\
\Gamma_4 &= -(e_1 - e_2) R_1 (e_1 - e_2)^T, \quad \Gamma_5 = -(e_2 - e_4) R_2 (e_2 - e_4)^T.
\end{aligned}$$

Therefore

$$\Delta V(k) \leq \zeta^T(k) \left(\sum_{i=1}^{5} \Gamma_i\right) \zeta(k) = \zeta^T(k) \Phi \zeta(k), \tag{11.25}$$

where

$$\Phi = \begin{bmatrix} \Phi_{1ij} & R_1 & \Phi_{2ij} & 0 & \Phi_{3ij} \\ * & Q_2 - Q_1 - R_1 - R_2 & 0 & R_2 & 0 \\ * & * & \Phi_{4ij} & 0 & \Phi_{5ij} \\ * & * & * & -Q_2 - R_2 & 0 \\ * & * & * & * & \Phi_{6ij} \end{bmatrix},$$

$$\begin{aligned}
\Phi_{1ij} &= A_{ij}^T P A_{ij} - P + Q_1 - R_1 + Q_3 + \left(A_{ij} - I\right)^T \tilde{R} \left(A_{ij} - I\right), \\
\Phi_{2ij} &= A_{ij}^T P A_{di} + (A_{ij} - I)^T \tilde{R} A_{di}, \quad \Phi_{3ij} = A_{ij}^T P B_{wi} + \left(A_{ij} - I\right)^T \tilde{R} B_{wi}, \\
\Phi_{4ij} &= A_{di}^T P A_{di} - Q_3 + A_{di}^T \tilde{R} A_{di}, \quad \Phi_{5ij} = A_{di}^T P B_{wi} + A_{di}^T \tilde{R} B_{wi}, \\
\Phi_{6ij} &= B_{wi}^T P B_{wi} + B_{wi}^T \tilde{R} B_{wi}.
\end{aligned}$$

Consider the following inequalities:

$$\left[\sum_{i=1}^{p}\sum_{j=1}^{c}\left(\underline{\gamma}_{ij}\underline{h}_{ij}+\overline{\gamma}_{ij}\overline{h}_{ij}\right)-1\right]M=0, \tag{11.26}$$

$$-\sum_{i=1}^{p}\sum_{j=1}^{c}\left(1-\underline{\gamma}_{ij}\right)\left(\underline{h}_{ij}-\overline{h}_{ij}\right)W_{ij}\geq 0, \tag{11.27}$$

$$-\sum_{i=1}^{p}\sum_{j=1}^{c}\underline{\gamma}_{ij}\left(\underline{h}_{ij}-\overline{h}_{ij}\right)X_{ij}\geq 0, \tag{11.28}$$

$$-\sum_{i=1}^{p}\sum_{j=1}^{c}\underline{\gamma}_{ij}\left(\underline{h}_{ij}-\overline{h}_{ij}\right)U_{ij}\geq 0, \tag{11.29}$$

$$-\sum_{i=1}^{p}\sum_{j=1}^{c}\underline{\gamma}_{ij}\left(\underline{h}_{ij}-\overline{h}_{ij}\right)Y_{ij}\geq 0, \tag{11.30}$$

where M is a matrix with appropriate dimensions, W_{ij}, N_{ij}, X_{ij}, U_{ij} and Y_{ij} are positive and symmetric matrices with appropriate dimensions. From (11.25) to (11.30), we can get

$$\begin{aligned}
\Delta V(k) \leq & \sum_{i=1}^{p}\sum_{j=1}^{c}\left(\underline{\gamma}_{ij}\underline{h}_{ij}+\left(1-\underline{\gamma}_{ij}\right)\bar{h}_{ij}\right)\zeta^{T}(k)\,\Phi\zeta(k) \\
& -\sum_{i=1}^{p}\sum_{j=1}^{c}\left(1-\underline{\gamma}_{ij}\right)\left(\underline{h}_{ij}-\overline{h}_{ij}\right)\zeta^{T}(k)\,W_{ij}\zeta(k) \\
& +\left[\sum_{i=1}^{p}\sum_{j=1}^{c}\left(\underline{\gamma}_{ij}\underline{h}_{ij}+\left(1-\underline{\gamma}_{ij}\right)\overline{h}_{ij}\right)-1\right]\zeta^{T}(k)\,M\zeta(k) \\
= & \,\zeta^{T}(k)\left[\sum_{i=1}^{p}\sum_{j=1}^{c}\left(\overline{h}_{ij}\Phi-(\underline{h}_{ij}-\overline{h}_{ij})W_{ij}+\overline{h}_{ij}M\right)-M\right]\zeta(k) \\
& +\zeta^{T}(k)\left[\sum_{i=1}^{p}\sum_{j=1}^{c}\underline{\gamma}_{ij}\left(\underline{h}_{ij}-\overline{h}_{ij}\right)\left(\Phi+W_{ij}+M\right)\right]\zeta(k).
\end{aligned}$$

By (11.19)–(11.20) and under zero input $w(k)=0$, it follows that $\Delta V(k)<0$, thus system (11.18) is asymptotically stable. Under zero initial condition, consider the following index:

$$
\begin{aligned}
J &= \sum_{k=0}^{\infty}\left[z^T(k)\,z(k) - \gamma^2 w^T(k)\,w(k) + \Delta V(k)\right] \\
&\le \sum_{k=0}^{\infty}\zeta^T(k)\left[\sum_{i=1}^{p}\sum_{j=1}^{c}\left(\overline{h}_{ij}\bar{\Phi} - \left(\underline{h}_{ij} - \overline{h}_{ij}\right)W_{ij} + \overline{h}_{ij}M\right) - M\right]\zeta(k) \\
&\quad + \sum_{k=0}^{\infty}\zeta^T(k)\left[\sum_{i=1}^{p}\sum_{j=1}^{c}\underline{\gamma}_{ij}\left(\underline{h}_{ij} - \overline{h}_{ij}\right)\left(\bar{\Phi} + W_{ij} + M\right)\right]\zeta(k) \\
&\quad - \sum_{k=0}^{\infty}\sum_{i=1}^{p}\sum_{j=1}^{c}\underline{\gamma}_{ij}\left(\underline{h}_{ij} - \overline{h}_{ij}\right)x^T(k+1)\,X_{ij}x(k+1) \\
&\quad - \sum_{k=0}^{\infty}\sum_{i=1}^{p}\sum_{j=1}^{c}\underline{\gamma}_{ij}\left(\underline{h}_{ij} - \overline{h}_{ij}\right)z^T(k)\,Y_{ij}z(k) \\
&\quad - \sum_{k=0}^{\infty}\sum_{i=1}^{p}\sum_{j=1}^{c}\underline{\gamma}_{ij}\left(\underline{h}_{ij} - \overline{h}_{ij}\right)\Delta x^T(k)\,U_{ij}\Delta x(k) \\
&= \sum_{\tau=0}^{\infty}\zeta^T(k)\left[\sum_{i=1}^{p}\sum_{j=1}^{c}\left(\overline{h}_{ij}\bar{\Phi} - \left(\underline{h}_{ij} - \overline{h}_{ij}\right)W_{ij} + \overline{h}_{ij}M\right) - M\right]\zeta(k) \\
&\quad + \sum_{\tau=0}^{\infty}\zeta^T(k)\left[\sum_{i=1}^{p}\sum_{j=1}^{c}\underline{\gamma}_{ij}\left(\underline{h}_{ij} - \overline{h}_{ij}\right)\left(\bar{\Phi} + W_{ij}\right.\right. \\
&\qquad\qquad \left.\left. + \, M - \Lambda_{1ij} - \Lambda_{2ij} - \Lambda_{3ij}\right)\right]\zeta(k).
\end{aligned}
\tag{11.31}
$$

Thus, $J < 0$ in (11.31) can be guaranteed by

$$
\sum_{i=1}^{p}\sum_{j=1}^{c}\left(\overline{h}_{ij}\bar{\Phi} - \left(\underline{h}_{ij} - \overline{h}_{ij}\right)W_{ij} + \overline{h}_{ij}M\right) - M < 0, \tag{11.32}
$$

and $\bar{\Phi} + W_{ij} + M - \Lambda_{1ij} - \Lambda_{2ij} - \Lambda_{3ij} > 0$ (which yields from condition (11.20)) for all $x \neq 0$. Considering $\underline{h}_{ij}$ in (11.15), $\overline{h}_{ij}$ in (11.16), and the equalities in (11.17), we express the following set of inequalities, which are equivalent to the set of inequalities in (11.32),

$$
\begin{aligned}
&\sum_{\tau=1}^{\lambda}\sum_{i_1=1}^{2}\sum_{i_2=1}^{2}\cdots\sum_{i_n=1}^{2}\prod_{r=1}^{n}v_{ri_r\tau l}\left(x_r(k)\right) \\
&\quad \times\left(\sum_{i=1}^{p}\sum_{j=1}^{c}\left(\overline{\varepsilon}\bar{\Phi} + \left(\overline{\varepsilon} - \underline{\varepsilon}\right)W_{ij} + \overline{\varepsilon}M\right) - M\right) < 0.
\end{aligned}
\tag{11.33}
$$

Obviously, the set of inequalities in (11.33) is satisfied by condition (11.19). Hence, $\|z\|_2 < \gamma \|w\|_2$ as $J < 0$. That is to say, if the conditions (11.19)–(11.20) hold, for all nonzero $w(k) \in \ell_2[0, \infty)$, we can obtain $J < 0$, which means $\|z\|_2 < \gamma \|w\|_2$. According to Definition 11.3, the system (11.18) is asymptotically stable with H_∞ performance γ. The proof is completed. □

Remark 11.5 A sufficient criterion of the stability of the system in (11.18) is provided in Theorem 11.4. The H_∞ performance is considered in the presented systems. The criterion in Theorem 11.4 is helpful for further controller design in the following part.

11.3.2 Fault-Tolerant Control

Next, the existence conditions of the controller are proposed based on the conditions in Theorem 11.4.

Theorem 11.6 *Considering the closed-loop system in (11.18) under imperfect premise matching, system (11.18) is asymptotically stable and has H_∞ performance γ, if there exist matrices $\bar{P} > 0$, $\bar{Q}_1 > 0$, $\bar{Q}_2 > 0$, $\bar{Q}_3 > 0$, $\bar{R}_1 > 0$, $\bar{R}_2 > 0$, $\bar{V}_0 > 0$, $\bar{V}_1 > 0$, $\bar{X}_{ij} > 0$, $\bar{Y}_{ij} > 0$, $\bar{U}_{ij} > 0$, $\bar{W}_{ij} > 0$, $\bar{M}$ and $\bar{K}_j$ with appropriate dimensions, for $i = 1, 2, \ldots, p$, $j = 1, 2, \ldots, c$, such that the following conditions hold:*

$$
\begin{aligned}
\Xi_{ij} &< 0, \quad \forall i_1, i_2, \ldots, i_n, \tau, i, j, && (11.34)\\
\Psi_{ij} &> 0, \quad \forall i, j, && (11.35)\\
\bar{X}_{ij} - \bar{P} &> 0, \quad \forall i, j, && (11.36)\\
\bar{Y}_{ij} - I &> 0, \quad \forall i, j, && (11.37)\\
\bar{U}_{ij} - d_m^2 \bar{R}_1 - d^2 \bar{R}_2 &> 0, \quad \forall i, j, && (11.38)
\end{aligned}
$$

where

$$
\Xi_{ij} = \begin{bmatrix} \Xi_{1ij} & \Xi_{2ij} & \Xi_{3ij} \\ * & \Xi_{4ij} & 0_{4\times 3} \\ * & * & \Xi_{5ij} \end{bmatrix}, \quad \Psi_{ij} = \begin{bmatrix} \Psi_{1ij} & \Psi_{2ij} & \Psi_{3ij} \\ * & \Psi_{4ij} & 0_{4\times 3} \\ * & * & \Psi_{5ij} \end{bmatrix},
$$

with

$$
\Xi_{1ij} = \begin{bmatrix} \Xi_{111ij} & \Xi_{112ij} & \Xi_{113ij} & \Xi_{114ij} & \Xi_{115ij} \\ * & \Xi_{122ij} & \Xi_{123ij} & \Xi_{124ij} & \Xi_{125ij} \\ * & * & \Xi_{133ij} & \Xi_{134ij} & \Xi_{135ij} \\ * & * & * & \Xi_{144ij} & \Xi_{145ij} \\ * & * & * & * & \Xi_{155ij} \end{bmatrix},
$$

$$
\Xi_{2ij} = \begin{bmatrix} \Xi_{211ij} & 0_{5\times 1} & \Xi_{213ij} & 0_{5\times 1} \end{bmatrix}, \quad \tilde{\varepsilon} = \bar{\varepsilon} - \underline{\varepsilon}, \quad \breve{\varepsilon} = \bar{\varepsilon} - \frac{1}{pc},
$$

$$\Xi_{3ij} = \begin{bmatrix} \Xi_{311ij} & 0_{5\times 1} & \Xi_{313ij} \end{bmatrix}, \quad \Xi_{4ij} = \begin{bmatrix} \Xi_{411ij} & 0_{2\times 2} \\ * & \Xi_{422ij} \end{bmatrix},$$

$$\Xi_{5ij} = \begin{bmatrix} d_m^2 \bar{R}_1 + d^2 \bar{R}_2 - 2\bar{P} & \bar{\varepsilon}^{\frac{1}{4}} B_i M_{a0} \beta_1 & 0 \\ * & -\bar{V}_0 & 0 \\ * & * & \bar{V}_0 - 2I \end{bmatrix},$$

$$\Xi_{111ij} = \bar{\varepsilon}(-\bar{P} + \bar{Q}_1 - \bar{R}_1 + \bar{Q}_3) + \tilde{\varepsilon}\bar{W}_{1ij} + \breve{\varepsilon}\bar{M}_1,$$

$$\Xi_{113ij} = \tilde{\varepsilon}\bar{W}_{3ij} + \breve{\varepsilon}\bar{M}_3, \quad \Xi_{114ij} = \tilde{\varepsilon}\bar{W}_{4ij} + \breve{\varepsilon}\bar{M}_4, \quad \Xi_{115ij} = \tilde{\varepsilon}\bar{W}_{11ij} + \breve{\varepsilon}\bar{M}_{11},$$

$$\Xi_{122ij} = \bar{\varepsilon}(\bar{Q}_2 - \bar{Q}_1 - \bar{R}_1 - \bar{R}_2) + \tilde{\varepsilon}\bar{W}_{5ij} + \breve{\varepsilon}\bar{M}_5, \quad \Xi_{123ij} = \tilde{\varepsilon}\bar{W}_{6ij} + \breve{\varepsilon}\bar{M}_6,$$

$$\Xi_{124ij} = \bar{\varepsilon}\bar{R}_2 + \tilde{\varepsilon}\bar{W}_{7ij} + \breve{\varepsilon}\bar{M}_7, \quad \Xi_{125ij} = \tilde{\varepsilon}\bar{W}_{12ij} + \breve{\varepsilon}\bar{M}_{12},$$

$$\Xi_{133ij} = \tilde{\varepsilon}\bar{W}_{8ij} + \breve{\varepsilon}\bar{M}_8 - \bar{\varepsilon}\bar{Q}_3, \quad \Xi_{134ij} = \tilde{\varepsilon}\bar{W}_{9ij} + \breve{\varepsilon}\bar{M}_9,$$

$$\Xi_{135ij} = \tilde{\varepsilon}\bar{W}_{13ij} + \breve{\varepsilon}\bar{M}_{13}, \quad \Xi_{144ij} = \bar{\varepsilon}\left(-\bar{Q}_2 - \bar{R}_2\right) + \tilde{\varepsilon}\bar{W}_{10ij} + \breve{\varepsilon}\bar{M}_{10},$$

$$\Xi_{145ij} = \tilde{\varepsilon}\bar{W}_{14ij} + \breve{\varepsilon}\bar{M}_{14}, \quad \Xi_{155ij} = -\bar{\varepsilon}\gamma^2 I + \tilde{\varepsilon}\bar{W}_{15ij} + \breve{\varepsilon}\bar{M}_{15},$$

$$\Xi_{211ij} = \begin{bmatrix} \sqrt{\bar{\varepsilon}}(A_i\bar{P} + B_i M_{a0}\bar{K}_j) & 0 & \sqrt{\bar{\varepsilon}}A_{di}\bar{P} & 0 & \sqrt{\bar{\varepsilon}}B_{wi} \end{bmatrix}^T,$$

$$\Xi_{313ij} = \begin{bmatrix} \sqrt{3}\bar{\varepsilon}^{\frac{1}{4}}\bar{K}_j & 0_{1\times 4} \end{bmatrix}^T, \quad \Xi_{112ij} = \bar{\varepsilon}\bar{R}_1 + \tilde{\varepsilon}\bar{W}_{2ij} + \breve{\varepsilon}\bar{M}_2,$$

$$\Xi_{213ij} = \begin{bmatrix} \sqrt{\bar{\varepsilon}}\left(E_i\bar{P} + G_i M_{a0}\bar{K}_j\right) & 0 & \sqrt{\bar{\varepsilon}}E_{di}\bar{P} & 0 & \sqrt{\bar{\varepsilon}}D_{wi} \end{bmatrix}^T,$$

$$\Xi_{311ij} = \begin{bmatrix} \sqrt{\bar{\varepsilon}}\left(A_i\bar{P} + B_i M_{a0}\bar{K}_j - \bar{P}\right) & 0 & \sqrt{\bar{\varepsilon}}A_{di}\bar{P} & 0 & \sqrt{\bar{\varepsilon}}B_{wi} \end{bmatrix}^T,$$

$$\Xi_{411ij} = \begin{bmatrix} -\bar{P} & \bar{\varepsilon}^{\frac{1}{4}} B_i M_{a0}\beta_1 \\ * & -\bar{V}_0 \end{bmatrix}, \quad \Xi_{422ij} = \begin{bmatrix} -I & \bar{\varepsilon}^{\frac{1}{4}} G_i M_{a0}\beta_1 \\ * & -\bar{V}_0 \end{bmatrix},$$

$$\Psi_{1ij} = \begin{bmatrix} \Psi_{111ij} & \bar{R}_1 + \bar{W}_{2ij} + \bar{M}_2 & \bar{W}_{3ij} + \bar{M}_3 & \bar{W}_{4ij} + \bar{M}_4 & \bar{W}_{11ij} + \bar{M}_{11} \\ * & \Psi_{122ij} & \bar{W}_{6ij} + \bar{M}_6 & \bar{R}_2 + \bar{W}_{7ij} + \bar{M}_7 & \bar{W}_{12ij} + \bar{M}_{12} \\ * & * & \Psi_{133ij} & \bar{W}_{9ij} + \bar{M}_9 & \bar{W}_{13ij} + \bar{M}_{13} \\ * & * & * & \Psi_{144ij} & \bar{W}_{14ij} + \bar{M}_{14} \\ * & * & * & * & \Psi_{155ij} \end{bmatrix},$$

$$\Psi_{2ij} = \begin{bmatrix} \Psi_{211ij} & 0_{5\times 1} & \Psi_{213ij} & 0_{5\times 1} \end{bmatrix}, \quad \Psi_{3ij} = \begin{bmatrix} \Psi_{311ij} & 0_{5\times 1} & \Psi_{313ij} \end{bmatrix},$$

$$\Psi_{4ij} = \begin{bmatrix} \Psi_{411ij} & 0_{2\times 2} \\ * & \Psi_{422ij} \end{bmatrix}, \quad \Psi_{5ij} = \begin{bmatrix} 2\bar{P} + \bar{R} - \bar{U}_{ij} & B_i M_{a0}\beta_1 & 0 \\ * & \bar{V}_1 & 0 \\ * & * & 2I - \bar{V}_1 \end{bmatrix},$$

$$\Psi_{111ij} = -\bar{P} + \bar{Q}_1 - \bar{R}_1 + \bar{Q}_3 + \bar{W}_{1ij} + \bar{M}_1, \quad \Psi_{133ij} = \bar{W}_{8ij} + \bar{M}_8 - \bar{Q}_3,$$

$$\Psi_{122ij} = \bar{Q}_2 - \bar{Q}_1 - \bar{R}_1 - \bar{R}_2 + \bar{W}_{5ij} + \bar{M}_5,$$

$$\Psi_{144ij} = -\bar{Q}_2 - \bar{R}_2 + \bar{W}_{10ij} + \bar{M}_{10}, \quad \Psi_{155ij} = -\gamma^2 I + \bar{W}_{15ij} + \bar{M}_{15},$$

$$\Psi_{211ij} = \begin{bmatrix} A_i\bar{P} + B_i M_{a0}\bar{K}_j & 0 & A_{di}\bar{P} & 0 & B_{wi} \end{bmatrix}^T,$$

$$\Psi_{213ij} = \begin{bmatrix} E_i\bar{P} + G_i M_{a0}\bar{K}_j & 0 & E_{di}\bar{P} & 0 & D_{wi} \end{bmatrix}^T,$$

$$\Psi_{313ij} = \begin{bmatrix} \sqrt{3}\bar{K}_j & 0_{1\times 4} \end{bmatrix}^T,$$

$$\Psi_{311ij} = \begin{bmatrix} A_i\bar{P} + B_i M_{a0}\bar{K}_j - \bar{P} & 0 & A_{di}\bar{P} & 0 & B_{wi} \end{bmatrix}^T,$$

$$\Psi_{411ij} = \begin{bmatrix} 3\bar{P} - \bar{X}_{ij} & B_i M_{a0}\beta_1 \\ * & \bar{V}_1 \end{bmatrix}, \quad \Psi_{422ij} = \begin{bmatrix} 3I - \bar{Y}_{ij} & G_i M_{a0}\beta_1 \\ * & \bar{V}_1 \end{bmatrix},$$

$$\bar{M} = \begin{bmatrix} \bar{M}_1 & \bar{M}_2 & \bar{M}_3 & \bar{M}_4 & \bar{M}_{11} \\ * & \bar{M}_5 & \bar{M}_6 & \bar{M}_7 & \bar{M}_{12} \\ * & * & \bar{M}_8 & \bar{M}_9 & \bar{M}_{13} \\ * & * & * & \bar{M}_{10} & \bar{M}_{14} \\ * & * & * & * & \bar{M}_{15} \end{bmatrix}, \quad \bar{W}_{ij} = \begin{bmatrix} \bar{W}_{1ij} & \bar{W}_{2ij} & \bar{W}_{3ij} & \bar{W}_{4ij} & \bar{W}_{11ij} \\ * & \bar{W}_{5ij} & \bar{W}_{6ij} & \bar{W}_{7ij} & \bar{W}_{12ij} \\ * & * & \bar{W}_{8ij} & \bar{W}_{9ij} & \bar{W}_{13ij} \\ * & * & * & \bar{W}_{10ij} & \bar{W}_{14ij} \\ * & * & * & * & \bar{W}_{15ij} \end{bmatrix}.$$

Furthermore, the controller gain matrix of the stable feedback controller in the form of (11.9) is given as

$$K_j = \bar{K}_j \bar{P}^{-1}.$$

Then, the fault-tolerant controller is given as

$$u^F(k) = M_a u(k) = \sum_{j=1}^{c} m_j(x(k)) M_a \bar{K}_j \bar{P}^{-1} x(k).$$

Proof Firstly, for the condition in (11.34), by using Schur complement and utilizing $-\tilde{R}^{-1} \le \tilde{R} - 2I$, $-V_0^{-1} \le V_0 - 2I$, $V_1^{-1} \ge 2I - V_1$, the following LMIs hold

$$\bar{\Xi}_{ij} = \begin{bmatrix} \bar{\Xi}_{1ij} & \bar{\Xi}_{2ij} \\ * & \bar{\Xi}_{3ij} \end{bmatrix} < 0, \tag{11.39}$$

where

$$\bar{\Xi}_{1ij} = \begin{bmatrix} \bar{\Xi}_{111ij} & \Xi_{112ij} & \Xi_{113ij} & \Xi_{114ij} & \Xi_{115ij} \\ * & \Xi_{122ij} & \Xi_{123ij} & \Xi_{124ij} & \Xi_{125ij} \\ * & * & \Xi_{133ij} & \Xi_{134ij} & \Xi_{135ij} \\ * & * & * & \Xi_{144ij} & \Xi_{145ij} \\ * & * & * & * & \Xi_{155ij} \end{bmatrix},$$

$$\bar{\Xi}_{2ij} = \begin{bmatrix} \Xi_{211ij} & \bar{\Xi}_{212ij} & \bar{\Xi}_{213ij} \end{bmatrix},$$

$$\begin{aligned} \bar{\Xi}_{3ij} = \mathrm{diag}\{&-\bar{P} + \sqrt{\bar{\varepsilon}} B_i M_{a0} V_0^{-1} \beta_1^2 M_{a0}^T B_i^T, -I + \sqrt{\bar{\varepsilon}} G_i M_{a0} V_0^{-1} \beta_1^2 M_{a0}^T G_i^T, \\ &\bar{R} - 2\bar{P} + \sqrt{\bar{\varepsilon}} B_i M_{a0} V_0^{-1} \beta_1^2 M_{a0}^T B_i^T\} \end{aligned}$$

$$\bar{\Xi}_{111ij} = \bar{\varepsilon}\left(-\bar{P} + \bar{Q}_1 - \bar{R}_1 + \bar{Q}_3\right) + \tilde{\varepsilon}\bar{W}_{1ij} + \breve{\varepsilon}\bar{M}_1 + 3\sqrt{\bar{\varepsilon}}\bar{K}_j^T V_0 \bar{K}_j,$$

$$\bar{\Xi}_{212ij} = \begin{bmatrix} \sqrt{\bar{\varepsilon}}\left(E_i \bar{P} + G_i M_{a0} \bar{K}_j\right) & 0 & \sqrt{\bar{\varepsilon}} E_{di} \bar{P} & 0 & \sqrt{\bar{\varepsilon}} D_{wi} \end{bmatrix}^T,$$

$$\bar{\Xi}_{213ij} = \begin{bmatrix} \sqrt{\bar{\varepsilon}}\left(A_i \bar{P} + B_i M_{a0} \bar{K}_j - \bar{P}\right) & 0 & \sqrt{\bar{\varepsilon}} A_{di} \bar{P} & 0 & \sqrt{\bar{\varepsilon}} B_{wi} \end{bmatrix}^T.$$

Define the following nonsingular matrices:

$$\begin{aligned} &\bar{P} = P^{-1}, \quad \bar{Q}_1 = P^{-1} Q_1 P^{-1}, \quad \bar{Q}_2 = P^{-1} Q_2 P^{-1}, \quad \bar{Q}_3 = P^{-1} Q_3 P^{-1}, \\ &\bar{K}_j = K_j P^{-1}, \quad \bar{R}_1 = P^{-1} R_1 P^{-1}, \quad \bar{R}_2 = P^{-1} R_2 P^{-1}, \end{aligned}$$

$$\begin{aligned}
\bar{X}_{ij} &= P^{-1}X_{ij}P^{-1}, \quad \bar{R} = d_m^2\bar{R}_1 + d^2\bar{R}_2, \quad \bar{U}_{ij} = P^{-1}U_{ij}P^{-1}, \quad \bar{V}_0 = V_0,\\
\bar{V}_1 &= V_1, \quad \bar{Y}_{ij} = Y_{ij}, \quad \bar{R}_1 = P^{-1}R_1P^{-1}, \quad \bar{R}_2 = P^{-1}R_2P^{-1},\\
\bar{W}_{1ij} &= P^{-1}W_{1ij}P^{-1}, \quad \bar{W}_{2ij} = P^{-1}W_{2ij}P^{-1}, \quad \bar{W}_{3ij} = P^{-1}W_{3ij}P^{-1},\\
\bar{W}_{4ij} &= P^{-1}W_{4ij}P^{-1}, \quad \bar{W}_{5ij} = P^{-1}W_{5ij}P^{-1}, \quad \bar{W}_{6ij} = P^{-1}W_{6ij}P^{-1},\\
\bar{W}_{7ij} &= P^{-1}W_{7ij}P^{-1}, \quad \bar{W}_{8ij} = P^{-1}W_{8ij}P^{-1}, \quad \bar{W}_{9ij} = P^{-1}W_{9ij}P^{-1},\\
\bar{W}_{10ij} &= P^{-1}W_{10ij}P^{-1}, \quad \bar{W}_{11ij} = P^{-1}W_{11ij}, \quad \bar{W}_{12ij} = P^{-1}W_{12ij},\\
\bar{W}_{13ij} &= P^{-1}W_{13ij}, \quad \bar{W}_{14ij} = P^{-1}W_{14ij}, \quad \bar{W}_{15ij} = W_{15ij},\\
\bar{M}_1 &= P^{-1}M_1P^{-1}, \quad \bar{M}_2 = P^{-1}M_2P^{-1}, \quad \bar{M}_3 = P^{-1}M_3P^{-1},\\
\bar{M}_4 &= P^{-1}M_4P^{-1}, \quad \bar{M}_5 = P^{-1}M_5P^{-1}, \quad \bar{M}_6 = P^{-1}M_6P^{-1},\\
\bar{M}_7 &= P^{-1}M_7P^{-1}, \quad \bar{M}_8 = P^{-1}M_8P^{-1}, \quad \bar{M}_9 = P^{-1}M_9P^{-1},\\
\bar{M}_{10} &= P^{-1}M_{10}P^{-1}, \quad \bar{M}_{11} = P^{-1}M_{11}, \quad \bar{M}_{12} = P^{-1}M_{12},\\
\bar{M}_{13} &= P^{-1}M_{13}, \quad \bar{M}_{14} = P^{-1}M_{14}, \quad \bar{M}_{15} = M_{15}.
\end{aligned}$$

Then, pre- and post-multiplying (11.39) by diag$\{P, P, P, P, I, P, I, P\}$ and its transpose, respectively. Notice $-P\tilde{R}^{-1}P \leq \tilde{R} - 2P$. Then, we have

$$\begin{aligned}
\hat{\Xi}_{ij} &= \begin{bmatrix} \tilde{\Xi}_{1ij} & \tilde{\Xi}_{2ij} \\ * & \tilde{\Xi}_{3ij} \end{bmatrix} + 3\sqrt{\bar{\varepsilon}} \begin{bmatrix} K_j^T \\ 0_{7\times 1} \end{bmatrix} V_0 \begin{bmatrix} K_j^T \\ 0_{7\times 1} \end{bmatrix}^T \\
&\quad + \sqrt{\bar{\varepsilon}} \begin{bmatrix} 0_{5\times 1} \\ PB_iM_{a0} \\ 0_{2\times 1} \end{bmatrix} \left(V_0^{-1}\beta_1^2\right) \begin{bmatrix} 0_{5\times 1} \\ PB_iM_{a0} \\ 0_{2\times 1} \end{bmatrix}^T \\
&\quad + \sqrt{\bar{\varepsilon}} \begin{bmatrix} 0_{6\times 1} \\ G_iM_{a0} \\ 0 \end{bmatrix} \left(V_0^{-1}\beta_1^2\right) \begin{bmatrix} 0_{6\times 1} \\ G_iM_{a0} \\ 0 \end{bmatrix}^T \\
&\quad + \sqrt{\bar{\varepsilon}} \begin{bmatrix} 0_{7\times 1} \\ PB_iM_{a0} \end{bmatrix} \left(V_0^{-1}\beta_1^2\right) \begin{bmatrix} 0_{7\times 1} \\ PB_iM_{a0} \end{bmatrix}^T \\
&= \begin{bmatrix} \hat{\Xi}_{1ij} & \tilde{\Xi}_{2ij} \\ * & \hat{\Xi}_{3ij} \end{bmatrix} < 0,
\end{aligned}$$

where

$$\begin{aligned}
\tilde{\Xi}_{1ij} &= \begin{bmatrix} \tilde{\Xi}_{111ij} & \tilde{\Xi}_{112ij} & \tilde{\Xi}_{113ij} & \tilde{\Xi}_{114ij} & \tilde{\Xi}_{115ij} \\ * & \tilde{\Xi}_{122ij} & \tilde{\Xi}_{123ij} & \tilde{\Xi}_{124ij} & \tilde{\Xi}_{125ij} \\ * & * & \tilde{\Xi}_{133ij} & \tilde{\Xi}_{134ij} & \tilde{\Xi}_{135ij} \\ * & * & * & \tilde{\Xi}_{144ij} & \tilde{\Xi}_{145ij} \\ * & * & * & * & \tilde{\Xi}_{155ij} \end{bmatrix}, \\
\tilde{\Xi}_{2ij} &= \begin{bmatrix} \tilde{\Xi}_{211ij} & \tilde{\Xi}_{212ij} & \tilde{\Xi}_{213ij} \end{bmatrix}, \\
\tilde{\Xi}_{111ij} &= \bar{\varepsilon}(-P + Q_1 - R_1 + Q_3) + \tilde{\varepsilon}W_{1ij} + \breve{\varepsilon}M_1,
\end{aligned}$$

$$\begin{aligned}
\tilde{\Xi}_{3ij} &= \mathrm{diag}\{-P, -I, -P\tilde{R}^{-1}P\}, \quad \tilde{\Xi}_{112ij} = \bar{\varepsilon}R_1 + \tilde{\varepsilon}W_{2ij} + \check{\varepsilon}M_2, \\
\tilde{\Xi}_{122ij} &= \bar{\varepsilon}\left(Q_2 - Q_1 - R_1 - R_2\right) + \tilde{\varepsilon}W_{5ij} + \check{\varepsilon}M_5, \\
\tilde{\Xi}_{114ij} &= \tilde{\varepsilon}W_{4ij} + \check{\varepsilon}M_4, \quad \tilde{\Xi}_{115ij} = \tilde{\varepsilon}W_{11ij} + \check{\varepsilon}M_{11}, \\
\tilde{\Xi}_{113ij} &= \tilde{\varepsilon}W_{3ij} + \check{\varepsilon}M_3, \quad \tilde{\Xi}_{123ij} = \tilde{\varepsilon}W_{6ij} + \check{\varepsilon}M_6, \\
\tilde{\Xi}_{124ij} &= \bar{\varepsilon}R_2 + \tilde{\varepsilon}W_{7ij} + \check{\varepsilon}M_7, \quad \tilde{\Xi}_{125ij} = \tilde{\varepsilon}W_{12ij} + \check{\varepsilon}M_{12}, \\
\tilde{\Xi}_{133ij} &= \tilde{\varepsilon}W_{8ij} + \check{\varepsilon}M_8 - \bar{\varepsilon}Q_3, \quad \tilde{\Xi}_{134ij} = \tilde{\varepsilon}W_{9ij} + \check{\varepsilon}M_9, \\
\tilde{\Xi}_{135ij} &= \tilde{\varepsilon}W_{13ij} + \check{\varepsilon}M_{13}, \quad \tilde{\Xi}_{145ij} = \tilde{\varepsilon}W_{14ij} + \check{\varepsilon}M_{14}, \\
\tilde{\Xi}_{144ij} &= \bar{\varepsilon}\left(-Q_2 - R_2\right) + \tilde{\varepsilon}W_{10ij} + \check{\varepsilon}M_{10}, \\
\tilde{\Xi}_{211ij} &= \left[\sqrt{\bar{\varepsilon}}P\left(A_i + B_iM_{a0}K_j\right) \quad 0 \quad \sqrt{\bar{\varepsilon}}PA_{di} \quad 0 \quad \sqrt{\bar{\varepsilon}}PB_{wi}\right]^T, \\
\tilde{\Xi}_{212ij} &= \left[\sqrt{\bar{\varepsilon}}\left(E_i + G_iM_{a0}K_j\right) \quad 0 \quad \sqrt{\bar{\varepsilon}}E_{di} \quad 0 \quad \sqrt{\bar{\varepsilon}}D_{wi}\right]^T, \\
\tilde{\Xi}_{213ij} &= \left[\sqrt{\bar{\varepsilon}}P\left(A_i + B_iM_{a0}\bar{K}_j - I\right) \quad 0 \quad \sqrt{\bar{\varepsilon}}PA_{di} \quad 0 \quad \sqrt{\bar{\varepsilon}}PB_{wi}\right]^T, \\
\hat{\Xi}_{1ij} &= \begin{bmatrix} \tilde{\Xi}_{111ij} + 3\sqrt{\bar{\varepsilon}}K_j^TV_0K_j & \tilde{\Xi}_{112ij} & \tilde{\Xi}_{113ij} & \tilde{\Xi}_{114ij} & \tilde{\Xi}_{115ij} \\ * & \tilde{\Xi}_{122ij} & \tilde{\Xi}_{123ij} & \tilde{\Xi}_{124ij} & \tilde{\Xi}_{125ij} \\ * & * & \tilde{\Xi}_{133ij} & \tilde{\Xi}_{134ij} & \tilde{\Xi}_{135ij} \\ * & * & * & \tilde{\Xi}_{144ij} & \tilde{\Xi}_{145ij} \\ * & * & * & * & \tilde{\Xi}_{155ij} \end{bmatrix}, \\
\tilde{\Xi}_{155ij} &= -\bar{\varepsilon}\gamma^2I + \tilde{\varepsilon}W_{15ij} + \check{\varepsilon}M_{15}, \\
\hat{\Xi}_{3ij} &= \mathrm{diag}\{-P + \sqrt{\bar{\varepsilon}}PB_iM_{a0}V_0^{-1}\beta_1^2M_{a0}^TB_i^TP^T, \\
&\quad -I + \sqrt{\bar{\varepsilon}}G_iM_{a0}V_0^{-1}\beta_1^2M_{a0}^TG_i^T, \\
&\quad -P\tilde{R}^{-1}P + \sqrt{\bar{\varepsilon}}PB_iM_{a0}V_0^{-1}\beta_1^2M_{a0}^TB_i^TP^T\}.
\end{aligned}$$

By (11.12)–(11.13) and Lemma 1.7 in [106] , and applying the inequality $2a^Tb \leq a^Ta + b^Tb$ for arbitrary diagonal matrix $0 < V_0 < I$, it follows that

$$\begin{aligned}
\tilde{\Xi}_{ij} = &\begin{bmatrix} \tilde{\Xi}_{1ij} & \tilde{\Xi}_{2ij} \\ * & \tilde{\Xi}_{3ij} \end{bmatrix} + \sqrt{\bar{\varepsilon}}\begin{bmatrix} K_j^T \\ 0_{7\times1} \end{bmatrix} L_a^T\left[0_{1\times5} \; M_{a0}^TB_iP \; 0_{1\times2}\right] \\
&+ \sqrt{\bar{\varepsilon}}\begin{bmatrix} K_j^T \\ 0_{7\times1} \end{bmatrix} L_a^T\left[0_{1\times6} \; M_{a0}^TG_i \; 0\right] + \sqrt{\bar{\varepsilon}}\begin{bmatrix} K_j^T \\ 0_{7\times1} \end{bmatrix} L_a^T \\
&\times\left[0_{1\times7} \; M_{a0}^TB_iP\right]\left(\sqrt{\bar{\varepsilon}}\begin{bmatrix} K_j^T \\ 0_{7\times1} \end{bmatrix} L_a^T\left[0_{1\times5} \; M_{a0}^TB_iP \; 0_{1\times2}\right]\right. \\
&+ \sqrt{\bar{\varepsilon}}\begin{bmatrix} K_j^T \\ 0_{7\times1} \end{bmatrix} L_a^T\left[0_{1\times6} \; M_{a0}^TG_i \; 0\right] \\
&\left.+ \sqrt{\bar{\varepsilon}}\begin{bmatrix} K_j^T \\ 0_{7\times1} \end{bmatrix} L_a^T\left[0_{1\times7} \; M_{a0}^TB_iP\right]\right)^T \leq \hat{\Xi}_{ij} < 0. \qquad (11.40)
\end{aligned}$$

Furthermore, let the following matrices be portioned as:

$$M = \begin{bmatrix} M_1 & M_2 & M_3 & M_4 & M_{11} \\ * & M_5 & M_6 & M_7 & M_{12} \\ * & * & M_8 & M_9 & M_{13} \\ * & * & * & M_{10} & M_{14} \\ * & * & * & * & M_{15} \end{bmatrix}, \quad W_{ij} = \begin{bmatrix} W_{1ij} & W_{2ij} & W_{3ij} & W_{4ij} & W_{11ij} \\ * & W_{5ij} & W_{6ij} & W_{7ij} & W_{12ij} \\ * & * & W_{8ij} & W_{9ij} & W_{13ij} \\ * & * & * & W_{10ij} & W_{14ij} \\ * & * & * & * & W_{15ij} \end{bmatrix}. \tag{11.41}$$

Then based on the set of inequality in (11.40), and by Schur complement, one can obtain that

$$\left(\bar{\varepsilon}\bar{\Phi} + \left(\bar{\varepsilon} - \underline{\varepsilon}\right) W_{ij} + \bar{\varepsilon}M\right) - \frac{1}{pc}M < 0, \quad \forall i_1, i_2, \ldots, i_n, k, i, j.$$

Thus, one can have

$$\sum_{i=1}^{p}\sum_{j=1}^{c} \left(\bar{\varepsilon}\bar{\Phi} + \left(\bar{\varepsilon} - \underline{\varepsilon}\right) W_{ij} + \bar{\varepsilon}M\right) - M < 0, \quad \forall i_1, i_2, \ldots, i_n, k,$$

which meets the condition (11.19) in Theorem 11.4.

Secondly, for the set of inequality in (11.35), similarly, by using Schur complement, the following LMIs hold

$$\bar{\Psi}_{ij} = \begin{bmatrix} \bar{\Psi}_{1ij} & \bar{\Psi}_{2ij} \\ * & \bar{\Psi}_{3ij} \end{bmatrix} < 0, \tag{11.42}$$

where

$$\bar{\Psi}_{1ij} = \begin{bmatrix} \bar{\Psi}_{111ij} & \bar{W}_{2ij} + \bar{M}_2 + \bar{R}_1 & \bar{W}_{3ij} + \bar{M}_3 & \bar{W}_{4ij} + \bar{M}_4 & \bar{W}_{11ij} + \bar{M}_{11} \\ * & \Psi_{122ij} & \bar{W}_{6ij} + \bar{M}_6 & \bar{W}_{7ij} + \bar{M}_7 + \bar{R}_2 & \bar{W}_{12ij} + \bar{M}_{12} \\ * & * & \Psi_{133ij} & \bar{W}_{9ij} + \bar{M}_9 & \bar{W}_{13ij} + \bar{M}_{13} \\ * & * & * & \Psi_{144ij} & \bar{W}_{14ij} + \bar{M}_{14} \\ * & * & * & * & \Psi_{155ij} \end{bmatrix},$$

$$\begin{aligned}
\bar{\Psi}_{2ij} &= \begin{bmatrix} \Psi_{211ij} & \bar{\Psi}_{212ij} & \bar{\Psi}_{213ij} \end{bmatrix}, \\
\bar{\Psi}_{3ij} &= \mathrm{diag}\{3\bar{P} - \bar{X}_{ij} - B_i M_{a0} V_1^{-1} \beta_1^2 M_{a0}^T B_i^T, \\
&\qquad 3I - \bar{Y}_{ij} - G_i M_{a0} V_1^{-1} \beta_1^2 M_{a0}^T G_i^T, \\
&\qquad 2\bar{P} + \bar{R} - \bar{U}_{ij} - B_i M_{a0} V_1^{-1} \beta_1^2 M_{a0}^T B_i^T\}, \\
\bar{\Psi}_{111ij} &= -\bar{P} + \bar{Q}_1 - \bar{R}_1 + \bar{Q}_3 + \bar{W}_{1ij} + \bar{M}_1 - 3\bar{K}_j^T V_1 \bar{K}_j, \\
\bar{\Psi}_{212ij} &= \begin{bmatrix} E_i\bar{P} + G_i M_{a0}\bar{K}_j & 0 & E_{di}\bar{P} & 0 & D_{wi} \end{bmatrix}^T, \\
\bar{\Psi}_{213ij} &= \begin{bmatrix} A_i\bar{P} + B_i M_{a0}\bar{K}_j - \bar{P} & 0 & A_{di}\bar{P} & 0 & B_{wi} \end{bmatrix}^T.
\end{aligned}$$

Pre- and post-multiplying (11.39) by diag$\{P, P, P, P, I, P, I, P\}$ and its transpose, respectively, we have

$$\begin{aligned}\hat{\Psi}_{ij} &= \begin{bmatrix} \tilde{\Psi}_{1ij} & \tilde{\Psi}_{2ij} \\ * & \tilde{\Psi}_{3ij} \end{bmatrix} + 3\begin{bmatrix} -K_j^T \\ 0_{7\times 1} \end{bmatrix}(-V_1)\begin{bmatrix} -K_j^T \\ 0_{7\times 1} \end{bmatrix}^T \\ &\quad + \begin{bmatrix} 0_{5\times 1} \\ PB_iM_{a0} \\ 0_{2\times 1} \end{bmatrix}\left(-V_1^{-1}\beta_1^2\right)\begin{bmatrix} 0_{5\times 1} \\ PB_iM_{a0} \\ 0_{2\times 1} \end{bmatrix}^T \\ &\quad + \begin{bmatrix} 0_{6\times 1} \\ G_iM_{a0} \\ 0 \end{bmatrix}\left(-V_1^{-1}\beta_1^2\right)\begin{bmatrix} 0_{6\times 1} \\ G_iM_{a0} \\ 0 \end{bmatrix}^T \\ &\quad + \begin{bmatrix} 0_{7\times 1} \\ PB_iM_{a0} \end{bmatrix}\left(-V_1^{-1}\beta_1^2\right)\begin{bmatrix} 0_{7\times 1} \\ PB_iM_{a0} \end{bmatrix}^T \\ &= \begin{bmatrix} \hat{\Psi}_{1ij} & \tilde{\Psi}_{2ij} \\ * & \hat{\Psi}_{3ij} \end{bmatrix} > 0,\end{aligned}$$

where

$$\begin{aligned}\tilde{\Psi}_{1ij} &= \begin{bmatrix} \tilde{\Psi}_{111ij} & W_{2ij}+M_2+R_1 & W_{3ij}+M_3 & W_{4ij}+M_4 & W_{11ij}+M_{11} \\ * & \tilde{\Psi}_{122ij} & W_{6ij}+M_6 & W_{7ij}+M_7+R_2 & W_{12ij}+M_{12} \\ * & * & \tilde{\Psi}_{133ij} & W_{9ij}+M_9 & W_{13ij}+M_{13} \\ * & * & * & \tilde{\Psi}_{144ij} & W_{14ij}+M_{14} \\ * & * & * & * & \tilde{\Psi}_{155ij} \end{bmatrix}, \\ \tilde{\Psi}_{2ij} &= \begin{bmatrix} \tilde{\Psi}_{211ij} & \tilde{\Psi}_{212ij} & \tilde{\Psi}_{213ij} \end{bmatrix}, \\ \tilde{\Psi}_{3ij} &= \text{diag}\{3P - X_{ij}, 3I - Y_{ij}, 2P + \tilde{R} - U_{ij}\}, \\ \tilde{\Psi}_{111ij} &= -P + Q_1 - R_1 + Q_3 + W_{1ij} + M_1, \\ \tilde{\Psi}_{122ij} &= Q_2 - Q_1 - R_1 - R_2 + W_{5ij} + M_5, \\ \tilde{\Psi}_{133ij} &= W_{8ij} + M_8 - Q_3, \quad \tilde{\Psi}_{144ij} = -Q_2 - R_2 + W_{10ij} + M_{10}, \\ \tilde{\Psi}_{155ij} &= -\gamma^2 I + W_{15ij} + M_{15}, \\ \tilde{\Psi}_{211ij} &= \begin{bmatrix} P\left(A_i + B_iM_{a0}K_j\right) & 0 & PA_{di} & 0 & PB_{wi} \end{bmatrix}^T, \\ \hat{\Psi}_{1ij} &= \begin{bmatrix} \hat{\Psi}_{11ij} & \hat{\Psi}_{12ij} \\ * & \hat{\Psi}_{13ij} \end{bmatrix}, \quad \hat{\Psi}_{13ij} = \begin{bmatrix} \tilde{\Psi}_{144ij} & W_{14ij}+M_{14} \\ * & \tilde{\Psi}_{155ij} \end{bmatrix}, \\ \hat{\Psi}_{11ij} &= \begin{bmatrix} \tilde{\Psi}_{111ij} - 3K_j^TV_1K_j & W_{2ij}+M_2+R_1 & W_{3ij}+M_3 \\ * & \tilde{\Psi}_{122ij} & W_{6ij}+M_6 \\ * & * & \tilde{\Psi}_{133ij} \end{bmatrix}, \\ \hat{\Psi}_{12ij} &= \begin{bmatrix} W_{4ij}+M_4 & W_{11ij}+M_{11} \\ W_{7ij}+M_7+R_2 & W_{12ij}+M_{12} \\ W_{9ij}+M_9 & W_{13ij}+M_{13} \end{bmatrix},\end{aligned}$$

$$\begin{aligned}
\tilde{\Psi}_{212ij} &= \left[E_i + G_i M_{a0} K_j \ 0 \ E_{di} \ 0 \ D_{wi} \right]^T, \\
\tilde{\Psi}_{213ij} &= \left[P(A_i + B_i M_{a0} \bar{K}_j - I) \ 0 \ P A_{di} \ 0 \ P B_{wi} \right]^T, \\
\hat{\Psi}_{3ij} &= \mathrm{diag}\{3P - X_{ij} - P B_i M_{a0} V_1^{-1} \beta_1^2 M_{a0}^T B_i^T P^T, \\
&\qquad 3I - Y_{ij} - G_i M_{a0} V_1^{-1} \beta_1^2 M_{a0}^T G_i^T, \\
&\qquad 2P + \tilde{R} - U_{ij} - P B_i M_{a0} V_1^{-1} \beta_1^2 M_{a0}^T B_i^T P^T\}.
\end{aligned}$$

By (11.12)–(11.13) and Lemma 1.7, and applying the inequality $-2a^T b \geq -a^T a - b^T b$ for arbitrarily diagonal matrix $0 < V_1 < I$, it follows that

$$\begin{aligned}
\tilde{\Psi}_{ij} &= \begin{bmatrix} \tilde{\Psi}_{1ij} & \tilde{\Psi}_{2ij} \\ * & \tilde{\Psi}_{3ij} \end{bmatrix} + \begin{bmatrix} -K_j^T \\ 0_{7\times1} \end{bmatrix} (-L_a^T) \left[0_{1\times5} \ M_{a0}^T B_i P \ 0_{1\times2} \right] \\
&\quad + \begin{bmatrix} -K_j^T \\ 0_{7\times1} \end{bmatrix} (-L_a^T) \left[0_{1\times6} \ M_{a0}^T G_i \ 0 \right] \\
&\quad + \begin{bmatrix} -K_j^T \\ 0_{7\times1} \end{bmatrix} (-L_a^T) \left[0_{1\times7} \ M_{a0}^T B_i P \right] \\
&\quad + \left(\begin{bmatrix} -K_j^T \\ 0_{7\times1} \end{bmatrix} (-L_a^T) \left[0_{1\times5} \ M_{a0}^T B_i P \ 0_{1\times2} \right] \right. \\
&\qquad + \begin{bmatrix} -K_j^T \\ 0_{7\times1} \end{bmatrix} (-L_a^T) \left[0_{1\times6} \ M_{a0}^T G_i \ 0 \right] \\
&\qquad \left. + \begin{bmatrix} -K_j^T \\ 0_{7\times1} \end{bmatrix} (-L_a^T) \left[0_{1\times7} \ M_{a0}^T B_i P \right] \right)^T \\
&\geq \hat{\Psi}_{ij} > 0.
\end{aligned} \tag{11.43}$$

For all i, j, pre- and post-multiplying (11.36) by P^T and its transpose, we can obtain $X_{ij} - P > 0$, it follows that

$$P \left(X_{ij} - P \right)^{-1} P \geq 3P - X_{ij}.$$

Similarly, from (11.37) and (11.38), one can get

$$\begin{aligned}
\left(Y_{ij} - I \right)^{-1} &\geq 3I - Y_{ij}, \\
P \left(U_{ij} - \tilde{R} \right)^{-1} P &\geq 2P - U_{ij} + \tilde{R}.
\end{aligned}$$

Thus, considering (11.36)–(11.38), from (11.43) we have

$$\begin{aligned}
\check{\Psi}_{ij} &= \begin{bmatrix} \tilde{\Psi}_{1ij} & \tilde{\Psi}_{2ij} \\ * & \check{\Psi}_{3ij} \end{bmatrix} > \tilde{\Psi}_{ij} > 0, \\
\check{\Psi}_{3ij} &= \mathrm{diag}\{P \left(X_{ij} - P \right)^{-1} P, P \left(Y_{ij} - I \right)^{-1} P, P \left(U_{ij} - \tilde{R} \right)^{-1} P\}.
\end{aligned}$$

Then, considering the matrices in (11.41), and by Schur complement, one can reach the condition in (11.20). The whole proof is completed. □

11.4 Simulation Results

Example 11.7 In this section, a numerical example is used to illustrate the effectiveness of the controller design method. Consider an IT2 fuzzy model with 3 rules in the following format:

Plant Rule i: IF $x_1(k)$ is M_1^i, THEN

$$x(k+1) = A_i x(k) + B_i u^F(k) + A_{di} x(k - d(k)) + B_{wi} w(k), \quad i = 1, 2, 3, \tag{11.44}$$

where

$$A_1 = \begin{bmatrix} f_1 & 1.30 \\ -0.08 & 1.4 \end{bmatrix}, \quad A_2 = \begin{bmatrix} f_2 & -1.50 \\ -0.08 & -1.4 \end{bmatrix}, \quad A_3 = \begin{bmatrix} f_3 & 1.30 \\ -0.08 & 1.4 \end{bmatrix},$$
$$A_{d1} = \begin{bmatrix} 0 & -0.015 \\ -0.64 * f_1 & -0.020 \end{bmatrix}, \quad A_{d2} = \begin{bmatrix} 0 & -0.011 \\ -0.47 * f_2 & -0.030 \end{bmatrix},$$
$$A_{d3} = \begin{bmatrix} 0 & -0.014 \\ -0.92 * f_3 & -0.020 \end{bmatrix},$$
$$B_1 = \begin{bmatrix} g_1 \\ -9.0300 \end{bmatrix}, \quad B_2 = \begin{bmatrix} g_2 \\ -9.06 \end{bmatrix}, \quad B_3 = \begin{bmatrix} g_3 \\ -9.06 \end{bmatrix},$$
$$B_{w1} = \begin{bmatrix} -0.003 \\ 0.01 \end{bmatrix}, \quad B_{w2} = \begin{bmatrix} -0.001 \\ 0.01 \end{bmatrix}^T, \quad B_{w3} = \begin{bmatrix} -0.002 \\ 0.01 \end{bmatrix}^T,$$
$$E_1 = \begin{bmatrix} 0.214 & -0.128 \end{bmatrix}, \quad E_2 = \begin{bmatrix} 0.120 & -0.120 \end{bmatrix},$$
$$E_3 = \begin{bmatrix} 0.214 & -0.128 \end{bmatrix}, \quad E_{d1} = \begin{bmatrix} 0.00 & -0.1020 \end{bmatrix},$$
$$E_{d2} = \begin{bmatrix} 0.00 & -0.1022 \end{bmatrix}, \quad E_{d3} = \begin{bmatrix} 0.00 & -0.1024 \end{bmatrix},$$
$$G_1 = G_3 = 0.214, \quad G_2 = 0.120, \quad D_{w1} = D_{w2} = D_{w3} = -0.001,$$

in which $f(\sigma(x_1)) = -0.02\sqrt{\sigma(k)}$, $g(\sigma(x_1)) = (-1)^{\sigma(x_1)}(0.0019\sigma(x_1) + 3.23)$, $\sigma(x_1) \in [121, 225]$, with

$$f_1 = f_{\max} = f(121) = -0.0220, \quad f_2 = f_{\min} = f(225) = -0.0300,$$
$$f_3 = f(144) = -0.0240, \quad g_1 = g_{\min} = g(255) = -3.6575,$$
$$g_2 = g_{\max} = g(224) = 3.6556, \quad g_3 = g(121) = -3.4599.$$

Because of the uncertain parameter σ existing in the nonlinear function $f(\sigma)$, the type-1 T–S fuzzy model approach cannot be applied completely. However, the IT2 fuzzy model is used successfully to represent the nonlinear system.

Table 11.1 LMFs and UMFs for the plant

LMFs for the plant
$\underline{\mu}_{M_1^1}(x_1) = 0.8 - \left(0.8/\left(1 + e^{(-(x_1+80)/15)}\right)\right)$
$\underline{\mu}_{M_1^3}(x_1) = 0.8/\left(1 + e^{(-(x_1+80)/15)}\right)$
$\underline{\mu}_{M_1^2}(x_1) = 1 - \overline{\mu}_{M_1^1}(x_1) - \overline{\mu}_{M_1^3}(x_1)$
UMFs for the plant
$\overline{\mu}_{M_1^1}(x_1) = 1 - \left(1/\left(1 + e^{(-(x_1+80)/15)}\right)\right)$
$\overline{\mu}_{M_1^3}(x_1) = 1/\left(1 + e^{(-(x_1+80)/15)}\right)$
$\overline{\mu}_{M_1^2}(x_1) = 1 - \underline{\mu}_{M_1^1}(x_1) - \underline{\mu}_{M_1^3}(x_1)$

Table 11.2 LMFs and UMFs for the controller

LMFs for the controller	UMFs for the controller
$\underline{\mu}_{N_1^1}(x_1) = e^{(-x_1^2/4000)}$	$\overline{\mu}_{N_1^1}(x_1) = e^{(-x_1^2/0.5)}$
$\underline{\mu}_{N_1^2}(x_1) = 1 - \overline{\mu}_{N_1^1}(x_1)$	$\overline{\mu}_{N_1^2}(x_1) = 1 - \underline{\mu}_{N_1^1}(x_1)$

For system (11.44), the LMFs and UMFs are defined in Table 11.1, and we choose $\underline{\alpha}_i = 1, \overline{\alpha}_i = 1 - \underline{\alpha}_i, x_1(k) \in [-80, 80]$. The state $x_1(k)$ is divided into 10 equal-size sub-states (i.e., $\tau = 1, 2, \ldots, 10$), and define

$$\upsilon_{11\tau}(x_1) = 1 - \frac{x_1 - \underline{x}_{1,\tau}}{\overline{x}_1 - \underline{x}_{1,\tau}}, \quad \upsilon_{12\tau}(x_1) = 1 - \upsilon_{11\tau}(x_1),$$
$$\underline{x}_{1,\tau} = 16(\tau - 6), \quad \overline{x}_{1,\tau} = 16(\tau - 5).$$

According to (11.2) and (11.8), the scale invariants are obtained as

$$\underline{\varepsilon}_{ij1\tau} = \underline{\theta}_i\left(\underline{x}_{1,\tau}\right)\underline{m}_j\left(\underline{x}_{1,\tau}\right), \quad \underline{\varepsilon}_{ij2\tau} = \underline{\theta}_i\left(\overline{x}_{1,\tau}\right)\underline{m}_j\left(\overline{x}_{1,\tau}\right),$$
$$\overline{\varepsilon}_{ij1\tau} = \overline{\theta}_i\left(\underline{x}_{1,\tau}\right)\overline{m}_j\left(\underline{x}_{1,\tau}\right), \quad \overline{\varepsilon}_{ij2\tau} = \overline{\theta}_i\left(\overline{x}_{1,\tau}\right)\overline{m}_j\left(\overline{x}_{1,\tau}\right), \quad \tau = 1, 2, \ldots, 10.$$

Besides, we choose the LMFs and UMFs of the controller in Table 11.2, with $\underline{\beta}_i = 0.5$ and $\overline{\beta}_i = 1 - \underline{\beta}_i$. Assume that the time-varying delay $2 \leq d(k) \leq 4$ and the initial state $x_0 = \left[-50\ 40\right]^T$. Then the state response of the open-loop system is obtained in Fig. 11.1, which shows that the open-loop system in (11.44) is not stable.

To analyze the stability of the closed-loop system in (11.44) with actuator failure, it is supposed that system (11.44) has partial failure between $\underline{m}_{ai} = 0.138$ and $\overline{m}_{ai} = 0.162$. For Theorem 11.6, by using the MATLAB Control Toolbox, the obtained control gains are

$$K_1 = \left[-0.0095\ -0.1817\right], \quad K_2 = \left[-0.0081\ 0.1077\right],$$

and H_∞ performance level $\gamma = 60.5110$. Under the controller $u^F = M_a K_j$, the state trajectories of closed-loop system (11.44) are shown in Fig. 11.2. Thus, the effectiveness of the proposed control method has been successfully verified.

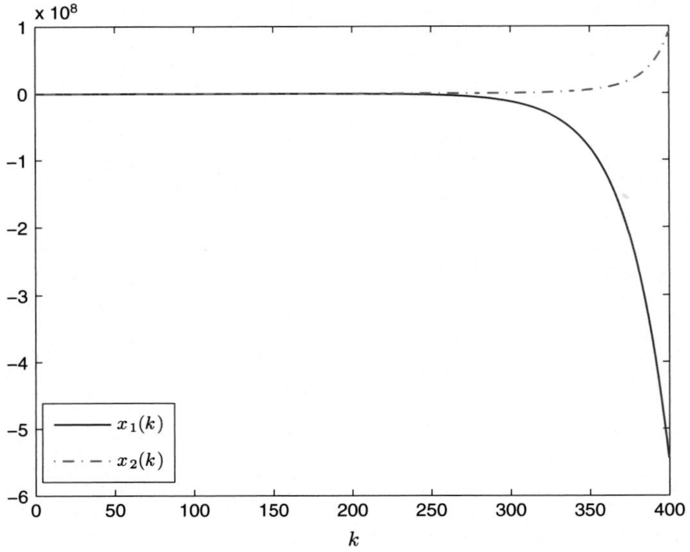

Fig. 11.1 States of the open-loop system

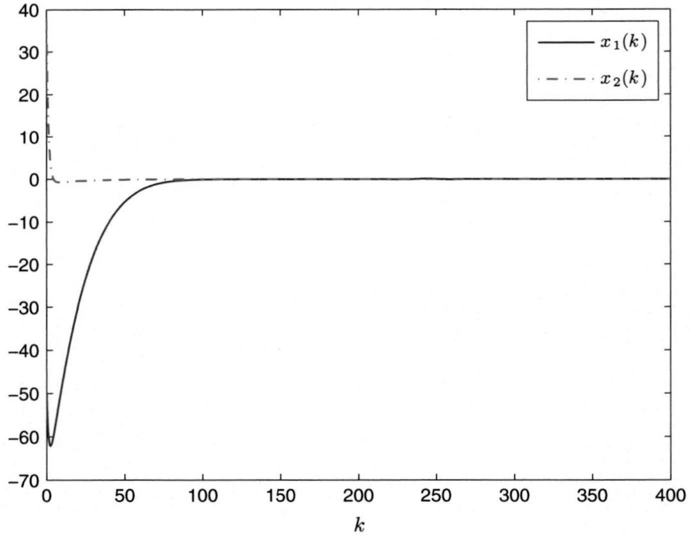

Fig. 11.2 States of the close-loop system with actuator failure

11.5 Conclusion

This chapter has considered the problem of fault-tolerant control for discrete-time IT2 fuzzy time delay system with actuator faults under imperfect premise matching. The time-varying delay and actuator failure have been first time taken into account for the IT2 fuzzy discrete-time systems. In this chapter, the fuzzy system and the IT2 controller do not share the same LMFs and UMFs, and the number of the fuzzy rules. By developing some new techniques, a new type fault-tolerant controller has been designed to guarantee that the closed-loop system is asymptotically stable under the actuator failures. The existence conditions of the fault-tolerant controller can be expressed by a convex optimization problem. Finally, a numerical example has been provided to demonstrate the effectiveness of the proposed results. In future work, the filtering problem or the fault detection problem will be considered based on the IT2 FMB systems with time delays and some other significant factors.

Chapter 12
Output-Feedback Control of Interval Type-2 Fuzzy-Model-Based Systems

This chapter focuses on designing a novel reliable static output-feedback controller for discrete-time IT2 FMB systems with mixed H_2/H_∞ performance. Firstly, the discrete-time IT2 FMB systems with sensor failure and the IT2 fuzzy controller under imperfect premise matching are constructed for control design objective. The mixed H_2/H_∞ performance index is established. Secondly, a sufficient condition of reliable stability is derived by applying the Lyapunov stability theory. Based on the condition, the desired IT2 fuzzy static output-feedback controller is designed under the sensor failure known case and unknown case, respectively. Some simulation results are provided to demonstrate the effectiveness of the proposed results.

12.1 Problem Formulation

Firstly, we introduce the IT2 fuzzy sets for further characterizing the membership functions in the fuzzy model systems of discrete form. Considering the premise variable of the plant, which is represented by p-rules T–S fuzzy model, let $M_{i\alpha}$ denotes an IT2 fuzzy set of ith rule for $i = 1, 2, \ldots, p$ and $\alpha = 1, 2, \ldots, \Theta$ (Θ is a positive integer). Define $f_\alpha(x(k))$ as the measurable premise variable, where $x(k)$ is the system state variable with k being the sampling time of discrete systems. Then, the firing strength of the ith rule corresponds to the interval sets $\Phi_i(x(k)) = \left[\underline{\phi}_i(x(k)), \overline{\phi}_i(x(k))\right]$, where $\underline{\phi}_i(x(k)) = \prod_{\alpha=1}^{\Theta} \underline{\mu}_{M_{i\alpha}}(f_\alpha(x(k))) \geq 0$, $\overline{\phi}_i(x(k)) = \prod_{\alpha=1}^{\Theta} \overline{\mu}_{M_{i\alpha}}(f_\alpha(x(k))) \geq 0$, $0 \leq \underline{\phi}_i(x(k)) \leq \overline{\phi}_i(x(k)) \leq 1$, $0 \leq \underline{\mu}_{M_{i\alpha}}(f_\alpha(x(k))) \leq \overline{\mu}_{M_{i\alpha}}(f_\alpha(x(k))) \leq 1$, $\underline{\mu}_{M_{i\alpha}}(f_\alpha(x(k)))$ and $\overline{\mu}_{M_{i\alpha}}(f_\alpha(x(k)))$ are the LMFs and UMFs, respectively. $\underline{\phi}_i(x(k))$ and $\overline{\phi}_i(x(k))$ are the lower and upper grade of membership, respectively.

Then, considering the premise variable of the controller with c fuzzy rules, which is under imperfect premise matching. Let $N_{j\beta}$ denote an IT2 fuzzy set of jth rule for $j = 1, 2, \ldots, c$ and $\beta = 1, 2, \ldots, \Omega$ (Ω is a positive integer). Define $g_\beta(x(k))$ as

© Springer Science+Business Media Singapore 2016
H. Li et al., *Analysis and Synthesis for Interval Type-2
Fuzzy-Model-Based Systems*, DOI 10.1007/978-981-10-0593-0_12

the measurable premise variable. Then the firing strength of the jth rule corresponds to the interval sets $\Psi_j(x(k)) = \left[\underline{\psi}_j(x(k)), \overline{\psi}_j(x(k))\right]$, where

$$\begin{aligned}
\underline{\psi}_j(x(k)) &= \prod_{\beta=1}^{\Omega} \underline{\mu}_{N_{j\beta}}\left(g_\beta(x(k))\right) \geq 0, \\
\overline{\psi}_j(x(k)) &= \prod_{\beta=1}^{\Omega} \overline{\mu}_{N_{j\beta}}\left(g_\beta(x(k))\right) \geq 0, \\
0 \leq \underline{\psi}_i&(x(k)) \leq \overline{\psi}_i(x(k)) \leq 1, \\
0 \leq \underline{\mu}_{N_{j\beta}}&\left(g_\beta(x(k))\right) \leq \overline{\mu}_{N_{j\beta}}\left(g_\beta(x(k))\right) \leq 1,
\end{aligned}$$

and $\underline{\mu}_{N_{j\beta}}\left(g_\beta(x(k))\right)$ and $\overline{\mu}_{N_{j\beta}}\left(g_\beta(x(k))\right)$ are the LMFs and UMFs, respectively. $\underline{\psi}_j(x(k))$ and $\overline{\psi}_j(x(k))$ are the lower and upper grade of membership, respectively. Based on the IT2 fuzzy sets introduced above, a p-rule discrete-time IT2 T–S fuzzy model [94] for describing a nonlinear plant is of the following form:

♦ **Plant Form**:

Plant Rule i: IF $f_1(x(k))$ is $M_{i1}, \ldots,$ and $f_\alpha(x(k))$ is $M_{i\alpha}, \ldots,$ and $f_\Theta(x(k))$ is $M_{i\Theta}$, THEN

$$\left\{\begin{aligned}
x(k+1) &= A_i x(k) + B_i u(k) + B_{wi} w(k), \\
y(k) &= C_i x(k) + D_{wi} w(k), \\
z(k) &= E_i x(k) + G_i u(k) + G_{wi} w(k),
\end{aligned}\right. \tag{12.1}$$

where $x(k) \in \mathbf{R}^n$ denotes the system state variable, $y(k) \in \mathbf{R}^s$ denotes the measured output, $z(k) \in \mathbf{R}^m$ denotes the controlled output, $u(k) \in \mathbf{R}^q$ is the control input, $w(k) \in \mathbf{R}^r$ is assumed to be an exogenous disturbance belonging to $\ell_2[0, \infty)$. The vector-valued initial function is defined as $\chi(k)$. A_i, B_i, B_{wi}, C_i, D_{wi}, E_i, G_i and G_{wi} are known appropriate dimensioned system matrices. Utilizing the bounds of the membership function from Preliminaries, the discrete-time IT2 T–S fuzzy system in (12.1) can be formulated as:

$$\left\{\begin{aligned}
x(k+1) &= \sum_{i=1}^{p} \phi_i(x(k))\left(A_i x(k) + B_i u(k) + B_{wi} w(k)\right), \\
y(k) &= \sum_{i=1}^{p} \phi_i(x(k))\left(C_i x(k) + D_{wi} w(k)\right), \\
z(k) &= \sum_{i=1}^{p} \phi_i(x(k))\left(E_i x(k) + G_i u(k) + G_{wi} w(k)\right),
\end{aligned}\right. \tag{12.2}$$

where for $i = 1, 2, \ldots, p$,

$$\phi_i (x (k)) = \underline{\alpha}_i (x (k)) \underline{\phi}_i (x (k)) + \overline{\alpha}_i (x (k)) \overline{\phi}_i (x (k)) \geq 0,$$

and $\left\{\underline{\alpha}_i (x (k)), \overline{\alpha}_i (x (k))\right\} \in [0, 1]$, $\underline{\alpha}_i (x (k)) + \overline{\alpha}_i (x (k)) = 1$. ($\underline{\alpha}_i (x (k))$ and $\overline{\alpha}_i (x (k))$ denote existent nonlinear weighting functions that are not necessary to be known in real applications); $\phi_i (x (k))$ is the grade of membership of the embedded membership function.

The IT2 fuzzy static output-feedback controller with c rules for the system (12.2) is of the following form:

♦ **Controller Form**:

Rule j: IF $g_1 (x (k))$ is $N_{j1}, \ldots,$ and $g_\beta (x (k))$ is $N_{j\beta}, \ldots,$ and $g_\Omega (x (k))$ is $N_{j\Omega}$, THEN

$$u (k) = K_j y^F (k), \tag{12.3}$$

where $K_j \in \mathbf{R}^{q\times s}$ is the feedback gain matrix to be determined. The IT2 T–S fuzzy controller in (12.3) can be defined as:

$$u (k) = \sum_{j=1}^{c} \psi_j (x (k)) K_j y^F (k), \tag{12.4}$$

where for $j = 1, 2, \ldots, c$,

$$\psi_j (x (k)) = \frac{\underline{\beta}_j (x (k)) \underline{\psi}_j (x (k)) + \overline{\beta}_j (x (k)) \overline{\psi}_j (x (k))}{\sum_{\kappa=1}^{c} \left(\underline{\beta}_\kappa (x (k)) \underline{\psi}_\kappa (x (k)) + \overline{\beta}_\kappa (x (k)) \overline{\psi}_\kappa (x (k))\right)} \geq 0,$$

in which $\underline{\beta}_j (x (k))$ and $\overline{\beta}_j (x (k))$ are predefined functions satisfying

$$\left\{\underline{\beta}_j (x (k)), \overline{\beta}_j (x (k))\right\} \in [0, 1], \quad \underline{\beta}_j (x (k)) + \overline{\beta}_j (x (k)) = 1,$$

and $\psi_j (x (k))$ is the grade of membership of the embedded membership function. From the details in (12.2) and (12.4), we have

$$\sum_{i=1}^{p} \phi_i (x (k)) = \sum_{j=1}^{c} \psi_j (x (k)) = \sum_{i=1}^{p} \sum_{j=1}^{c} \phi_i (x (k)) \psi_j (x (k)) = 1.$$

We adopt the following model of sensor failure from [51]:

$$y^F (k) = \zeta y (k), \tag{12.5}$$

where $\zeta = \text{diag}\{\zeta_1, \zeta_2, \ldots, \zeta_s\}$ and $0 \leq \underline{\zeta}_\iota \leq \zeta_\iota \leq \overline{\zeta}_\iota \leq 1$ $(\iota = 1, 2, \ldots, s)$. The variables ζ_ι quantify the failures of the sensor.

Remark 12.1 In the above model of sensor failure, there exist three cases of the feedback signal in sensor. When $\underline{\zeta}_\iota = 1$, it corresponds to the normal case $y^F(k) = y(k)$. When $\overline{\zeta}_\iota = 0$, it covers the outage case [190]. When $\underline{\zeta}_\iota \neq 0$ and $\overline{\zeta}_\iota \neq 1$, it corresponds to the partial failure case.

In order to design the reliable controller, let

$$\begin{aligned}
&\underline{\zeta} = \text{diag}\left\{\underline{\zeta}_1, \underline{\zeta}_2, \ldots, \underline{\zeta}_s\right\}, \quad \overline{\zeta} = \text{diag}\left\{\overline{\zeta}_1, \overline{\zeta}_2, \ldots, \overline{\zeta}_s\right\}, \\
&\hat{\lambda} = \text{diag}\left\{\hat{\lambda}_1, \hat{\lambda}_2, \ldots, \hat{\lambda}_s\right\}, \quad \check{\lambda} = \text{diag}\left\{\check{\lambda}_1, \check{\lambda}_2, \ldots, \check{\lambda}_s\right\}, \\
&\bar{\lambda} = \text{diag}\left\{\bar{\lambda}_1, \bar{\lambda}_2, \ldots, \bar{\lambda}_s\right\},
\end{aligned}$$

where $\hat{\lambda}_\iota = \left(\underline{\zeta}_\iota + \overline{\zeta}_\iota\right)/2$, $\check{\lambda}_\iota = \left(\overline{\zeta}_\iota - \underline{\zeta}_\iota\right)/2$. Thus, one can obtain

$$\zeta = \hat{\lambda} + \bar{\lambda}, \quad |\bar{\lambda}| \leq \check{\lambda}.$$

Hence, it follows from (12.2), (12.4) and (12.5) that the closed-loop IT2 FMB control system is represented as:

$$\begin{cases}
x(k+1) = \sum_{i=1}^{p}\sum_{j=1}^{c} \phi_i(x(k))\,\psi_j(x(k))\left[\bar{A}_{ij}x(k) + \bar{B}_{wij}w(k)\right], \\
z(k) \quad = \sum_{i=1}^{p}\sum_{j=1}^{c} \phi_i(x(k))\,\psi_j(x(k))\left[\bar{E}_{ij}x(k) + \bar{G}_{wij}w(k)\right],
\end{cases} \tag{12.6}$$

where

$$\bar{A}_{ij} = A_i + B_i K_j \zeta C_i, \quad \bar{B}_{wij} = B_{wi} + B_i K_j \zeta D_{wi}, \tag{12.7}$$

$$\bar{E}_{ij} = E_i + G_i K_j \zeta C_i, \quad \bar{G}_{wij} = G_{wi} + G_i K_j \zeta D_{wi}. \tag{12.8}$$

In addition, to consider the performances of the system in (12.6), we introduce the following definitions:

Definition 12.2 Considering the disturbance-free system ($w(k) \equiv 0$) in (12.6), the corresponding H_2 performance cost function is defined as

$$J_2 = \sum_{k=0}^{\infty} z^T(k)\,z(k). \tag{12.9}$$

Definition 12.3 Considering the system with disturbance input in (12.6), if the output $z(k)$ of system (12.6) and a prescribed level of disturbance attenuation $\gamma > 0$ under the zero initial condition satisfy

$$\|z\|_2 < \gamma \|w\|_2, \quad \forall 0 \neq w \in \ell_2[0, \infty), \tag{12.10}$$

in which

$$\|z\|_2 = \sqrt{\sum_{k=0}^{\infty} z^T(k) z(k)},$$

then system (12.6) is said to be with γ-disturbance attenuation.

Definition 12.4 The IT2 fuzzy controller in (12.4) is said to be a reliable mixed H_2/H_∞ fuzzy static output-feedback controller for IT2 FMB system (12.2) if the closed-loop system (12.2) is reliable stable and satisfies the Definitions in 12.2 and 12.3.

In this work, we consider two cases of sensor failure matrix ζ, namely, the known failure and the unknown failure. The primary aim of this study is to design a fuzzy static output-feedback controller in the form of (12.4) under the two cases of ζ such that the closed-loop system with sensor failure in (12.6) is asymptotically stable and has the mixed H_2/H_∞ performance for all $\iota = 1, 2, \ldots, s$.

12.2 System Transformation

This section mainly processes the system transformation from the closed-loop system in (12.6) for the control design objective. The FOU [93] and the state space of interest in model system are both considered for system transformation.

To deal with the parameter uncertainties in closed-loop system (12.6), we use the reconstructed membership functions expressed by the LMFs and UMFs to transform the system model for further analysis. Moreover, the state space of interest is considered for less conservativeness. Concretely, according to [94], the state space of interest and the FOU are both divided for the further stability analysis of the IT2 FMB control system in (12.6).

1. The state space Θ is partitioned θ connected sub-state spaces denoted as Θ_τ ($\tau = 1, 2, \ldots, \theta$), such that $\Theta = \bigcup_{\tau=1}^{\theta} \Theta_\tau$.
2. The FOU is divided into $\vartheta + 1$ sub-FOUs. For $\upsilon = 1, 2, \ldots, \vartheta + 1$, the LMFs and UMFs in the υth sub-FOU are defined as follows for $\forall i, j, \upsilon, \tau$:

$$\underline{h}_{ijv}(x(k)) = \sum_{\tau=1}^{\theta}\sum_{i_1=1}^{2}\sum_{i_2=1}^{2}\cdots\sum_{i_n=1}^{2}\prod_{a=1}^{n}\underline{\epsilon}_{iji_1i_2\ldots i_nv\tau}\varrho_{ai_av\tau}(x_a(k)), \tag{12.11}$$

$$\overline{h}_{ijv}(x(k)) = \sum_{\tau=1}^{\theta}\sum_{i_1=1}^{2}\sum_{i_2=1}^{2}\cdots\sum_{i_n=1}^{2}\prod_{a=1}^{n}\overline{\epsilon}_{iji_1i_2\ldots i_nv\tau}\varrho_{ai_av\tau}(x_a(k)), \tag{12.12}$$

where $\underline{\epsilon}_{iji_1i_2\ldots i_nv\tau}$ and $\overline{\epsilon}_{iji_1i_2\ldots i_nv\tau}$ are constant scalars to be designed, and $0 \leq \underline{\epsilon}_{iji_1i_2\ldots i_nv\tau} \leq \overline{\epsilon}_{iji_1i_2\ldots i_nv\tau} \leq 1$. For $x(k) \in \Theta_\tau$, $\tau = 1, 2, \ldots, \theta$, and $a, b = 1, 2, \ldots, n$, it holds that $0 \leq \varrho_{ai_bv\tau}(x_a(k)) \leq 1$ and $\varrho_{a1v\tau}(x_a(k)) + \varrho_{a2v\tau}(x_a(k)) = 1$ ($i_a, i_b = 1, 2$); and $\varrho_{ai_bv\tau}(x_a(k)) = 0$ if else. Then, it follows that for $v = 1, 2, \ldots, \vartheta + 1$,

$$\sum_{\tau=1}^{\theta}\sum_{i_1=1}^{2}\sum_{i_2=1}^{2}\cdots\sum_{i_n=1}^{2}\prod_{a=1}^{n}\varrho_{ai_av\tau}(x_a(k)) = 1. \tag{12.13}$$

Hence, for the stability analysis of the considered system in next section, we rewrite the IT2 fuzzy system in (12.6) as follows:

$$\begin{cases} x(k+1) = \sum_{i=1}^{p}\sum_{j=1}^{c} h_{ij}(x(k))\left[\bar{A}_{ij}x(k) + \bar{B}_{wij}w(k)\right], \\ z(k) \quad = \sum_{i=1}^{p}\sum_{j=1}^{c} h_{ij}(x(k))\left[\bar{E}_{ij}x(k) + \bar{G}_{wij}w(k)\right], \end{cases} \tag{12.14}$$

where

$$\begin{aligned} h_{ij}(x(k)) &= \phi_i(x(k))\psi_j(x(k)) \\ &= \sum_{v=1}^{\vartheta+1}\sigma_{ijv}(x(k))\left(\underline{\rho}_{ijv}(x(k))\underline{h}_{ijv}(x(k)) + \overline{\rho}_{ijv}(x(k))\overline{h}_{ijv}(x(k))\right), \end{aligned}$$

and $\sum_{i=1}^{p}\sum_{j=1}^{c} h_{ij}(x(k)) = 1$. The following two functions

$$0 \leq \underline{\rho}_{ijv}(x(k)) \leq \overline{\rho}_{ijv}(x(k)) \leq 1$$

satisfy that $\underline{\rho}_{ijv}(x(k)) + \overline{\rho}_{ijv}(x(k)) = 1$, which are not necessary to be known. $\sigma_{ijv}(x(k)) = 1$ if the membership function $h_{ijv}(x(k))$ is within the vth sub-FOU. Otherwise, $\sigma_{ijv}(x(k)) = 0$.

Based on the transformed system in (12.14), the stability analysis and controller synthesis can be tackled without the implementation of the IT2 T–S fuzzy model

(12.2). Moreover, we give the property

$$\sum_{i=1}^{p} \phi_i (x (k)) = \sum_{j=1}^{c} \psi_j (x (k)) = \sum_{i=1}^{p} \sum_{j=1}^{c} \phi_i (x (k)) \psi_j (x (k)) = \sum_{i=1}^{p} \sum_{j=1}^{c} h_{ij} (x (k)) = 1$$

for further study.

In next section, the reliable control scheme for nonlinear systems based on the transformed system (12.14) is provided, which means the reliable stability with mixed H_2/H_∞ performance can be achieved for the closed-loop system in (12.6).

12.3 Main Results

In this section, the reliable stability analysis and controller design under known sensor failure and unknown sensor failure are presented for IT2 FMB systems with mixed H_2/H_∞ performance. By applying Lyapunov stability theory, a sufficient criterion of reliable stability is derived for system (12.14). Based on the criterion, two reliable mixed H_2/H_∞ fuzzy static output-feedback controllers are designed.

12.3.1 Stability Analysis

On the basis of the transformed system in (12.14), considering the H_2 performance in (12.9) and H_∞ performance in (12.10), a sufficient condition of reliable stability with mixed H_2/H_∞ performance is given for the closed-loop system (12.14) in the following theorem:

Theorem 12.5 *Considering the system with sensor failure in (12.14), for a given scalar $\gamma > 0$, system (12.14) is reliable stable and has an H_∞ performance index γ, if there exist symmetric matrices $P > 0$, $R_{ij\upsilon} > 0$, $X_{ij\upsilon} > 0$, $Y_{ij\upsilon} > 0$, $U_{ij\upsilon} > 0$, $V_{ij\upsilon} > 0$, $(i = 1, 2, \ldots, p,\ j = 1, 2, \ldots, c,\ \upsilon = 1, 2, \ldots, \vartheta + 1)$, and S with appropriate dimensions satisfying the following inequalities for $i = 1, 2, \ldots, p$, $j = 1, 2, \ldots, c$, $\upsilon = 1, 2, \ldots, \vartheta + 1$, $\tau = 1, 2, \ldots, \theta$:*

$$\sum_{i=1}^{p} \sum_{j=1}^{c} \Xi_{ij} - S < 0, \quad \forall i_1, i_2, \ldots, i_n, \upsilon, \tau, \tag{12.15}$$

$$\Sigma_{2ij} + R_{ij\upsilon} + S > 0, \quad \forall i, j, \upsilon, \tag{12.16}$$

where

$$
\begin{aligned}
\Xi_{ij} &= \bar{\epsilon}_{iji_1i_2\ldots i_n\upsilon T}\Sigma_{1ij} + \left(\bar{\epsilon}_{iji_1i_2\ldots i_n\upsilon T} - \underline{\epsilon}_{iji_1i_2\ldots i_n\upsilon T}\right) R_{ij\upsilon} + \bar{\epsilon}_{iji_1i_2\ldots i_n\upsilon T} S,\\
\Sigma_{1ij} &= \begin{bmatrix} \Sigma_{11ij} & \bar{A}_{ij}^T \left(P + U_{ij\upsilon}\right) \bar{B}_{wij} + \bar{E}_{ij}^T \left(I + V_{ij\upsilon}\right) \bar{G}_{wij} \\ * & \bar{B}_{wij}^T \left(P + U_{ij\upsilon}\right) \bar{B}_{wij} + \bar{G}_{wij}^T \left(I + V_{ij\upsilon}\right) \bar{G}_{wij} - \gamma^2 I \end{bmatrix},\\
\Sigma_{2ij} &= \begin{bmatrix} \Sigma_{21ij} & \bar{A}_{ij}^T \left(P - X_{ij\upsilon}\right) \bar{B}_{wij} + \bar{E}_{ij}^T \left(I - Y_{ij\upsilon}\right) \bar{G}_{wij} \\ * & \bar{B}_{wij}^T \left(P - X_{ij\upsilon}\right) \bar{B}_{wij} + \bar{G}_{wij}^T \left(I - Y_{ij\upsilon}\right) \bar{G}_{wij} - \gamma^2 I \end{bmatrix},\\
\Sigma_{11ij} &= -P + \bar{A}_{ij}^T \left(P + U_{ij\upsilon}\right) \bar{A}_{ij} + \bar{E}_{ij}^T \left(I + V_{ij\upsilon}\right) \bar{E}_{ij},\\
\Sigma_{21ij} &= -P + \bar{A}_{ij}^T \left(P - X_{ij\upsilon}\right) \bar{A}_{ij} + \bar{E}_{ij}^T \left(I - Y_{ij\upsilon}\right) \bar{E}_{ij}.
\end{aligned}
$$

If the above conditions have a feasible solution, then the bound of H_2 performance cost function in (12.9) is determined by

$$
J_2^* = x_0^T P x_0, \tag{12.17}
$$

where x_0 is the initial state.

Proof Based on the closed-loop system with disturbance input in (12.14), considering the Lyapunov function $V(x(k)) = x^T(k) P x(k)$, applying Lemma 1.7 and introducing some slack matrices (S is an arbitrary symmetric matrix, and symmetric matrices $R_{ij\upsilon} > 0$, $U_{ij\upsilon} > 0$, and $X_{ij\upsilon} > 0$ with appropriate dimensions) based on the *S-procedure* [17], we have

$$
\begin{aligned}
\Delta V(x(k)) \le{}& \sum_{i=1}^{p}\sum_{j=1}^{c}\sum_{\upsilon=1}^{\vartheta+1} \sigma_{ij\upsilon}\left(\underline{\rho}_{ij\upsilon}\underline{h}_{ij\upsilon} + \bar{\rho}_{ij\upsilon}\bar{h}_{ij\upsilon}\right)\eta^T(k)\,\Sigma_{0ij}\eta(k)\\
&+ \left[\sum_{i=1}^{p}\sum_{j=1}^{c}\sum_{\upsilon=1}^{\vartheta+1} \sigma_{ij\upsilon}\left(\underline{\rho}_{ij\upsilon}\underline{h}_{ij\upsilon} + \bar{\rho}_{ij\upsilon}\bar{h}_{ij\upsilon}\right) - 1\right]\eta^T(k)\,S\eta(k)\\
&- \sum_{i=1}^{p}\sum_{j=1}^{c}\sum_{\upsilon=1}^{\vartheta+1} \sigma_{ij\upsilon}\left(1 - \underline{\rho}_{ij\upsilon}\right)\left(\underline{h}_{ij\upsilon} - \bar{h}_{ij\upsilon}\right)\eta^T(k)\,R_{ij\upsilon}\eta(k)\\
&+ \sum_{i=1}^{p}\sum_{j=1}^{c}\sum_{\upsilon=1}^{\vartheta+1} \sigma_{ij\upsilon}\bar{h}_{ij\upsilon}x^T(k+1)\,U_{ij\upsilon}x(k+1)\\
&- \sum_{i=1}^{p}\sum_{j=1}^{c}\sum_{\upsilon=1}^{\vartheta+1} \sigma_{ij\upsilon}\underline{\rho}_{ij\upsilon}\left(\underline{h}_{ij\upsilon} - \bar{h}_{ij\upsilon}\right)x^T(k+1)\,X_{ij\upsilon}x(k+1)
\end{aligned}
$$

$$
\begin{aligned}
&= \eta^T(k) \left\{ \sum_{i=1}^{p} \sum_{j=1}^{c} \sum_{\upsilon=1}^{\vartheta+1} \sigma_{ij\upsilon} \left[\overline{h}_{ij\upsilon} \tilde{\Sigma}_{1ij\upsilon} - \left(\underline{h}_{ij\upsilon} - \overline{h}_{ij\upsilon} \right) R_{ij\upsilon} + \overline{h}_{ij\upsilon} S \right] \right. \\
&\quad \left. - S \right\} \eta(k) + \eta^T(k) \sum_{i=1}^{p} \sum_{j=1}^{c} \sum_{\upsilon=1}^{\vartheta+1} \sigma_{ij\upsilon} \underline{\rho}_{ij\upsilon} \\
&\quad \times \left(\underline{h}_{ij\upsilon} - \overline{h}_{ij\upsilon} \right) \left(\tilde{\Sigma}_{2ij\upsilon} + R_{ij\upsilon} + S \right) \eta(k), \qquad (12.18)
\end{aligned}
$$

where $\eta(k) = \left[x^T(k) \ w^T(k) \right]^T$, and

$$
\begin{aligned}
\Sigma_{0ij} &= \begin{bmatrix} -P + \bar{A}_{ij}^T P \bar{A}_{ij} & \bar{A}_{ij}^T P \bar{B}_{wij} \\ * & \bar{B}_{wij}^T P \bar{B}_{wij} \end{bmatrix}, \\
\tilde{\Sigma}_{1ij} &= \begin{bmatrix} -P + \bar{A}_{ij}^T \left(P + U_{ij\upsilon} \right) \bar{A}_{ij} & \bar{A}_{ij}^T \left(P + U_{ij\upsilon} \right) \bar{B}_{wij} \\ * & \bar{B}_{wij}^T \left(P + U_{ij\upsilon} \right) \bar{B}_{wij} \end{bmatrix}, \\
\tilde{\Sigma}_{2ij} &= \begin{bmatrix} -P + \bar{A}_{ij}^T \left(P - X_{ij\upsilon} \right) \bar{A}_{ij} & \bar{A}_{ij}^T \left(P - X_{ij\upsilon} \right) \bar{B}_{wij} \\ * & \bar{B}_{wij}^T \left(P - X_{ij\upsilon} \right) \bar{B}_{wij} \end{bmatrix}.
\end{aligned}
$$

Firstly, considering the H_∞ performance in (12.10) and introducing some slack matrices (symmetric matrices $V_{ij\upsilon} > 0$ and $Y_{ij\upsilon} > 0$ with appropriate dimensions), under the zero initial condition, we have

$$
\begin{aligned}
J_\infty &= \sum_{k=0}^{\infty} \left[z^T(k) z(k) - \gamma^2 w^T(k) w(k) + \Delta V(x(k)) \right] \\
&\le \sum_{k=0}^{\infty} \left\{ \sum_{i=1}^{p} \sum_{j=1}^{c} \sum_{\upsilon=1}^{\vartheta+1} \sigma_{ij\upsilon} \left(\underline{\rho}_{ij\upsilon} \underline{h}_{ij\upsilon} + \overline{\rho}_{ij\upsilon} \overline{h}_{ij\upsilon} \right) \eta^T(k) \begin{bmatrix} \bar{E}_{ij}^T E_{ij} & \bar{E}_{ij}^T \bar{G}_{wij} \\ * & \bar{G}_{wij}^T \bar{G}_{wij} \end{bmatrix} \right. \\
&\quad \times \eta(k) + \Delta V(x(k)) + \sum_{i=1}^{p} \sum_{j=1}^{c} \sum_{\upsilon=1}^{\vartheta+1} \sigma_{ij\upsilon} \overline{h}_{ij\upsilon} z^T(k) V_{ij\upsilon} z(k) \\
&\quad \left. - \sum_{i=1}^{p} \sum_{j=1}^{c} \sum_{\upsilon=1}^{\vartheta+1} \sigma_{ij\upsilon} \underline{\rho}_{ij\upsilon} \left(\underline{h}_{ij\upsilon} - \overline{h}_{ij\upsilon} \right) z^T(k) Y_{ij\upsilon} z(k) \right\} \\
&= \sum_{k=0}^{\infty} \eta^T(k) \left\{ \sum_{i=1}^{p} \sum_{j=1}^{c} \sum_{\upsilon=1}^{\vartheta+1} \sigma_{ij\upsilon} \left[\overline{h}_{ij\upsilon} \Sigma_{1ij\upsilon} - \left(\underline{h}_{ij\upsilon} - \overline{h}_{ij\upsilon} \right) R_{ij\upsilon} + \overline{h}_{ij\upsilon} S \right] \right. \\
&\quad \left. - S \right\} \eta(k) + \sum_{k=0}^{\infty} \eta^T(k) \sum_{i=1}^{p} \sum_{j=1}^{c} \sum_{\upsilon=1}^{\vartheta+1} \sigma_{ij\upsilon} \underline{\rho}_{ij\upsilon} \left(\underline{h}_{ij\upsilon} - \overline{h}_{ij\upsilon} \right) \\
&\quad \times \left(\Sigma_{2ij\upsilon} + R_{ij\upsilon} + S \right) \eta(k). \qquad (12.19)
\end{aligned}
$$

Obviously, $J_\infty < 0$ in (12.19) can be obtained from the following two inequalities: $\Xi_{2ijv} + R_{ijv} + Q > 0$ (which is guaranteed by the condition in (12.16)), and

$$\sum_{i=1}^{p}\sum_{j=1}^{c}\sum_{v=1}^{\vartheta+1}\sigma_{ijv}\left(x\left(k\right)\right)\left[\overline{h}_{ijv}\Sigma_{1ijv} - \left(\underline{h}_{ijv} - \overline{h}_{ijv}\right)R_{ijv} + \overline{h}_{ijv}S\right] - S < 0. \tag{12.20}$$

Noticing that only one $\sigma_{ijv}\left(x\left(k\right)\right) = 1$ for each fixed value of i and j at any time instant and $\sum_{v=1}^{\vartheta+1}\sigma_{ijv}\left(x\left(k\right)\right) = 1$. The inequality in (12.20) is satisfied by the following inequality:

$$\sum_{i=1}^{p}\sum_{j=1}^{c}\left[\overline{h}_{ijv}\Sigma_{1ijv} - \left(\underline{h}_{ijv} - \overline{h}_{ijv}\right)R_{ijv} + \overline{h}_{ijv}S\right] - S < 0. \tag{12.21}$$

Considering $\underline{h}_{ijv}$ in (12.11), $\overline{h}_{ijv}$ in (12.12), and the equation in (12.13), we can obtain the following inequality, which is equivalent to the inequality in (12.21),

$$\sum_{\tau=1}^{\theta}\sum_{i_1=1}^{2}\sum_{i_2=1}^{2}\cdots\sum_{i_n=1}^{2}\prod_{a=1}^{n}\varrho_{ai_a v\tau}\left(x_a\left(k\right)\right)\left(\sum_{i=1}^{p}\sum_{j=1}^{c}\Xi_{ij} - S\right) < 0. \tag{12.22}$$

Thus, the inequality in (12.22) is satisfied by the condition in (12.15). Therefore, $\|z\|_2 < \gamma\|w\|_2$ as $J_\infty < 0$, which means for all nonzero $w\left(k\right) \in \ell_2\left[0, \infty\right)$, the conditions in Theorem 12.5 can guarantee that the system in (12.14) is asymptotically stable with an H_∞ performance index γ.

Moreover, under the disturbance-free case, it can be easily obtained $\Delta V\left(x\left(k\right)\right) < 0$ from (12.18), which means the system in (12.14) is asymptotically stable. Then, considering the H_2 performance cost function in (12.9) and the inequality in (12.19), we have

$$\begin{aligned} J_2 = \sum_{k=0}^{\infty} z^T\left(k\right)z\left(k\right) &\leq -\sum_{k=0}^{\infty}\Delta V\left(x\left(k\right)\right) = V\left(x(0)\right) - V\left(x(\infty)\right) \\ &\leq V\left(x(0)\right) = x^T\left(0\right)Px\left(0\right) = J_2^*. \end{aligned}$$

The proof is completed. □

12.3.2 Output-Feedback Control

In this subsection, the reliable IT2 fuzzy controller is designed based on the conditions in Theorem 12.5. The failure parameter of the sensor is considered with two cases, in which the sensor failure parameter matrix is known or unknown. The controller design results of two cases are given in the following part.

First case: Reliable controller design under known sensor failure parameter.

Firstly, assume that the sensor failure parameter matrix is known, the reliable mixed IT2 fuzzy controller is designed in the following theorem.

Theorem 12.6 *Considering the system with sensor failure in (12.14), for a given sensor failure diagonal matrix ζ and a scalar $\gamma > 0$, system (12.14) is reliable stable and has an H_∞ performance index γ, if there exist symmetric matrices $P > 0$, $R_{1ijv} > 0$, $R_{2ijv} > 0$, $R_{3ijv} > 0$, $X_{ijv} > 0$, $Y_{ijv} > 0$, $U_{ijv} > 0$, $V_{ijv} > 0$, $(i = 1, 2, \ldots, p,\ j = 1, 2, \ldots, c,\ v = 1, 2, \ldots, \vartheta + 1)$, S_1, S_3, and arbitrary matrix S_2 with appropriate dimensions, such that the following LMIs hold for $i = 1, 2, \ldots, p$, $j = 1, 2, \ldots, c$, $v = 1, 2, \ldots, \vartheta + 1$, $\tau = 1, 2, \ldots, \theta$:*

$$\begin{bmatrix} \hat{\Gamma} & \hat{\epsilon}\bar{\Pi}_1 \\ * & \Lambda_1 \end{bmatrix} < 0, \quad \forall i, j, i_1, i_2, \ldots, i_n, v, \tau, \tag{12.23}$$

$$\begin{bmatrix} \check{\Gamma} & \bar{\Pi}_1 \\ * & \Lambda_2 \end{bmatrix} > 0, \quad \forall i, j, v, \tag{12.24}$$

$$P - X_{ijv} < 0, \quad \forall i, j, v, \tag{12.25}$$

$$I - Y_{ijv} < 0, \quad \forall i, j, v, \tag{12.26}$$

where

$$\hat{\Gamma} = \begin{bmatrix} \hat{\Gamma}_1 & \tilde{\epsilon}R_{2ijv} + \breve{\epsilon}S_2 \\ * & \hat{\Gamma}_2 \end{bmatrix}, \quad \check{\Gamma} = \begin{bmatrix} \check{\Gamma}_1 & R_{1ijv} + S_1 \\ * & \check{\Gamma}_2 \end{bmatrix}, \quad \bar{\Pi}_1 = \begin{bmatrix} \bar{A}_{ij}^T & \bar{E}_{ij}^T \\ \bar{B}_{wij}^T & \bar{G}_{wij}^T \end{bmatrix},$$

$$\Lambda_1 = \text{diag}\left\{U_{ijv} + P - 2I, U_{ijv} - I\right\}, \quad \hat{\Gamma}_1 = -\hat{\epsilon}^2 P + \tilde{\epsilon}R_{1ijv} + \breve{\epsilon}S_1,$$

$$\Lambda_2 = \text{diag}\left\{X_{ijv} - P - 2I, Y_{ijv} - 3I\right\}, \quad \hat{\Gamma}_2 = -\hat{\epsilon}^2\gamma^2 I + \tilde{\epsilon}R_{3ijv} + \breve{\epsilon}S_3,$$

$$\check{\Gamma}_1 = -P + R_{1ijv} + S_1, \quad \check{\Gamma}_2 = -\gamma^2 I + R_{2ijv} + S_2, \quad \hat{\epsilon} = \sqrt{\bar{\epsilon}_{iji_1i_2\ldots i_n v\tau}},$$

$$\tilde{\epsilon} = \bar{\epsilon}_{iji_1i_2\ldots i_n v\tau} - \underline{\epsilon}_{iji_1i_2\ldots i_n v\tau}, \quad \breve{\epsilon} = \bar{\epsilon}_{iji_1i_2\ldots i_n v\tau} - \frac{1}{pc},$$

and $\bar{A}_{ij}$, $\bar{B}_{wij}$, $\bar{E}_{ij}$ and $\bar{G}_{wij}$ are defined in (12.7) and (12.8). If the above conditions have a feasible solution, then the matrices K_j for the desired controller in the form of (12.3) can be obtained from the solution. Moreover, the H_2 performance cost function bound is determined by

$$J_2^* = x_0^T P x_0.$$

Proof For $U_{ijv} + P > 0$, the following inequality holds:

$$\left(U_{ijv} + P - I\right)^T \left(U_{ijv} + P\right)^{-1} \left(U_{ijv} + P - I\right) \geq 0.$$

Thus, we have

$$-\left(U_{ijv}+P\right)^{-1} \leq U_{ijv}+P-2I. \tag{12.27}$$

Similarly, for $V_{ijv}+I>0$, we have

$$-\left(V_{ijv}+I\right)^{-1} \leq V_{ijv}-3I. \tag{12.28}$$

Therefore, from the condition in (12.23), the following inequality holds:

$$\forall i_1, i_2, \ldots, i_n, i, j, v, \tau,$$

$$\begin{bmatrix} \hat{\Gamma}_1 \; \breve{\epsilon} R_{2ijv}+\breve{\epsilon} S_2 & \hat{\epsilon}\bar{A}_{ij}^T & \hat{\epsilon}\bar{E}_{ij}^T \\ * \quad\quad \hat{\Gamma}_2 & \hat{\epsilon}\bar{B}_{wij}^T & \hat{\epsilon}\bar{G}_{wij}^T \\ * \quad\quad * & -\left(U_{ijv}+P\right)^{-1} & 0 \\ * \quad\quad * & * & -\left(V_{ijv}+I\right)^{-1} \end{bmatrix} < 0. \tag{12.29}$$

Define the following matrices:

$$R_{ijv}=\begin{bmatrix} R_{1ijv} & R_{2ijv} \\ * & R_{3ijv} \end{bmatrix}, \quad S=\begin{bmatrix} S_1 & S_2 \\ * & S_3 \end{bmatrix}. \tag{12.30}$$

Then, based on the inequality in (12.29), according to Schur complement, one can obtain that the following inequality holds:

$$\Xi_{ij}-\frac{1}{pc}S<0, \quad \forall\, i_1, i_2, \ldots, i_n, i, j, v, \tau,$$

which can derive the condition (12.15) in Theorem 12.5. Similar to the above proof, it can be seen that the condition (12.24) together with the conditions in (12.25) and (12.26) can guarantee the condition (12.15) in Theorem 12.5 holds. The proof is completed. □

Remark 12.7 From Theorem 12.6, under the known sensor failure case, the existence condition of the desired controller in the form of (12.3) is provided, which can guarantee the reliable stability for the closed-loop system in (12.14) with mixed H_2/H_∞ performance. However, in some practical applications, the sensor failure is often unknown, which may destroy the stability of the system unpredictably. Therefore, it is necessary to design a reliable controller to tolerate the unknown sensor failure in the process. In the following part, the desired controller is designed under the sensor failure unknown case.

Second case: Reliable controller design under known sensor failure parameter.

Assuming that the sensor failure parameter matrix is unknown, based on Theorem 12.6, the reliable mixed H_2/H_∞ IT2 controller is designed in the following theorem.

Theorem 12.8 *Considering the system with unknown sensor failure in (12.14), for a given scalar $\gamma > 0$, system (12.14) is reliable stable and satisfies the H_∞ performance index γ, if there exist symmetric matrices $P > 0$, $R_{1ij\upsilon} > 0$, $R_{2ij\upsilon} > 0$, $R_{3ij\upsilon} > 0$, $X_{ij\upsilon} > 0$, $Y_{ij\upsilon} > 0$, $U_{ij\upsilon} > 0$, $V_{ij\upsilon} > 0$, $(i = 1, 2, \ldots, p, \ j = 1, 2, \ldots, c, \ \upsilon = 1, 2, \ldots, \vartheta + 1)$, S_1, S_3, arbitrary matrix S_2 with appropriate dimensions, and a scalar $\delta > 0$ satisfying the following inequalities:*

$$
\begin{bmatrix} \hat{\Gamma} & \hat{\epsilon}\tilde{\Pi}_1 & \hat{\epsilon}\Pi_2 \\ * & \Lambda_1 & \Pi_3 \\ * & * & -\Lambda_0 \end{bmatrix} < 0, \quad \forall i, j, i_1, i_2, \ldots, i_n, \upsilon, \tau, \tag{12.31}
$$

$$
\begin{bmatrix} \check{\Gamma} & \tilde{\Pi}_1 & \Pi_2 \\ * & \Lambda_2 & -\Pi_3 \\ * & * & \Lambda_0 \end{bmatrix} > 0, \quad \forall i, j, \upsilon, \tag{12.32}
$$

$$
P - X_{ij\upsilon} < 0, \quad \forall i, j, \upsilon, \tag{12.33}
$$

$$
I - Y_{ij\upsilon} < 0, \quad \forall i, j, \upsilon, \tag{12.34}
$$

where

$$
\tilde{\Pi}_1 = \begin{bmatrix} \tilde{A}_{ij}^T & \tilde{E}_{ij}^T \\ \tilde{B}_{wij}^T & \tilde{G}_{wij}^T \end{bmatrix}, \quad \Pi_2 = \begin{bmatrix} \delta C_i^T & \delta C_i^T & 0 & 0 \\ \delta D_{wi}^T & \delta D_{wi}^T & 0 & 0 \end{bmatrix},
$$

$$
\Pi_3 = \begin{bmatrix} 0 & 0 & B_i K_j \check{\lambda} & 0 \\ 0 & 0 & 0 & G_i K_j \check{\lambda} \end{bmatrix}, \quad \Lambda_0 = \mathrm{diag}\{\delta I, \delta I, \delta I, \delta I\},
$$

$$
\tilde{A}_{ij} = A_i + B_i K_j \hat{\lambda} C_i, \quad \tilde{B}_{wij} = B_{wi} + B_i K_j \hat{\lambda} D_{wi},
$$

$$
\tilde{E}_{ij} = E_i + G_i K_j \hat{\lambda} C_i, \quad \tilde{G}_{wij} = G_{wi} + G_i K_j \hat{\lambda} D_{wi},
$$

and $\hat{\Gamma}$, $\check{\Gamma}$, Λ_1 and Λ_2 are defined in Theorem 12.6. If the above conditions have a feasible solution, then the controller gain K_j in the form of (12.3) can be obtained from the solution. Moreover, the H_2 performance cost function bound is determined by

$$
J_2^* = x^T(0) P x(0).
$$

Proof For the condition in (12.31), considering a scalar $\delta > 0$ and Lemma 1.8, one can obtain that

$$
\begin{bmatrix} \hat{\Gamma} & \hat{\epsilon}\tilde{\Pi}_1 \\ * & \Lambda_1 \end{bmatrix} = \begin{bmatrix} \hat{\Gamma} & \hat{\epsilon}\tilde{\Pi}_1 \\ * & \Lambda_1 \end{bmatrix} + \begin{bmatrix} 0 & 0 & \hat{\epsilon}\left(B_i K_j \bar{\lambda} C_i\right)^T & \hat{\epsilon}\left(G_i K_j \bar{\lambda} C_i\right)^T \\ * & 0 & \hat{\epsilon}\left(B_i K_j \bar{\lambda} D_{wi}\right)^T & \hat{\epsilon}\left(G_i K_j \bar{\lambda} D_{wi}\right)^T \\ * & * & 0 & 0 \\ * & * & * & 0 \end{bmatrix}
$$

$$
\begin{aligned}
&= \begin{bmatrix} \hat{\Gamma} & \hat{\epsilon}\tilde{\Pi}_1 \\ * & \Lambda_1 \end{bmatrix} + \begin{bmatrix} \hat{\epsilon}C_i^T & \hat{\epsilon}C_i^T \\ \hat{\epsilon}D_{wi}^T & \hat{\epsilon}D_{wi}^T \\ 0 & 0 \\ 0 & 0 \end{bmatrix} \left(\begin{bmatrix} \bar{\lambda} & 0 \\ 0 & \bar{\lambda} \end{bmatrix}^T \begin{bmatrix} 0 & 0 & (B_i K_j)^T & 0 \\ 0 & 0 & 0 & (G_i K_j)^T \end{bmatrix} \right) \\
&+ \left(\begin{bmatrix} 0 & 0 & (B_i K_j)^T & 0 \\ 0 & 0 & 0 & (G_i K_j)^T \end{bmatrix}^T \begin{bmatrix} \bar{\lambda} & 0 \\ 0 & \bar{\lambda} \end{bmatrix} \right) \begin{bmatrix} \hat{\epsilon}C_i^T & \hat{\epsilon}C_i^T \\ \hat{\epsilon}D_{wi}^T & \hat{\epsilon}D_{wi}^T \\ 0 & 0 \\ 0 & 0 \end{bmatrix}^T \\
&\le \begin{bmatrix} \hat{\Gamma} & \hat{\epsilon}\tilde{\Pi}_1 \\ * & \Lambda_1 \end{bmatrix} + \delta \begin{bmatrix} \hat{\epsilon}C_i^T & \hat{\epsilon}C_i^T \\ \hat{\epsilon}D_{wi}^T & \hat{\epsilon}D_{wi}^T \\ 0 & 0 \\ 0 & 0 \end{bmatrix} \begin{bmatrix} \hat{\epsilon}C_i^T & \hat{\epsilon}C_i^T \\ \hat{\epsilon}D_{wi}^T & \hat{\epsilon}D_{wi}^T \\ 0 & 0 \\ 0 & 0 \end{bmatrix}^T \\
&+ \delta^{-1} \left(\begin{bmatrix} 0 & 0 & (B_i K_j)^T & 0 \\ 0 & 0 & 0 & (G_i K_j)^T \end{bmatrix}^T \begin{bmatrix} \check{\lambda} & 0 \\ 0 & \check{\lambda} \end{bmatrix} \right) \\
&\times \left(\begin{bmatrix} \check{\lambda} & 0 \\ 0 & \check{\lambda} \end{bmatrix}^T \begin{bmatrix} 0 & 0 & (B_i K_j)^T & 0 \\ 0 & 0 & 0 & (G_i K_j)^T \end{bmatrix} \right). \qquad (12.35)
\end{aligned}
$$

Thus, according to Schur complement, it can be obtained that (12.29) holds from (12.31), which satisfies (12.23) in Theorem 12.6. Also, by the same approach applying to (12.24), one can obtain that (12.32) satisfies (12.24) in Theorem 12.6.

$$
\begin{aligned}
\begin{bmatrix} \check{\Gamma} & \bar{\Pi}_1 \\ * & \Lambda_2 \end{bmatrix} &= \begin{bmatrix} \check{\Gamma} & \tilde{\Pi}_1 \\ * & \Lambda_2 \end{bmatrix} + \begin{bmatrix} 0 & 0 & (B_i K_j \bar{\lambda} C_i)^T & (G_i K_j \bar{\lambda} C_i)^T \\ * & 0 & (B_i K_j \bar{\lambda} D_{wi})^T & (G_i K_j \bar{\lambda} D_{wi})^T \\ * & * & 0 & 0 \\ * & * & * & 0 \end{bmatrix} \\
&= \begin{bmatrix} \check{\Gamma} & \tilde{\Pi}_1 \\ * & \Lambda_2 \end{bmatrix} - \begin{bmatrix} C_i^T & C_i^T \\ D_{wi}^T & D_{wi}^T \\ 0 & 0 \\ 0 & 0 \end{bmatrix} \left(\begin{bmatrix} -\bar{\lambda} & 0 \\ 0 & -\bar{\lambda} \end{bmatrix}^T \begin{bmatrix} 0 & 0 & (B_i K_j)^T & 0 \\ 0 & 0 & 0 & (G_i K_j)^T \end{bmatrix} \right) \\
&- \left(\begin{bmatrix} 0 & 0 & (B_i K_j)^T & 0 \\ 0 & 0 & 0 & (G_i K_j)^T \end{bmatrix}^T \begin{bmatrix} -\bar{\lambda} & 0 \\ 0 & -\bar{\lambda} \end{bmatrix} \right) \begin{bmatrix} C_i^T & C_i^T \\ D_{wi}^T & D_{wi}^T \\ 0 & 0 \\ 0 & 0 \end{bmatrix}^T \\
&\ge \begin{bmatrix} \check{\Gamma} & \tilde{\Pi}_1 \\ * & \Lambda_2 \end{bmatrix} - \delta \begin{bmatrix} C_i^T & C_i^T \\ D_{wi}^T & D_{wi}^T \\ 0 & 0 \\ 0 & 0 \end{bmatrix} \begin{bmatrix} C_i^T & C_i^T \\ D_{wi}^T & D_{wi}^T \\ 0 & 0 \\ 0 & 0 \end{bmatrix}^T
\end{aligned}
$$

$$-\delta^{-1}\left(\begin{bmatrix} 0\,0\,(B_iK_j)^T & 0 \\ 0\,0 \quad 0 & (G_iK_j)^T \end{bmatrix}^T \begin{bmatrix} -\check{\lambda} & 0 \\ 0 & -\check{\lambda} \end{bmatrix}\right)$$
$$\times\left(\begin{bmatrix} -\check{\lambda} & 0 \\ 0 & -\check{\lambda} \end{bmatrix}^T \begin{bmatrix} 0\,0\,(B_iK_j)^T & 0 \\ 0\,0 \quad 0 & (G_iK_j)^T \end{bmatrix}\right).$$

Therefore, all the conditions in Theorem 12.8 are satisfied by the criteria in Theorem 12.5. This completes the proof. □

Remark 12.9 Theorems 12.6 and 12.8 provide the sufficient conditions for the existence of the reliable mixed H_2/H_∞ fuzzy IT2 controller in the form of (12.4), respectively. When LMIs (12.23)–(12.24) and (12.31)–(12.32) are feasible, each H_2 performance cost function is bounded by J_2^*. Actually, the upper bound of cost function (12.17) depends on the initial state x_0. In [201], x_0 is assumed to be a zero mean random variable satisfying $\mathbf{E}\left\{x_0x_0^T\right\} = I$ to remove the dependence. According to this assumption, the cost bound (12.17) is expressed as $J_2 = \mathbf{E}\left\{J_2\right\} \le \mathbf{E}\left\{x_0^T P x_0\right\} = \text{trace}\left\{P\right\} = J_2^*$. In this chapter, we use $\mathcal{X}_0\mathcal{X}_0^T = x^T(0)\,x(0)$ to remove the dependence, and an optimal H_2 performance cost function bound is described, which results in the following corollary.

Corollary 12.10 *Consider the closed-loop system in (12.14) associated with an H_2 performance cost function in (12.9). Suppose that the optimization problem*

$$\min\ \tilde{J}_2^* = \text{trace}\left(\mathcal{Z}_0\right) \tag{12.36}$$

subject to (12.23) and (12.24) (or (12.31) and (12.32)), and

$$\begin{bmatrix} -\mathcal{Z}_0 & \mathcal{X}_0 P \\ * & -P \end{bmatrix} < 0, \tag{12.37}$$

has a feasible solution, where trace $(\cdot)$ *denotes the trace of a matrix, symmetric matrix $\mathcal{Z}_0 > 0$, then the IT2 fuzzy controller in (12.4) is an optimal reliable mixed H_2/H_∞ controller, which guarantees the minimization of the H_2 performance cost function bound (12.17) for system (12.14), where $\mathcal{X}_0\mathcal{X}_0^T = x_0^T x_0$.*

Proof Since (12.23) and (12.24) ((12.31) and (12.32)) have been given the proof in Theorem 12.6 (Theorem 12.8), we just proof (12.37) in the following. Recalling trace (P_1P_2) = trace (P_2P_1), from (12.37), one can obtain that $\mathcal{X}_0 P P^{-1}\left(\mathcal{X}_0 P\right)^T = \mathcal{X}_0^T P \mathcal{X}_0 < \mathcal{Z}_0$, thus,

$$x_0^T P x_0 = \text{trace}\left(x_0^T P x_0\right) = \text{trace}\left(P x_0^T x_0\right) = \text{trace}\left(P \mathcal{X}_0\mathcal{X}_0^T\right) < \text{trace}\left(\mathcal{Z}_0\right).$$

It follows from (12.17) that $J_2^* < \tilde{J}_2^*$. Then, the minimization of $\tilde{J}_2^*$ implies the minimization of the H_2 performance cost function bound for the system in (12.14). This completes the proof. □

12.4 Simulation Results

In this section, we provide the simulation results from a numerical example to verify the effectiveness of the control design scheme. Firstly, the desired static output-feedback controller design method under sensor failure known case is used to testify the availability for the reliable mixed $H_2/\mathcal{H}_\infty$ performance of the system via Theorem 12.6. Then, considering a disturbance-free system with unknown sensor failure, the desired H_2 performance controller is obtained via Corollary 12.10.

Example 12.11 We give an IT2 fuzzy system of discrete-time form representing a nonlinear system, which is with two uncertain parameters a and b. For simplicity, the following 3-rule IT2 fuzzy model is employed to describe the nonlinear system with the sampling period $T = 0.1$ s:

Plant Rule i : IF $a(x_1(k))$ is M_{i1}, THEN

$$\begin{cases} x(k+1) = A_i x(k) + B_i u(k) + B_{wi} w(k), \\ z(k) = E_i x(k) + G_i u(k) + G_{wi} w(k), \\ y(k) = C_i x(k) + D_{wi} w(k), \end{cases} \tag{12.38}$$

where

$$A_1 = \begin{bmatrix} 0.18 & -a_{\min} \\ 0.08 & -a_{\min} - 0.5 \end{bmatrix}, \quad A_2 = \begin{bmatrix} 0.18 & -a_{\text{avg}} \\ 0.08 & -a_{\text{avg}} - 0.5 \end{bmatrix}, \quad B_{w1} = \begin{bmatrix} -0.13 \\ 0.4 \end{bmatrix},$$

$$A_3 = \begin{bmatrix} 0.18 & -a_{\max} \\ 0.08 & -a_{\max} - 0.5 \end{bmatrix}, \quad B_1 = \begin{bmatrix} 2a_{\min} - 0.05 \\ 0.13 \end{bmatrix}, \quad B_2 = \begin{bmatrix} 2a_{\text{avg}} - 0.05 \\ 0.26 \end{bmatrix},$$

$$B_3 = \begin{bmatrix} 2a_{\max} - 0.05 \\ 0.16 \end{bmatrix}, \quad B_{w2} = \begin{bmatrix} -0.11 \\ 0.2 \end{bmatrix}, \quad B_{w3} = \begin{bmatrix} -0.12 \\ 0.1 \end{bmatrix},$$

$$E_1 = \begin{bmatrix} -0.214 & 0.128 \end{bmatrix}, \quad E_2 = \begin{bmatrix} -0.120 & 0.120 \end{bmatrix},$$

$$E_3 = \begin{bmatrix} -0.214 & 0.128 \end{bmatrix}, \quad G_1 = -0.214, \quad G_2 = -0.120, \quad G_3 = -0.214,$$

$$G_{w1} = G_{w2} = G_{w3} = 0.01, \quad C_1 = \begin{bmatrix} -0.03 & 0.020 \\ -0.01 & b_{\max} \end{bmatrix}, \quad C_2 = \begin{bmatrix} -0.02 & 0.018 \\ -0.01 & b_{\text{avg}} \end{bmatrix},$$

$$C_3 = \begin{bmatrix} -0.01 & 0.012 \\ -0.01 & b_{\min} \end{bmatrix}, \quad D_{w1} = D_{w2} = D_{w3} = \begin{bmatrix} 0.01 & -0.02 \end{bmatrix}^T.$$

Assuming that $x_1 \in [-80, 80]$, the uncertain parameters a and b satisfy $a_{\min} = 0.1 \le a(x_1) \le a_{\max} = 0.2$ and $b_{\min} = 0.012 \le b(x_1) \le b_{\max} = 0.025$, respectively. Thus, $a_{\text{avg}} = (a_{\min} + a_{\max})/2$ and $b_{\text{avg}} = (b_{\min} + b_{\max})/2$. The LMFs and UMFs of the plant and the static output-feedback controller are defined in Table 12.1 while $\phi_i(x(k))$, i.e., the grade of membership of the embedded membership functions is determined by the following weighting functions:

Table 12.1 LMFs and UMFs of the plant and the controller

LMFs of the pant	UMFs of the pant
$\underline{\mu}_{M_1^1}(x_1) = 0.8 - 0.8/e^{\left(-\frac{x_1+80}{15}\right)}$	$\overline{\mu}_{M_1^1}(x_1) = 1 - 1/e^{\left(-\frac{x_1+80}{15}\right)}$
$\underline{\mu}_{M_1^3}(x_1) = 0.8/e^{\left(-\frac{x_1-80}{15}\right)}$	$\overline{\mu}_{M_1^3}(x_1) = 1/e^{\left(-\frac{x_1-80}{15}\right)}$
$\underline{\mu}_{M_1^2}(x_1) = 1 - \overline{\mu}_{M_1^1}(x_1) - \overline{\mu}_{M_1^3}(x_1)$	$\overline{\mu}_{M_1^2}(x_1) = 1 - \underline{\mu}_{M_1^1}(x_1) - \underline{\mu}_{M_1^3}(x_1)$
LMFs of the controller	UMFs of the controller
$\underline{\mu}_{N_1^1}(x_1) = e^{\left(-\frac{x_1^2}{4000}\right)}$	$\overline{\mu}_{N_1^1}(x_1) = \underline{\mu}_{N_1^1}(x_1)$
$\underline{\mu}_{N_1^2}(x_1) = \underline{\mu}_{N_1^1}(x_1)$	$\overline{\mu}_{N_1^2}(x_1) = \underline{\mu}_{N_1^2}(x_1)$

$$
\begin{aligned}
\underline{\alpha}_1(x_1) &= \frac{1}{2}\sin^2(x_1), \quad \underline{\alpha}_3(x_1) = \frac{1}{2}\sin^2(x_1), \\
\underline{\alpha}_2(x_1) &= \frac{1}{\overline{\phi}_2(x_1) - \underline{\phi}_2(x_1)}\left[\overline{\phi}_1(x_1) - \underline{\alpha}_1(x_1)\left(\overline{\phi}_1(x_1) - \underline{\phi}_1(x_1) + \right)\right. \\
&\quad \left. + \overline{\phi}_3(x_1) - \underline{\alpha}_3(x_1)\left(\overline{\phi}_3(x_1) - \underline{\phi}_3(x_1) + \right) + \overline{\phi}_2(x_1) - 1\right],
\end{aligned}
$$

and $\overline{\alpha}_i(x_1) = 1 - \underline{\alpha}_i(x_1)$. We choose $\underline{\beta}_j(x_1) = \overline{\beta}_j(x_1) = 0.5$ to determine the actual membership functions of the plant and the controller, respectively.

Considering the computational burden, we use only one sub-FOU (i.e., $\upsilon = 1$) and divide the state x_1 into 100 equal-size sub-states (i.e., $\tau = 1, 2, \ldots, 100$), where the upper and lower bounds of τth state $x_1^{\upsilon,\tau}$ in the sub-FOU υ are defined as $\underline{x}_1^{\upsilon,\tau} = 1.6(\tau - 51)$, $\overline{x}_1^{\upsilon,\tau} = 1.6(\tau - 50)$. Then the constant scalars in the form of (12.11) and (12.12) are determined by

$$
\begin{aligned}
\underline{\epsilon}_{ij11\tau} &= \underline{\phi}_i\left(\underline{x}_1^{\upsilon,\tau}\right)\underline{\psi}_j\left(\underline{x}_1^{\upsilon,\tau}\right), \quad \underline{\epsilon}_{ij21\tau} = \underline{\phi}_i\left(\overline{x}_1^{\upsilon,\tau}\right)\underline{\psi}_j\left(\overline{x}_1^{\upsilon,\tau}\right), \\
\overline{\epsilon}_{ij11\tau} &= \overline{\phi}_i\left(\underline{x}_1^{\upsilon,\tau}\right)\overline{\psi}_j\left(\underline{x}_1^{\upsilon,\tau}\right), \quad \overline{\epsilon}_{ij21\tau} = \overline{\phi}_i\left(\overline{x}_1^{\upsilon,\tau}\right)\overline{\psi}_j\left(\overline{x}_1^{\upsilon,\tau}\right).
\end{aligned}
$$

Moreover, the LMFs and UMFs $\underline{h}_{ij1}$ and $\overline{h}_{ij1}$ are defined by choosing $\varrho_{111\tau}(x_1) = 1 - \left(x_1 - \underline{x}_1^{\upsilon,\tau}\right)/\left(\overline{x}_1^{\upsilon,\tau} - \underline{x}_1^{\upsilon,\tau}\right)$ and $\varrho_{121\tau}(x_1) = 1 - \varrho_{111\tau}(x_1)$, respectively. The state responses of the open-loop system in (12.38) based on the above parameters are plotted in Fig. 12.1, which shows that this system is not stable.

Firstly, in order to make a comparison between the sensor failure (known) case and the sensor normal case, we choose the sensor failure matrix $\zeta = \text{diag}\{0.2, 0.3\}$ and the normal one $\zeta = \text{diag}\{1, 1\}$, respectively, and give an H_∞ performance index $\gamma = 0.20$. Then, based on Theorem 12.6, the feasible solutions for controller gain matrices in system (12.38) with disturbance input are computed as follows:

(1) Sensor failure case

$$
K_1 = \begin{bmatrix} 162.6078 & 103.7725 \end{bmatrix}, \quad K_2 = \begin{bmatrix} 172.2039 & 103.9296 \end{bmatrix}. \tag{12.39}
$$

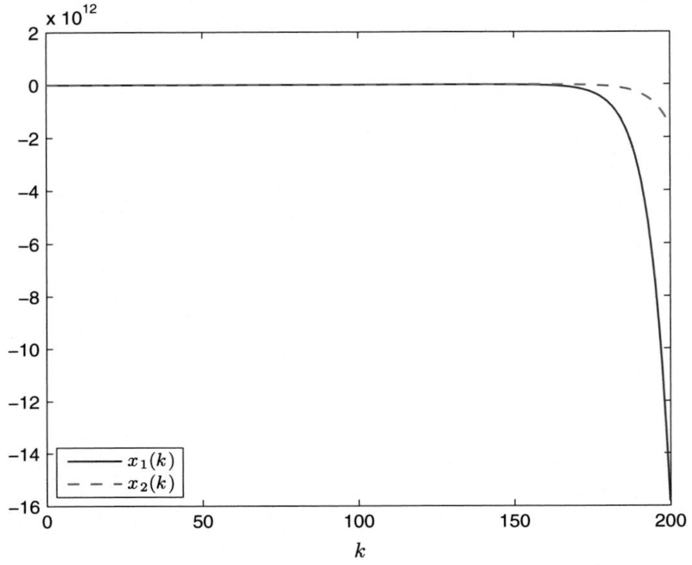

Fig. 12.1 States of the open-loop system

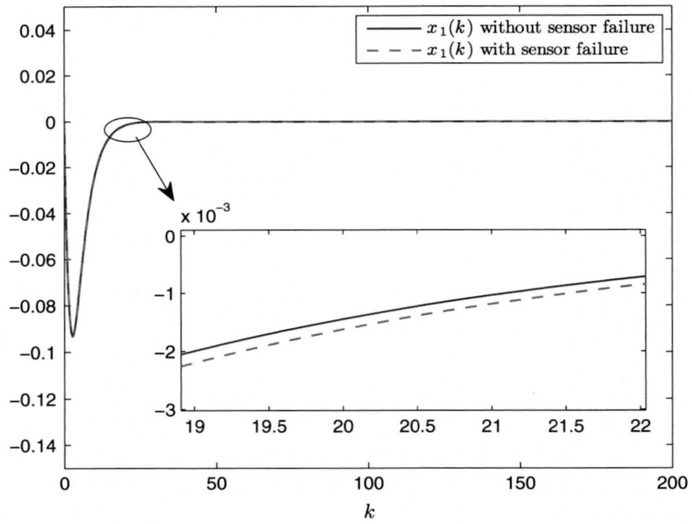

Fig. 12.2 State x_1 for the closed-loop system

(2) Sensor normal case

$$K_1 = \begin{bmatrix} 31.9047 & 32.3183 \end{bmatrix}, \quad K_2 = \begin{bmatrix} 34.7427 & 30.7439 \end{bmatrix}. \tag{12.40}$$

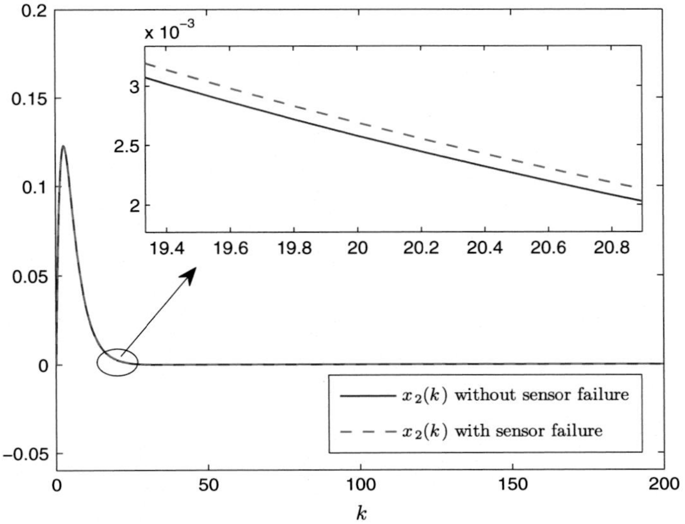

Fig. 12.3 State x_2 for the closed-loop system

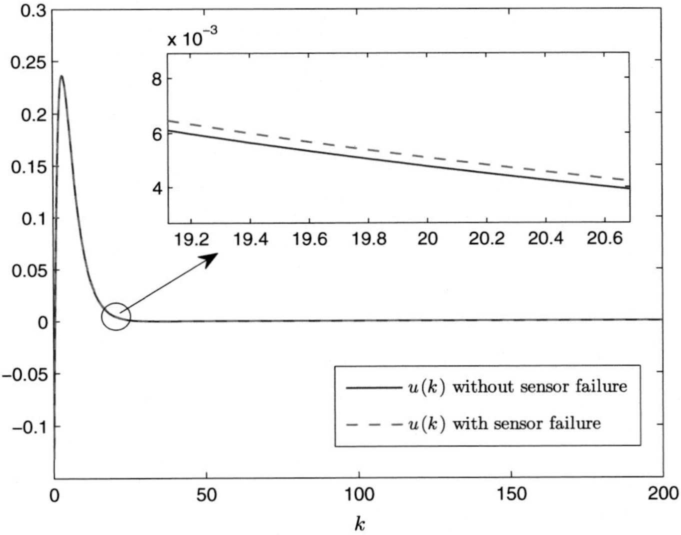

Fig. 12.4 Control input for the closed-loop system

Next, based on the controller gains in (12.39) and (12.40), we analyze the stability of the plant under zero initial state $x(0) = \begin{bmatrix} 0 & 0 \end{bmatrix}^T$. Considering the disturbance input $w(k) = 1/\left(2^k + 1\right)$, Figs. 12.2 and 12.3 shows the state responses of the closed-loop system under the above two cases. It can be seen that the states of both cases are stable while the states in the sensor failure case are slightly worse than those in the normal one. Figure 12.4 plots the control forces to the plant under the two cases. These figures illustrate that the nonlinear system in (12.38) with sensor failure can be controlled subject to uncertainties a and b under zero initial condition, and the failure in the sensor can be completely tolerated. The effectiveness of the proposed design method is validated.

Secondly, to further analyze the H_2 performance of the closed-loop system, we use Corollary 12.10 to obtain the desired IT2 fuzzy static output-feedback controller and guarantee an optimal H_2 performance cost function bound. Assume that the sensor failure is unknown between the bounds of $\underline{\zeta}=\text{diag}\{0.65, 0.72\}$ and $\overline{\zeta}=\text{diag}\{0.72, 0.85\}$. Considering the disturbance-free system in (12.38) under initial state $x(0) = \begin{bmatrix} 0.1 & 0.2 \end{bmatrix}^T$, and giving another H_∞ performance index $\gamma = 1.0$, by solving the convex optimization problem in (12.36), we obtain the following fuzzy static output-feedback controller matrices:

$$K_1 = \begin{bmatrix} 40.2662 & 74.1886 \end{bmatrix}, \quad K_2 = \begin{bmatrix} 6.7712 & 1.9683 \end{bmatrix}, \tag{12.41}$$

and the minimum H_2 performance cost function bound $\tilde{J}_2^* = 50$. Besides, the actual H_2 performance cost function is $J_2 = \sum_{k=0}^{200} z^T(k) z(k) = 40.6686$, which satisfies $J_2 < \tilde{J}_2^*$. This also verifies the effectiveness of the proposed approach. Furthermore, to observe the stability of the sensor failure system in (12.38), we assume the actual sensor failure matrix $\zeta=\text{diag}\{0.68, 0.84\}$ satisfies the above bound. Based on the controller matrices in (12.41), the simulation results are obtained in Figs. 12.5 and 12.6.

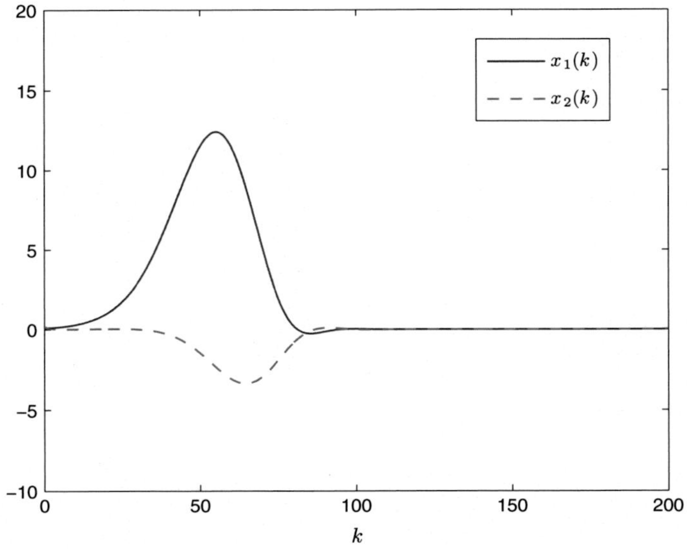

Fig. 12.5 States of the closed-loop system

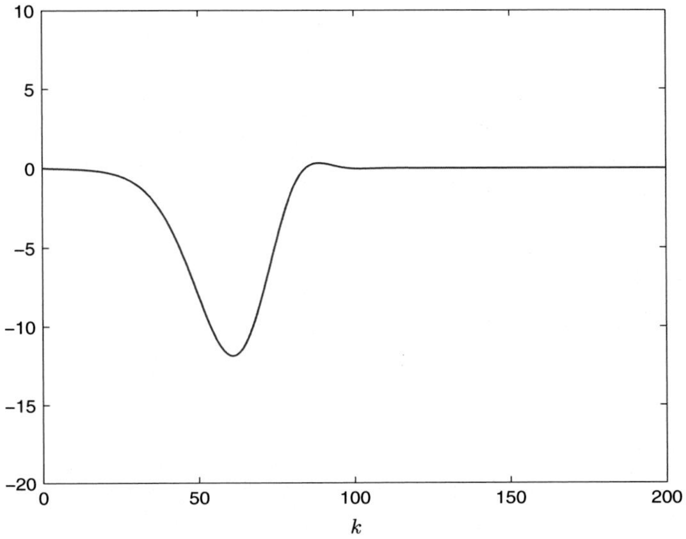

Fig. 12.6 Control input for the closed-loop system

Figure 12.5 shows the state responses and Fig. 12.6 depicts the corresponding control input of the closed-loop system with sensor failure. Figures 12.5 and 12.6 also present that the controlled system in (12.38) is reliable stability under the sensor failure unknown case.

12.5 Conclusion

In this chapter, the problem of reliable control for discrete-time IT2 FMB systems with sensor failure has been solved. The mixed H_2/H_∞ performance has been considered. The number of fuzzy rules and the membership functions for the static output-feedback controller are different from those of the plant. A sufficient criterion of reliable stability with mixed H_2/H_∞ performance has been given for the closed-loop system with sensor failure. The constraints of the static output-feedback controller parameters have been provided for sensor failure known case and sensor failure unknown case, which can guarantee the reliable stability of the plant with mixed H_2/H_∞ performance. Furthermore, the criteria of optimal H_2/H_∞ performance for the closed-loop system are proposed. A numerical example has been employed to verify the effectiveness of the proposed approach. In future work, the dynamic output-feedback control for IT2 FMB systems will be investigated by considering possible faults or errors occurring in the IT2 FMB systems.

Chapter 13
Output Tracking Control of Interval Type-2 Fuzzy-Model-Based Systems

13.1 Introduction

In this chapter, the guaranteed cost output tracking control problem for the discrete-time IT2 fuzzy system under imperfect premise matching is considered for the first time. The nonlinear systems subject to parameter uncertainties are modeled by the IT2 T–S fuzzy model approach, in which the LMFs and UMFs are introduced to represent and capture the uncertainties. The controller to be designed does not share the same premise variables as those of the system model, which makes the controller design more flexible. The guaranteed cost and tracking control are first considered simultaneously. Finally, some simulation results are given to show the effectiveness of the proposed method.

13.2 Problem Formulation

Consider the following IT2 T–S fuzzy system.

◆ **Plant Form**:

Rule i: IF $f_1(k)$ is $M_{i1}, \ldots,$ and $f_\alpha(k)$ is $M_{i\alpha}, \ldots,$ and $f_\delta(k)$ is $M_{i\delta}$, THEN,

$$\begin{cases} x(k+1) = A_i x(k) + B_i u(k) + L_i w(k), \\ \quad y(k) = C_i x(k) + D_i u(k), \end{cases} \tag{13.1}$$

where $M_{i\alpha}$ represents an IT2 fuzzy set of ith rule according to the known function $f_\alpha(k)$ for $i = 1, 2, \ldots, \sigma$ and $\alpha = 1, 2, \ldots, \delta, x(k) \in \mathbf{R}^n$ represents the system state variable, $u(k) \in \mathbf{R}^q$ is the control input, $w(k) \in \mathbf{R}^l$ is assumed to be a disturbance input, $y(k) \in \mathbf{R}^m$ stands for the measured output. A_i, B_i, C_i, D_i and L_i are known

© Springer Science+Business Media Singapore 2016
H. Li et al., *Analysis and Synthesis for Interval Type-2
Fuzzy-Model-Based Systems*, DOI 10.1007/978-981-10-0593-0_13

real constant matrices with appropriate dimensions. The following interval sets stand for the emission intensity of the ith rule:

$$\Phi_i(x(k)) = \left[\underline{\varphi}_i(x(k)), \overline{\varphi}_i(x(k))\right], \tag{13.2}$$

where

$$\begin{aligned}
\underline{\varphi}_i(x(k)) &= \prod_{\alpha=1}^{\delta} \underline{\mu}_{F_{i\alpha}}(f_\alpha(k)), \quad \overline{\varphi}_i(x(k)) = \prod_{\alpha=1}^{\delta} \overline{\mu}_{F_{i\alpha}}(f_\alpha(k)), \\
\overline{\mu}_{F_{i\alpha}}(f_\alpha(k)) &\geq \underline{\mu}_{F_{i\alpha}}(f_\alpha(k)) \geq 0, \quad \overline{\varphi}_i(x(k)) \geq \underline{\varphi}_i(x(k)) \geq 0, \\
1 &\geq \overline{\mu}_{F_{i\alpha}}(f_\alpha(k)) \geq 0, \quad 1 \geq \underline{\mu}_{F_{i\alpha}}(f_\alpha(k)) \geq 0,
\end{aligned}$$

$\underline{\varphi}_i(x(k))$ and $\overline{\varphi}_i(x(k))$ represent the lower grade of membership and the upper grade of membership, respectively. $\underline{\mu}_{F_{i\alpha}}(f_\alpha(k))$ and $\overline{\mu}_{F_{i\alpha}}(f_\alpha(k))$ represent the LMF and the UMF, respectively. Then the IT2 T–S fuzzy model of (13.1) can be written as:

$$\begin{cases}
x(k+1) = \sum_{i=1}^{\sigma} \varphi_i(x(k))\left(A_i x(k) + B_i u(k) + L_i w(k)\right), \\
y(k) = \sum_{i=1}^{\sigma} \varphi_i(x(k))\left(C_i x(k) + D_i u(k)\right),
\end{cases} \tag{13.3}$$

where

$$\begin{aligned}
\varphi_i(x(k)) &= \underline{\alpha}_i(x(k))\,\underline{\varphi}_i(x(k)) + \overline{\alpha}_i(x(k))\,\overline{\varphi}_i(x(k)) \geq 0, \\
1 &= \sum_{i=1}^{\sigma} \varphi_i(x(k)), \quad \underline{\alpha}_i(x(k)) \in [0, 1], \quad \overline{\alpha}_i(x(k)) \in [0, 1], \\
1 &= \underline{\alpha}_i(x(k)) + \overline{\alpha}_i(x(k)),
\end{aligned}$$

in which $\underline{\alpha}_i(x(k))$ and $\overline{\alpha}_i(x(k))$ rely on parameter uncertainties, respectively, which are not necessary to be known in the paper. $\varphi_i(x(k))$ is the grade of membership of the embedded membership function.

In order to track the controlled system, the reference model is introduced as follows:

$$\begin{cases}
x_r(k+1) = E x_r(k) + r(k), \\
y_r(k) = F x_r(k).
\end{cases} \tag{13.4}$$

Define the tracking error as $e(k) = y(k) - y_r(k)$.

Then, consider the fuzzy guaranteed cost state-feedback controller as follows:

♦ **Controller Form**:

Rule j: IF $g_1(k)$ is N_{j1} …, and $g_\beta(k)$ is $N_{j\beta}$, …, and $g_z(k)$ is N_{jz}, THEN,

$$u(k) = K_j x(k) + K_{rj} x_r(k), \tag{13.5}$$

where $N_{j\beta}$ ($j = 1, 2, \ldots, \lambda$ and $\beta = 1, 2, \ldots, z$) represents a fuzzy set of jth rule, K_j and K_{rj} are the controller gain, respectively. The following interval sets stand for the emission intensity of the jth rule:

$$\Psi_j(x(k)) = \left[\underline{\psi}_j(x(k)), \overline{\psi}_j(x(k))\right],$$

where

$$\underline{\psi}_j(x(k)) = \prod_{\beta=1}^{z} \underline{\mu}_{N_{j\beta}}\left(g_\beta(k)\right), \quad \overline{\psi}_j(x(k)) = \prod_{\beta=1}^{z} \overline{\mu}_{N_{j\beta}}\left(g_\beta(k)\right),$$
$$\overline{\mu}_{N_{j\beta}}\left(g_\beta(k)\right) \geq \underline{\mu}_{N_{j\beta}}\left(g_\beta(k)\right) \geq 0, \quad \overline{\psi}_j(x(k)) \geq \underline{\psi}_j(x(k)) \geq 0,$$
$$1 \geq \underline{\mu}_{N_{j\beta}}\left(g_\beta(k)\right) \geq 0, \quad 1 \geq \overline{\mu}_{N_{j\beta}}\left(g_\beta(k)\right) \geq 0,$$

in which $\underline{\mu}_{N_{j\beta}}\left(g_\beta(k)\right)$ stands for the LMF and $\overline{\mu}_{N_{j\beta}}\left(g_\beta(k)\right)$ stands for the UMF. $\underline{\psi}_j(x(k))$ and $\overline{\psi}_j(x(k))$ represent the lower and the upper grade of membership, respectively. Then the IT2 T–S fuzzy controller in (13.5) can be written as:

$$u(k) = \sum_{j=1}^{\lambda} \psi_j(x(k)) \left[K_j x(k) + K_{rj} x_r(k)\right], \tag{13.6}$$

where

$$\psi_j(x(k)) = \frac{\underline{\beta}_j(x(k))\,\underline{\psi}_j(x(k)) + \overline{\beta}_j(x(k))\,\overline{\psi}_j(x(k))}{\sum_{\check{k}=1}^{\lambda}\left(\underline{\beta}_{\check{k}}(x(k))\,\underline{\psi}_{\check{k}}(x(k)) + \overline{\beta}_{\check{k}}(x(k))\,\overline{\psi}_{\check{k}}(x(k))\right)} \geq 0,$$
$$1 = \sum_{j=1}^{\lambda} \psi_j(x(k)), \quad \underline{\beta}_j(x(k)) \in [0, 1], \quad \overline{\beta}_j(x(k)) \in [0, 1],$$
$$1 = \underline{\beta}_j(x(k)) + \overline{\beta}_j(x(k)),$$

in which $\underline{\beta}_j(x(k))$ and $\overline{\beta}_j(x(k))$ are predefined nonlinear functions. $\psi_j(x(k))$ represents the grade of membership of the embedded membership functions.

The guaranteed cost function [153] is defined as follows:

$$J = \sum_{k=0}^{\infty} \left[e^T (k) Qe (k) + u^T (k) Ru (k) \right], \tag{13.7}$$

where $Q > 0$ and $R > 0$.

In this section, considering the equations in (13.3) and (13.6), the closed-loop system can be described as follows:

$$\begin{cases} x (k+1) = \sum_{i=1}^{\sigma} \sum_{j=1}^{\lambda} \varphi_i (x (k)) \psi_j (x (k)) \left[A_{ij} x (k) + B_i K_{rj} x_r (k) + L_i w (k) \right], \\ y (k) = \sum_{i=1}^{\sigma} \sum_{j=1}^{\lambda} \varphi_i (x (k)) \psi_j (x (k)) \left[\left(C_i + D_i K_j \right) x (k) + D_i K_{rj} x_r (k) \right], \end{cases} \tag{13.8}$$

where $A_{ij} = A_i + B_i K_j$.

This chapter studies the stability of nonlinear systems with parameter uncertainties. For the sake of analyzing the stability, Ψ_k represents the sub-state spaces, $k = 1, 2, \ldots, \varrho$ and $\Psi = \cup_{\zeta=1}^{\varrho} \Psi_\zeta$. Considering the FOU, the LMFs and UMFs can be rewritten as follows [94]:

$$\begin{aligned} \underline{h}_{ij} (x (k)) &= \sum_{k=1}^{\varrho} \sum_{i_1=1}^{2} \sum_{i_2=1}^{2} \cdots \sum_{i_n=1}^{2} \prod_{r=1}^{n} v_{ri_r\zeta} (x_r (k)) \underline{\epsilon}_{iji_1i_2 \ldots i_n\zeta}, \\ \overline{h}_{ij} (x (k)) &= \sum_{k=1}^{\varrho} \sum_{i_1=1}^{2} \sum_{i_2=1}^{2} \cdots \sum_{i_n=1}^{2} \prod_{r=1}^{n} v_{ri_r\zeta} (x_r (k)) \underline{\epsilon}_{iji_1i_2 \ldots i_n\zeta}, \\ 1 &= \sum_{k=1}^{\varrho} \sum_{i_1=1}^{2} \sum_{i_2=1}^{2} \cdots \sum_{i_n=1}^{2} \prod_{r=1}^{n} v_{ri_r\zeta} (x_r (k)), \\ 0 \le v_{ri_r\zeta} &(x_r (k)) \le 1, \quad 0 \le \underline{\epsilon}_{iji_1i_2 \ldots i_n\zeta} \le \overline{\epsilon}_{iji_1i_2 \ldots i_n\zeta} \le 1, \end{aligned}$$

where $v_{r1\zeta} (x_r (k)) + v_{r2\zeta} (x_r (k)) = 1$, in which $r, \delta = 1, 2, \ldots, n$ and $k = 1, \ldots, q$ and $\underline{\epsilon}_{iji_1i_2 \ldots i_n\zeta}$ and $\overline{\epsilon}_{iji_1i_2 \ldots i_n\zeta}$ are constant scalars; for $i_r, i_s = 1, 2, \zeta = 1, 2, \ldots, q$, and $x (k) \in \Psi_\zeta$.

According to (13.4)–(13.8), defining the augmented state vector $\varsigma (k) = \left[x^T (k) \; x_r^T (k) \right]^T$, the IT2 fuzzy closed-loop system can be rewritten as:

$$\begin{cases} \varsigma (k+1) = \sum_{i=1}^{\sigma} \sum_{j=1}^{\lambda} h_{ij} (x (k)) \left[\overline{A}_{ij} \varsigma (k) + \overline{r} (k) \right], \\ e (k) = \sum_{i=1}^{\sigma} \sum_{j=1}^{\lambda} h_{ij} (x (k)) H_i \varsigma (k), \end{cases} \tag{13.9}$$

where

$$
\begin{aligned}
h_{ij}(x(k)) &\equiv \varphi_i(x(k))\,\psi_j(x(k)) \\
&= \underline{\gamma}_{ij}(x(k))\,\underline{h}_{ij}(x(k)) + \overline{\gamma}_{ij}(x(k))\,\overline{h}_{ij}x(k), \quad \forall i, j, \\
\sum_{i=1}^{\sigma}\sum_{j=1}^{\lambda} h_{ij}(x(k)) &= 1, \quad \underline{\gamma}_{ij}(x(k)) + \overline{\gamma}_{ij}(x(k)) = 1, \\
0 \le \underline{\gamma}_{ij}(x(k)) &\le \overline{\gamma}_{ij}(x(k)) \le 1, \quad \forall i, j, \\
\overline{A}_{ij} &= \begin{bmatrix} A_{ij} & B_i K_{rj} \\ 0 & E \end{bmatrix}, \quad \overline{r}(k) = \begin{bmatrix} L_i & 0 \\ 0 & I \end{bmatrix}\begin{bmatrix} w(k) \\ r(k) \end{bmatrix}, \\
H_i &= \begin{bmatrix} C_i + D_i K_j & -F + D_i K_{rj} \end{bmatrix}, \\
U_j &= \begin{bmatrix} K_j & K_{rj} \end{bmatrix},
\end{aligned}
\tag{13.10}
$$

in which the $\underline{\gamma}_{ij}(x(k))$ and $\overline{\gamma}_{ij}(x(k))$ are two unknown functions.

For brevity, symbols $\underline{\varphi}_i(x(k))$, $\overline{\varphi}_i(x(k))$, $\varphi_i(x(k))$, $\underline{\psi}_j(x(k))$, $\overline{\psi}_j(x(k))$, $\psi_j(x(k))$, $\underline{\gamma}_{ij}(x(k))$, $\overline{\gamma}_{ij}(x(k))$, $\underline{\epsilon}_{iji_1i_2\ldots i_nk}$, $\overline{\epsilon}_{iji_1i_2\ldots i_nk}$, $\underline{h}_{ij}(x(k))$, $\overline{h}_{ij}(x(k))$ and $h_{ij}(x(k))$ are denoted as $\underline{\varphi}_i$, $\overline{\varphi}_i$, φ_i, $\underline{\psi}_j$, $\overline{\psi}_j$, ψ_j, $\underline{\gamma}_{ij}$, $\overline{\gamma}_{ij}$, $\underline{\epsilon}$, $\overline{\epsilon}$, $\underline{h}_{ij}$, $\overline{h}_{ij}$ and h_{ij}, respectively. Moreover, from the details in the above content, we have $\sum_{i=1}^{\sigma}\varphi_i = \sum_{j=1}^{\lambda}\psi_j = \sum_{i=1}^{\sigma}\sum_{j=1}^{\lambda}\varphi_i\psi_j = \sum_{i=1}^{\sigma}\sum_{j=1}^{\lambda}h_{ij} = 1$.

13.3 Main Results

13.3.1 Stability Analysis

Theorem 13.1 *Consider the discrete-time IT2 fuzzy system (13.9) under imperfect premise matching and cost function (13.7). If there exist matrices* $P > 0$, $Q > 0$, $R > 0$, $V_{ij} > 0$, $G_{ij} > 0$, $N_{ij} > 0$, $W_{ij} > 0$ *and* M *with appropriate dimensions, for* $i = 1, 2, \ldots, \sigma$, $j = 1, 2, \ldots, \lambda$ *satisfying the following conditions:*

$$
\Theta_{ij} - \Xi_{ij} + \Pi_{ij} - \overline{A}_{ij}^T V_{ij}\overline{A}_{ij} + W_{ij} + M > 0, \quad \forall i, j, \tag{13.11}
$$

$$
\sum_{i=1}^{\sigma}\sum_{j=1}^{\lambda}\left[\overline{\epsilon}\left(\Theta_{ij} + \Pi_{ij}\right) + \left(\overline{\epsilon} - \underline{\epsilon}\right)W_{ij} + \overline{\epsilon}_1 M\right] < 0, \quad \forall i, j, \tag{13.12}
$$

where

$$\Theta_{ij} = \overline{A}_{ij}^T P \overline{A}_{ij} - P, \quad \Xi_{ij} = H_i^T G_{ij} H_i + U_j^T N_{ij} U_j,$$
$$\Pi_{ij} = H_i^T Q H_i + U_j^T R U_j, \quad V_{ij} = \mathrm{diag}\{V_{1ij}, V_{2ij}\}, \quad \overline{\epsilon}_1 = \overline{\epsilon} - \frac{1}{\sigma\lambda}.$$

Then, the closed-loop system is asymptotically stable and the cost function (13.7) satisfies the following bound

$$J \le J_0,$$

where $J_0 = \varsigma^T(0) P \varsigma(0) + \sum_{k=0}^{\infty} \overline{r}^T(k) P \overline{r}(k)$.

Proof Firstly, consider a Lyapunov function for system (13.9) as follows:

$$V(k) = \varsigma^T(k) P \varsigma(k). \tag{13.13}$$

By using the well-known upper bound, we can obtain

$$\begin{aligned}
\Delta V(k) &= \varsigma^T(k+1) P \varsigma(k+1) - \varsigma^T(k) P \varsigma(k) \\
&\le \sum_{i=1}^{\sigma} \sum_{j=1}^{\lambda} h_{ij} \varsigma^T(k) \overline{A}_{ij}^T P \overline{A}_{ij} \varsigma(k) - \varsigma^T(k) P \varsigma(k) + \overline{r}^T(k) P \overline{r}(k) \\
&\quad + e^T(k) Q e(k) + u^T(k) R u(k) \\
&= \sum_{i=1}^{\sigma} \sum_{j=1}^{\lambda} h_{ij} \varsigma^T(k) (H_i^T Q H_i + U_j^T R U_j) \varsigma(k).
\end{aligned} \tag{13.14}$$

Since V_{ij}, G_{ij}, N_{ij} and W_{ij} are positive definite matrices, according to [94], one can obtain

$$\begin{aligned}
\Delta V(k) \le & \sum_{i=1}^{\sigma} \sum_{j=1}^{\lambda} \left(\underline{\gamma}_{ij} \underline{h}_{ij} + \left(1 - \underline{\gamma}_{ij}\right) \bar{h}_{ij} \right) \varsigma^T(k) \Theta_{ij} \varsigma(k) + \overline{r}^T(k) P \overline{r}(k) \\
& - \sum_{i=1}^{\sigma} \sum_{j=1}^{\lambda} \left(1 - \underline{\gamma}_{ij}\right) \left(\underline{h}_{ij} - \overline{h}_{ij} \right) \varsigma^T(k) W_{ij} \varsigma(k) \\
& + \left[\sum_{i=1}^{\sigma} \sum_{j=1}^{\lambda} \left(\underline{\gamma}_{ij} \underline{h}_{ij} + \left(1 - \underline{\gamma}_{ij}\right) \overline{h}_{ij} \right) - 1 \right] \varsigma^T(k) M \varsigma(k) \\
& - \sum_{i=1}^{\sigma} \sum_{j=1}^{\lambda} \underline{\gamma}_{ij} \left(\underline{h}_{ij} - \overline{h}_{ij} \right) \varsigma^T(k) \overline{A}_{ij}^T V_{ij} \overline{A}_{ij} \varsigma(k) \\
& - \sum_{i=1}^{\sigma} \sum_{j=1}^{\lambda} \underline{\gamma}_{ij} \left(\underline{h}_{ij} - \overline{h}_{ij} \right) \varsigma^T(k) H_i^T G_{ij} H_i \varsigma(k)
\end{aligned}$$

$$
\begin{aligned}
&- \sum_{i=1}^{\sigma} \sum_{j=1}^{\lambda} \underline{\gamma}_{ij} \left(\underline{h}_{ij} - \overline{h}_{ij} \right) \varsigma^{T}(k) U_j^T N_{ij} U_j \varsigma(k) \\
&= \varsigma^{T}(k) \left[\sum_{i=1}^{\sigma} \sum_{j=1}^{\lambda} \underline{\gamma}_{ij} \left(\underline{h}_{ij} - \overline{h}_{ij} \right) \left(\Theta_{ij} + W_{ij} + M - \Xi_{ij} - \overline{A}_{ij}^T V_{ij} \overline{A}_{ij} \right) \right] \varsigma(k) \\
&+ \varsigma^{T}(k) \left[\sum_{i=1}^{\sigma} \sum_{j=1}^{\lambda} \left(\overline{h}_{ij} \Theta_{ij} - \left(\underline{h}_{ij} - \overline{h}_{ij} \right) W_{ij} + \overline{h}_{ij} M \right) - M \right] \varsigma(k) + \overline{r}^{T}(k) P \overline{r}(k).
\end{aligned}
$$

Then, by using the conditions (13.11)–(13.14) in Theorem 13.1, one can get

$$
\Delta V(k) \le - \left[e^{T}(k) Q e(k) + u^{T}(k) R u(k) \right] + \overline{r}^{T}(k) P \overline{r}(k).
$$

In the above inequality, Q and R are positive definite matrices, resulting in $\Delta V(k) \le \overline{r}^{T}(k) P \overline{r}(k)$. The "input to state stability" condition (ISSC) [188] is satisfied for this inequality and the system is asymptotically stable.

Then, the cost function

$$
\begin{aligned}
J &= \sum_{k=0}^{\infty} \left[e^{T}(k) Q e(k) + u^{T}(k) R u(k) \right] \\
&\le - \sum_{k=0}^{\infty} \Delta V(k) + \sum_{k=0}^{\infty} \overline{r}^{T}(k) P \overline{r}(k) \\
&= V(0) - V(\infty) + \sum_{k=0}^{\infty} \overline{r}^{T}(k) P \overline{r}(k) \\
&\le V(0) + \sum_{k=0}^{\infty} \overline{r}^{T}(k) P \overline{r}(k) = J_0,
\end{aligned}
$$

where $J_0 = \varsigma^{T}(0) P \varsigma(0) + \sum_{k=0}^{\infty} \overline{r}^{T}(k) P \overline{r}(k)$. The proof is completed. □

13.3.2 Output Tracking Control

Theorem 13.2 *Given the system (13.9) and the cost function (13.7), if there exists a matrix* $P > 0$, $X > 0$, $Q > 0$, $R > 0$, $\overline{V}_{ij} > 0$, $G_{ij} > 0$, $N_{ij} > 0$, $\overline{W}_{ij} > 0$ *and* $\overline{M}$ *with appropriate dimensions, for* $i = 1, 2, \ldots, \sigma$, $j = 1, 2, \ldots, \lambda$ *such that the following optimization problem has a solution* $(\alpha_{\min}, X_{\min}, J_{\min})$,

$$
\min \quad J = \alpha + \sum_{k=0}^{\infty} \beta_k I, \tag{13.15}
$$

$$s.t \begin{cases} \Omega_{ij} > 0, \quad \forall i, j, \\ \Delta_{ij} < 0, \quad \forall i, j, \\ G_{ij} - Q > 0, \quad \forall i, j, \\ V_{ij} - P > 0, \quad \forall i, j, \\ N_{ij} - R > 0, \quad \forall i, j, \\ \begin{bmatrix} -\alpha & \varsigma^T(0) \\ * & -X \end{bmatrix} < 0, \\ \begin{bmatrix} -\beta_k I & \overline{r}^T(k) \\ * & -X \end{bmatrix} < 0, \end{cases} \tag{13.16}$$

then the guaranteed cost function has a minimum upper bound and the optimal guaranteed cost output tracking controller is given as follows:

$$\begin{aligned} u(k) &= \sum_{j=1}^{\lambda} \psi_j \left[K_j x(k) + K_{rj} x_r(k) \right] \\ &= \sum_{j=1}^{\lambda} \psi_j \overline{U}_j X_{\min}^{-1} \varsigma(k), \end{aligned}$$

where

$$\Omega_{ij} = \begin{bmatrix} \Omega_{123ij} & \Omega_{4ij} \\ * & \Omega_{5ij} \end{bmatrix}, \quad \Omega_{123ij} = \begin{bmatrix} \Omega_{1ij} & \Omega_{2ij} \\ * & \Omega_{3ij} \end{bmatrix},$$

$$\Omega_{4ij} = \begin{bmatrix} P_1 A_i^T + \bar{K}_j^T B_i^T & 0 & P_1 C_i^T + \bar{K}_j^T D_i^T & \bar{K}_j^T \\ \bar{K}_{rj}^T B_i^T & P_2 E^T & -P_2 F^T + \bar{K}_{rj}^T D_i^T & \bar{K}_{rj}^T \end{bmatrix},$$

$$\Omega_{5ij} = \begin{bmatrix} 3P_1 - \overline{V}_{1ij} & 0 & 0 & 0 \\ * & 3P_2 - \overline{V}_{2ij} & 0 & 0 \\ * & * & 2I + Q - G_{ij} & 0 \\ * & * & * & 2I + R - N_{ij} \end{bmatrix},$$

$$\Delta_{ij} = \begin{bmatrix} \Delta_{123ij} & \Delta_{4ij} \\ * & \Delta_{5ij} \end{bmatrix}, \quad \Delta_{123ij} = \begin{bmatrix} \Delta_{1ij} & \Delta_{2ij} & \sqrt{\bar{\epsilon}} P_1 A_i^T + \sqrt{\bar{\epsilon}} \bar{K}_j^T B_i^T \\ * & \Delta_{3ij} & \sqrt{\bar{\epsilon}} \bar{K}_{rj}^T B_i^T \\ * & * & -P_1 \end{bmatrix},$$

$$\Delta_{4ij} = \begin{bmatrix} 0 & \sqrt{\bar{\epsilon}} P_1 C_i^T Q^T + \sqrt{\bar{\epsilon}} \bar{K}_j^T D_i^T Q^T & \sqrt{\bar{\epsilon}} \bar{K}_j^T R^T \\ \sqrt{\bar{\epsilon}} P_2 E^T & -\sqrt{\bar{\epsilon}} P_2 F^T Q^T + \sqrt{\bar{\epsilon}} \bar{K}_{rj}^T D_i^T Q^T & \sqrt{\bar{\epsilon}} \bar{K}_{rj}^T R^T \\ 0 & 0 & 0 \end{bmatrix},$$

$$\Delta_{5ij} = \begin{bmatrix} -P_2 & 0 & 0 \\ * & -Q & 0 \\ * & * & -R \end{bmatrix}, \quad \Delta_{2ij} = \left(\bar{\epsilon} - \underline{\epsilon}\right)\overline{W}_{2ij} + \bar{\epsilon}_1\overline{M}_2,$$

$$\begin{aligned}
\Omega_{1ij} &= -P_1 + \overline{W}_{1ij} + \overline{M}_1, \quad \Omega_{3ij} = -P_2 + \overline{W}_{3ij} + \overline{M}_3, \\
\Delta_{1ij} &= -\bar{\epsilon}P_1 + \left(\bar{\epsilon} - \underline{\epsilon}\right)\overline{W}_{1ij} + \bar{\epsilon}_1\overline{M}_1, \quad \Omega_{2ij} = \overline{W}_{2ij} + \overline{M}_2, \\
\Delta_{3ij} &= -\bar{\epsilon}P_2 + \left(\bar{\epsilon} - \underline{\epsilon}\right)\overline{W}_{3ij} + \bar{\epsilon}_1\overline{M}_3, \quad P = \text{diag}\{P_1^{-1}, P_2^{-1}\}, \\
\overline{U}_j &= \left[\, \bar{K}_j \ \bar{K}_{rj} \,\right], \quad X = P^{-1}, \\
K_j &= \bar{K}_j P_1^{-1}, \quad K_{rj} = \bar{K}_{rj} P_2^{-1}.
\end{aligned} \tag{13.17}$$

Proof Firstly, for the condition $\Omega_{ij} > 0$ in (13.16), by using $2I + Q - G_{ij} < \left(G_{ij} - Q\right)^{-1} = -\left(Q - G_{ij}\right)^{-1}$ and $2I + R - N_{ij} < \left(N_{ij} - R\right)^{-1} = -\left(R - N_{ij}\right)^{-1}$, the following LMI holds:

$$\overline{\Omega}_{ij} > \Omega_{ij} > 0, \tag{13.18}$$

where

$$\overline{\Omega}_{ij} = \begin{bmatrix} \overline{\Omega}_{123ij} & \overline{\Omega}_{4ij} \\ * & \overline{\Omega}_{5ij} \end{bmatrix},$$

$$\overline{\Omega}_{123ij} = \begin{bmatrix} \Omega_{1ij} & \Omega_{2ij} & P_1 A_i^T + \bar{K}_j^T B_i^T \\ * & \Omega_{3ij} & \bar{K}_{rj}^T B_i^T \\ * & * & 3P_1 - \overline{V}_{1ij} \end{bmatrix},$$

$$\overline{\Omega}_{4ij} = \begin{bmatrix} 0 & P_1 C_i^T + \bar{K}_j^T D_i^T & \bar{K}_j^T \\ P_2 E^T & -P_2 F^T + \bar{K}_{rj}^T D_i^T & \bar{K}_{rj}^T \\ 0 & 0 & 0 \end{bmatrix},$$

$$\overline{\Omega}_{5ij} = \begin{bmatrix} 3P_2 - \overline{V}_{2ij} & 0 & 0 \\ * & -\left(Q - G_{ij}\right)^{-1} & 0 \\ * & * & -\left(R - N_{ij}\right)^{-1} \end{bmatrix}.$$

According to Schur complement [210], (13.10) and (13.17), (13.18) can be rewritten as follows:

$$\overline{\Lambda} = \begin{bmatrix} \overline{\Lambda}_1 & \overline{\Lambda}_2 \\ * & \overline{\Lambda}_3 \end{bmatrix} > 0, \tag{13.19}$$

where

$$\overline{\Lambda}_1 = -P^{-1} + P^{-T}\left(-\Xi_{ij} + \Pi_{ij} + W_{ij} + M\right)P^{-1}$$

$$
\begin{aligned}
&= \begin{bmatrix} -P_1 + \overline{W}_{1ij} + \overline{M}_1 & \overline{W}_{2ij} + \overline{M}_2 \\ * & -P_2 + \overline{W}_{3ij} + \overline{M}_3 \end{bmatrix} \\
&\quad + \begin{bmatrix} P_1 C_i^T + \bar{K}_j^T D_i^T \\ -P_2 F^T + \bar{K}_{rj}^T D_i^T \end{bmatrix} \left(Q - G_{ij}\right) \begin{bmatrix} C_i P_1 + D_i \bar{K}_j & -F P_2 + D_i \bar{K}_{rj} \end{bmatrix} \\
&\quad + \overline{U}_j^T \left(R - N_{ij}\right) \overline{U}_j, \\
\overline{\Lambda}_2 &= P^{-T} \overline{A}_{ij}^T = \begin{bmatrix} P_1 A_i^T + \bar{K}_j^T B_i^T & 0 \\ \bar{K}_{rj}^T B_i^T & P_2 E^T \end{bmatrix}, \\
\overline{\Lambda}_3 &= 3P^{-T} - P^{-T} V_{ij} P^{-1} = \begin{bmatrix} 3P_1 - \overline{V}_{1ij} & 0 \\ * & 3P_2 - \overline{V}_{2ij} \end{bmatrix}.
\end{aligned}
$$

Define the following nonsingular matrices:

$$
\begin{aligned}
W_{ij} &= \begin{bmatrix} W_{1ij} & W_{2ij} \\ * & W_{3ij} \end{bmatrix}, \quad M = \begin{bmatrix} M_1 & M_2 \\ * & M_3 \end{bmatrix}, \\
\overline{W}_{1ij} &= P_1 W_{1ij} P_1, \quad \overline{W}_{2ij} = P_1 W_{2ij} P_2, \quad \overline{W}_{3ij} = P_2 W_{3ij} P_2, \\
\overline{M}_1 &= P_1 M_1 P_1, \quad \overline{M}_2 = P_1 M_2 P_2, \quad \overline{M}_3 = P_2 M_3 P_2, \\
\overline{V}_{1ij} &= P_1 V_{1ij} P_1, \quad \overline{V}_{2ij} = P_2 V_{2ij} P_2, \quad \bar{K}_j = K_j P_1, \quad \bar{K}_{rj} = K_{rj} P_2.
\end{aligned}
$$

Now, per-and post-multiplying (13.19) by $\operatorname{diag}\{P, P\}$ and its transpose, one can obtain

$$
\Lambda = \begin{bmatrix} \Lambda_1 & \Lambda_2 \\ * & \Lambda_3 \end{bmatrix} > 0,
$$

where

$$
\begin{aligned}
\Lambda_1 &= -P - \Xi_{ij} + \Pi_{ij} + W_{ij} + M, \\
\Lambda_2 &= \overline{A}_{ij}^T P, \quad \Lambda_3 = 3P - V_{ij}.
\end{aligned}
$$

By using $P\left(V_{ij} - P\right)^{-1} P \geq 2P - \left(V_{ij} - P\right) = 3P - V_{ij}$, we have

$$
\Lambda = \begin{bmatrix} \Lambda_1 & \Lambda_2 \\ * & P(V_{ij} - P)^{-1} P \end{bmatrix} > 0.
$$

According to Schur complement, the inequality (13.11) in Theorem 13.1 is satisfied.

Similarly, the condition $\Delta_{ij} < 0$ in (13.16) is rewritten as

$$
\bar{\Gamma} = \begin{bmatrix} \bar{\Gamma}_1 & \bar{\Gamma}_2 \\ * & \bar{\Gamma}_3 \end{bmatrix} < 0, \tag{13.20}
$$

where

$$
\begin{aligned}
\bar{\Gamma}_1 &= -\bar{\epsilon}P^{-1} + \bar{\epsilon}P^{-T}\Pi_{ij}P^{-1} + \left(\bar{\epsilon} - \underline{\epsilon}\right)\overline{W}_{ij} + \bar{\epsilon}_1\overline{M} \\
&= \begin{bmatrix} -\bar{\epsilon}P_1 + \left(\bar{\epsilon} - \underline{\epsilon}\right)\overline{W}_{1ij} + \bar{\epsilon}_1\overline{M}_1 & \left(\bar{\epsilon} - \underline{\epsilon}\right)\overline{W}_{2ij} + \bar{\epsilon}_1\overline{M}_2 \\ * & -\bar{\epsilon}P_2 + \left(\bar{\epsilon} - \underline{\epsilon}\right)\overline{W}_{3ij} + \bar{\epsilon}_1\overline{M}_3 \end{bmatrix} \\
&+ \begin{bmatrix} \sqrt{\bar{\epsilon}}P_1C_i^TQ^T + \sqrt{\bar{\epsilon}}\bar{K}_j^TD_i^TQ^T \\ -\sqrt{\bar{\epsilon}}P_2F^TQ^T + \sqrt{\bar{\epsilon}}\bar{K}_{rj}^TD_i^TQ^T \end{bmatrix} Q^{-1} \\
&\times \left[\sqrt{\bar{\epsilon}}QC_iP_1 + \sqrt{\bar{\epsilon}}QD_i\bar{K}_j \;\; -\sqrt{\bar{\epsilon}}QFP_2 + \sqrt{\bar{\epsilon}}QD_i\bar{K}_{rj}\right] \\
&+ \left(\bar{\epsilon}R\overline{U}_j\right)^T R^{-1}\bar{\epsilon}R\overline{U}_j, \\
\bar{\Gamma}_2 &= \sqrt{\bar{\epsilon}}P^{-T}\overline{A}_{ij}^T = \begin{bmatrix} \sqrt{\bar{\epsilon}}P_1A_i^T + \sqrt{\bar{\epsilon}}\bar{K}_j^TB_i^T & 0 \\ \sqrt{\bar{\epsilon}}\bar{K}_{rj}^TB_i^T & \sqrt{\bar{\epsilon}}P_2E^T \end{bmatrix}, \\
\bar{\Gamma}_3 &= -P^{-1} = \begin{bmatrix} -P_1 & 0 \\ * & -P_2 \end{bmatrix}.
\end{aligned}
$$

Pre- and post-multiply (13.20) by diag$\{P, P\}$ and its transpose, which yields

$$
\Gamma = \begin{bmatrix} \Gamma_1 & \Gamma_2 \\ * & \Gamma_3 \end{bmatrix} < 0,
$$

where

$$
\begin{aligned}
\Gamma_1 &= -\bar{\epsilon}P + \bar{\epsilon}\Pi_{ij} + \left(\bar{\epsilon} - \underline{\epsilon}\right)W_{ij} + \left(\bar{\epsilon} - \frac{1}{\sigma\lambda}\right)M, \\
\Gamma_2 &= \sqrt{\bar{\epsilon}}\overline{A}_{ij}^TP, \quad \Gamma_3 = -P.
\end{aligned}
$$

By means of Schur complement, the inequality (13.12) in Theorem 13.1 holds. Furthermore, the condition $\begin{bmatrix} -\beta_kI & \bar{r}^T(k) \\ * & -X \end{bmatrix} < 0$ is equivalent to $\bar{r}^T(k)X^{-1}\bar{r}(k) < \beta_kI$, and the condition $\begin{bmatrix} -\alpha & \varsigma^T(0) \\ * & -X \end{bmatrix} < 0$ is equivalent to $\varsigma^T(0)X^{-1}\varsigma(0) < \alpha$. Therefore, $J = \alpha + \sum_{k=0}^{\infty}\beta_kI$ implies the minimum value of optimal guaranteed cost. This completes the proof. □

Remark 13.3 In Theorem 13.2, a sufficient state-feedback controller design condition has been obtained. However, the state variables are usually unavailable in practical applications, which brings considerable difficulties to realize the strategy proposed in this chapter. Recently, a novel observer-based mode-dependent control scheme was proposed for systems with the actuator fault, input disturbances and sensor fault in [103], in which the state, disturbance and fault were assembled into the state of the new system and nice conditions to stabilize the resulting closed-loop system were obtained. The methodology used in [103] can be borrowed to deal with the unavailable state case with the external disturbance and fault in the framework of the IT2 T–S fuzzy model.

13.4 Simulation Results

Example 13.4 A numerical example is used to illustrate the effectiveness of proposed control scheme. Consider the following 3-rule IT2 fuzzy model:

Plant Rule i: IF $x_1(k)$ is M_{i1}, THEN,

$$\begin{cases} x(k+1) = A_i x(k) + B_i u(k) + L_i w(k), \\ y(k) = C_i x(k) + D_i u(k). \end{cases}$$

The reference model and reference input for the tracking control system are given as:

$$\begin{cases} x_r(k+1) = E x_r(k) + r(k), \\ y_r(k) = F x_r(k), \end{cases}$$

where

$$A_1 = \begin{bmatrix} -6.5019 & 4.9999 \\ 0.3144 & -2.5095 \end{bmatrix}, \quad A_2 = \begin{bmatrix} 0.0102 & 1.1097 \\ 0.10971 & 1.0109 \end{bmatrix},$$
$$A_3 = \begin{bmatrix} -3.0200 & 4.0201 \\ 3.9159 & -1.9101 \end{bmatrix}, \quad B_1 = \begin{bmatrix} 5.0261 & 5.0443 \end{bmatrix}^T,$$
$$B_2 = \begin{bmatrix} 5.0392 & 5.1251 \end{bmatrix}^T, \quad B_3 = \begin{bmatrix} 5.0948 & 5.0741 \end{bmatrix}^T,$$
$$C_1 = \begin{bmatrix} 22.5000 & -0.1094 \end{bmatrix}, \quad C_2 = \begin{bmatrix} -0.6000 & -0.9836 \end{bmatrix},$$
$$C_3 = \begin{bmatrix} 6.0000 & -1.5050 \end{bmatrix}, \quad L_1 = \begin{bmatrix} 3.0261 & 3.0443 \end{bmatrix}^T,$$
$$L_2 = \begin{bmatrix} 3.0392 & 3.1251 \end{bmatrix}^T, \quad L_3 = \begin{bmatrix} 3.0948 & 3.0741 \end{bmatrix}^T,$$
$$D_1 = -15.0261, \quad D_2 = -15.0392, \quad D_3 = 15.0948,$$
$$Q = 0.1, \quad R = 0.1, \quad E = -0.990, \quad F = 3.2,$$
$$r(k) = 0.04\cos(3.24k - 3.24), \quad w(k) = 0.04\cos(3.24k - 3.24).$$

Tables 13.1 and 13.2 show the LMFs and UMFs of the plant and the controller, respectively. Let $\underline{\beta}_i = \overline{\beta}_i = 0.5, x_1 \in [-81, 81], \overline{\alpha}_i = 1 - \underline{\alpha}_i$. The state x_1 is divided into 19 equal-size sub-states (i.e., $\zeta = 1, 2, \ldots, 19$).

Define

$$\underline{\alpha}_1 = \frac{1}{2}\sin^2 x_1, \quad \underline{\alpha}_3 = \frac{1}{2}\sin^2 x_1,$$
$$\underline{\alpha}_2 = \frac{1}{\overline{\varphi}_2 - \underline{\varphi}_2}\left(-1 + \left(\overline{\varphi}_1 - \underline{\alpha}_1\left(\overline{\varphi}_1 - \underline{\varphi}_1\right)\right.\right.$$

Table 13.1 LMFs and UMFs for the plant

LMFs for the plant
$\underline{\mu}_{F_1^1}(x_1) = 0.8 - \left(0.8/\left(1 + \mathrm{e}^{(-(x_1+81)/14)}\right)\right)$
$\underline{\mu}_{F_1^3}(x_1) = 0.8/\left(1 + \mathrm{e}^{(-(x_1-81)/14)}\right)$
$\underline{\mu}_{F_1^2}(x_1) = 1 - \overline{\mu}_{F_1^1}(x_1) - \overline{\mu}_{F_1^3}(x_1)$
UMFs for the plant
$\overline{\mu}_{F_1^1}(x_1) = 1 - \left(1/\left(1 + \mathrm{e}^{(-(x_1+81)/14)}\right)\right)$
$\overline{\mu}_{F_1^3}(x_1) = 1/\left(1 + \mathrm{e}^{(-(x_1-81)/14)}\right)$
$\overline{\mu}_{F_1^2}(x_1) = 1 - \underline{\mu}_{F_1^1}(x_1) - \underline{\mu}_{F_1^3}(x_1)$

Table 13.2 LMFs and UMFs for the controller

LMFs for the controller	UMFs for the controller
$\underline{\mu}_{N_1^1}(x_1) = \mathrm{e}^{(-x_1^2/5000)}$	$\overline{\mu}_{N_1^1}(x_1) = \mathrm{e}^{(-x_1^2/0.5)}$
$\underline{\mu}_{N_1^2}(x_1) = 1 - \overline{\mu}_{N_1^1}(x_1)$	$\overline{\mu}_{N_1^2}(x_1) = 1 - \underline{\mu}_{N_1^1}(x_1)$

$$
\begin{aligned}
&+ \overline{\varphi}_3 - \underline{\alpha}_3\left(\overline{\varphi}_3 - \underline{\varphi}_3\right) + \overline{\varphi}_2\Big)\Big), \\
&\upsilon_{11\varsigma}(x_1) = 1 - \frac{x_1 - \underline{x}_{1,\varsigma}}{\overline{x}_{1,\varsigma} - \underline{x}_{1,\varsigma}}, \quad \upsilon_{12\varsigma}(x_1) = 1 - \upsilon_{11\varsigma}(x_1), \\
&\underline{x}_{1,\varsigma} = \frac{162}{19}(\varsigma - 10), \quad \overline{x}_{1,\varsigma} = \frac{162}{19}(\varsigma - 9), \\
&\underline{\epsilon}_{ij1\varsigma} = \underline{\varphi}_i\left(\underline{x}_{1,\varsigma}\right)\underline{\psi}_j\left(\underline{x}_{1,\varsigma}\right), \quad \underline{\epsilon}_{ij2\varsigma} = \underline{\varphi}_i\left(\overline{x}_{1,\varsigma}\right)\underline{\psi}_j\left(\overline{x}_{1,\varsigma}\right), \\
&\overline{\epsilon}_{ij1\varsigma} = \overline{\varphi}_i\left(\underline{x}_{1,\varsigma}\right)\overline{\psi}_j\left(\underline{x}_{1,\varsigma}\right), \quad \overline{\epsilon}_{ij2\varsigma} = \overline{\varphi}_i\left(\overline{x}_{1,\varsigma}\right)\overline{\psi}_j\left(\overline{x}_{1,\varsigma}\right), \quad \forall\varsigma.
\end{aligned}
$$

For demonstration, the initial conditions are given as $x_0 = [0.2 \quad -0.2]^T$ and $x_{r0} = 0.2$. The optimization problem in Theorem 13.2 is solved through MATLAB Control Toolbox when minimizing the cost function J, and we can obtain the upper bound of the minimum cost $J_0 = 13.100452$.

Applying Theorem 13.2, we can obtain the optimal reliable guaranteed cost controller gains as follows:

$$
\begin{aligned}
K_1 &= \begin{bmatrix} 0.0001 & -0.2031 \end{bmatrix}, \quad K_2 = \begin{bmatrix} 0.0579 & -0.1804 \end{bmatrix}, \\
K_{r1} &= -0.0076, \quad K_{r2} = -0.0062.
\end{aligned}
$$

The state trajectories of the open-loop and closed-loop system are shown in Figs. 13.1 and 13.2, respectively. It can be observed that the unstable open-loop system becomes stable after the controller is designed for the system. Figure 13.3 plots the outputs of the closed-loop system and the reference model. It can be seen that the tracking performance of the designed closed-loop system performs well. Additionally, Fig. 13.4 plots the control input.

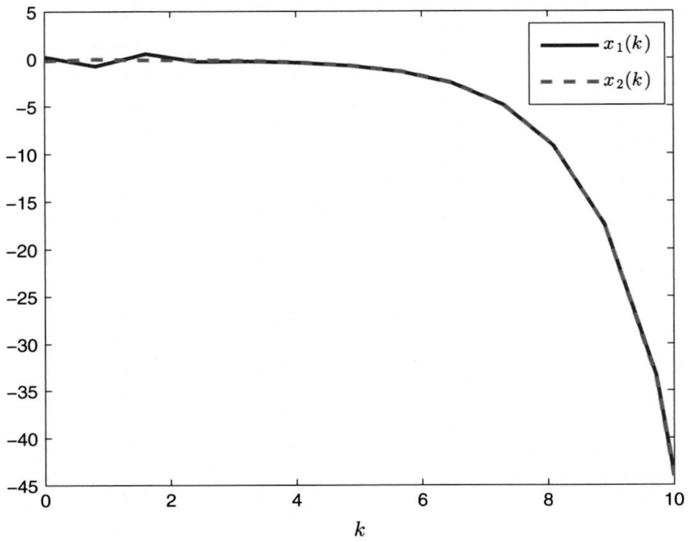

Fig. 13.1 States of the open-loop system

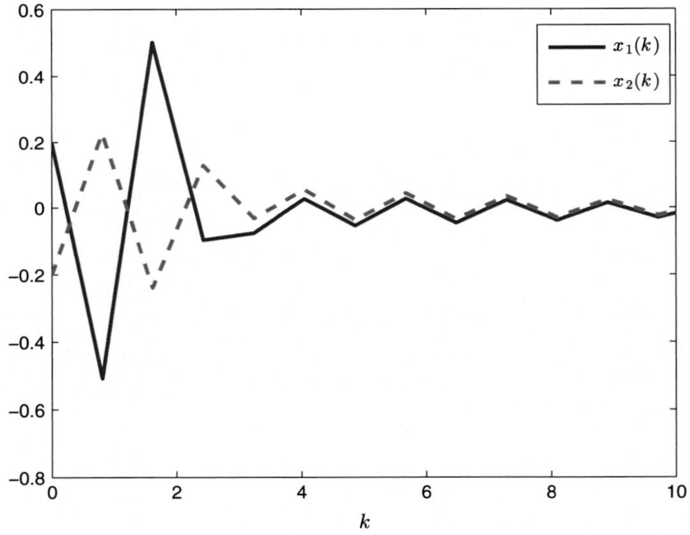

Fig. 13.2 States of the closed-loop system

Example 13.5 Taking into account the mass-spring-damping system shown in Fig. 6.1 and on the basis of Newton's law, we can obtain:

$$m\ddot{x} + F_f + Fs = u\,(t)\,,$$

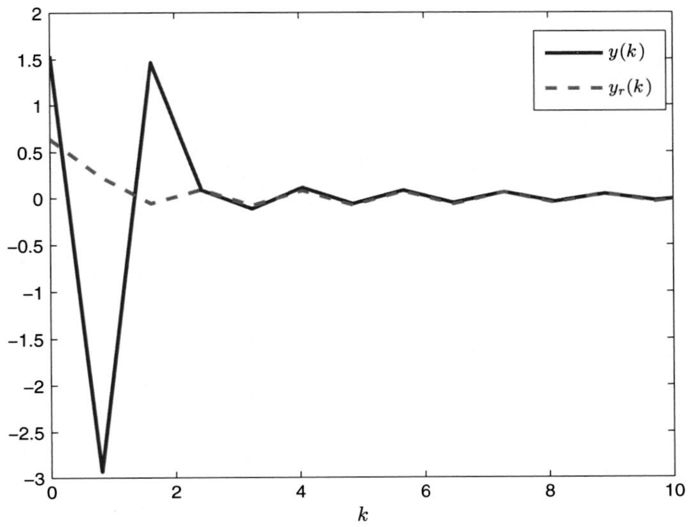

Fig. 13.3 Outputs $y(k)$ and $y_r(k)$ of the closed-loop system

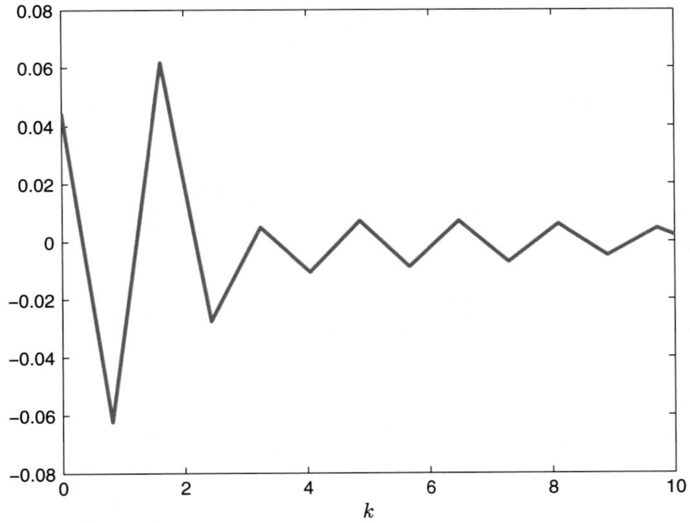

Fig. 13.4 Control input $u(k)$

where m represents the mass, F_f represents the friction force, F_s represents the restoring force of the spring and u represents the external control input. The friction force $F_f = c\dot{x}$ with $c > 0$ and the hardening spring force $F_s = k\left(1 + a^2x^2\right)x$ with constants k and a. Thus, the dynamic equation is as follows:

$$m\ddot{x} + c\dot{x} + kx + ka^2x^3 = u(t),$$

in which x represents the displacement from a reference point. Define $x(t) = \left[x_1^T(t)\ x_2^T(t)\right]^T = \left[x^T\ \dot{x}^T\right]^T$ and $\tilde{f} = \frac{-k-ka^2x_1^2(t)}{m}$. Let $x_1(t) \in [-3, 3]$, $m = 1$ kg, $c = 5$ N·m/s, $k_{\min} = 5$ N/m, $k_{\max} = 10$ N/m and $a = 2$ m^{-1}. Then, $\tilde{f}_{\max} = -5$ (i.e., the maximum value of $\tilde{f}$) with $k = 5$ and $x_1(t) = 0$. $\tilde{f}_{\min} = -370$ (i.e., the minimum value of $\tilde{f}$) with $k = 10$ and $x_1^2(t) = 9$. According to the membership function property $m_1(x_1(t)) + m_2(x_1(t)) = 1$, $\tilde{f}$ can be represented as

$$\tilde{f} = m_1(x_1(t))\,\tilde{f}_{\min} + m_2(x_1(t))\,\tilde{f}_{\max}.$$

Then, it can be found that

$$m_1(x_1(t)) = \frac{-\tilde{f} + \tilde{f}_{\max}}{\tilde{f}_{\max} - \tilde{f}_{\min}}, \quad m_2(x_1(t)) = \frac{\tilde{f} - \tilde{f}_{\min}}{\tilde{f}_{\max} - \tilde{f}_{\min}}.$$

According to the uncertain parameter k, the membership functions for IT2 fuzzy system can be obtained as follows:

$$\underline{m}_1(x_1(t)) = \frac{-\tilde{f}(t, k=5) + \tilde{f}_{\max}}{\tilde{f}_{\max} - \tilde{f}_{\min}}, \quad \bar{m}_2(x_1(t)) = \frac{\tilde{f}(t, k=5) - \tilde{f}_{\min}}{\tilde{f}(t, k=5)_{\max} - \tilde{f}_{\min}},$$
$$\bar{m}_1(x_1(t)) = \frac{-\tilde{f}(t, k=10) + \tilde{f}_{\max}}{\tilde{f}_{\max} - \tilde{f}_{\min}}, \quad \underline{m}_2(x_1(t)) = \frac{\tilde{f} - \tilde{f}(t, k=10)_{\min}}{\tilde{f}_{\max} - \tilde{f}_{\min}}.$$

Membership functions of the controller are given in Table 13.3.

Thus, we have the following continuous-time IT2 T–S fuzzy model for the mass-spring-damping system:

$$\dot{x} = \sum_{i=1}^{2} \varphi_i(x(t))\left[A_i x(t) + B_i u(t)\right],$$

where

Table 13.3 LMFs and UMFs of the controller

Lower bounds	Upper bounds
$\underline{u}_{N_{11}}(x_1) = e^{\left(-\frac{x_1^2}{0.5}\right)}$	$\overline{u}_{N_{11}}(x_1) = \underline{u}_{N_{11}}(x_1)$
$\underline{u}_{N_{12}}(x_1) = \underline{u}_{N_{11}}(x_1)$	$\overline{u}_{N_{12}}(x_1) = \underline{u}_{N_{12}}(x_1)$

$$A_1 = \begin{bmatrix} 0 & 1 \\ \tilde{f}_{\min} & -\frac{c}{m} \end{bmatrix}, \quad B_1 = \begin{bmatrix} 0 \\ \frac{1}{m} \end{bmatrix},$$
$$A_2 = \begin{bmatrix} 0 & 1 \\ \tilde{f}_{\max} & -\frac{c}{m} \end{bmatrix}, \quad B_2 = \begin{bmatrix} 0 \\ \frac{1}{m} \end{bmatrix}.$$

With the sampling time $T = 1$ s, via the method in [212], we can get the following discrete-time IT2 T–S fuzzy model for the mass-spring-damping system:

$$x(k+1) = \sum_{i=1}^{2} \varphi_i (x(k)) \left[A_i x(k) + B_i u(k)\right],$$

where

$$A_1 = \begin{bmatrix} 0.0824 & 0.0010 \\ -0.3517 & 0.0777 \end{bmatrix}, \quad B_1 = \begin{bmatrix} 0.0025 \\ 0.0010 \end{bmatrix},$$
$$A_2 = \begin{bmatrix} 0.3897 & 0.1003 \\ -0.5014 & -0.1118 \end{bmatrix}, \quad B_2 = \begin{bmatrix} 0.1221 \\ 0.1003 \end{bmatrix}.$$

The reference model is defined the same as that in Example 13.4. For demonstration, the external disturbance will be added into the system. Other relevant matrices are given as

$$L_1 = \begin{bmatrix} 0.0055 \\ 0.01399 \end{bmatrix}, \quad L_2 = \begin{bmatrix} -0.01776 \\ 0.0330 \end{bmatrix}, \quad F_1 = 0.0354,$$
$$C_1 = \begin{bmatrix} -01.0474 & -01.1704 \end{bmatrix}, \quad D_1 = 01.10794, \quad E = -0.3466, \quad R = 1,$$
$$C_2 = \begin{bmatrix} -01.0054 & -01.0881 \end{bmatrix}, \quad D_2 = 01.10463, \quad F = 0.3466, \quad Q = 1.$$

In this example, the number of sub-state is 20. The initial conditions are $x_0 = [0.0 \;\; 0.0]^T$ and $x_{r0} = 13.5$, respectively. Applying Theorem 13.2, the optimal reliable guaranteed cost controller gains can be calculated as follows:

$$K_1 = \begin{bmatrix} 0.4792 & 0.4991 \end{bmatrix}, \quad K_2 = \begin{bmatrix} 0.4883 & 0.5410 \end{bmatrix},$$
$$K_{r1} = 0.1532, \quad K_{r2} = 0.1632.$$

The state trajectories of the closed-loop system are shown in Fig. 13.5. It can be observed that the unstable open-loop system becomes stable after the controller is designed for the system. Figure 13.6 plots the outputs of the closed-loop system and the reference model. It can be seen that the tracking performance of the designed closed-loop system performs well. Additionally, Fig. 13.7 plots the control input.

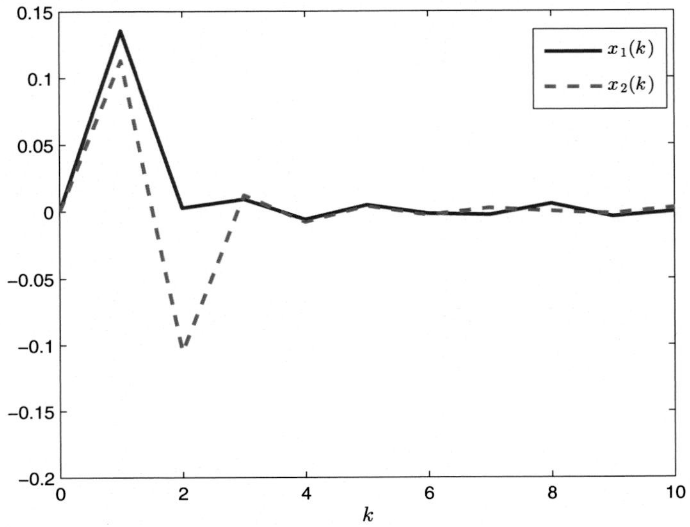

Fig. 13.5 States of the closed-loop system

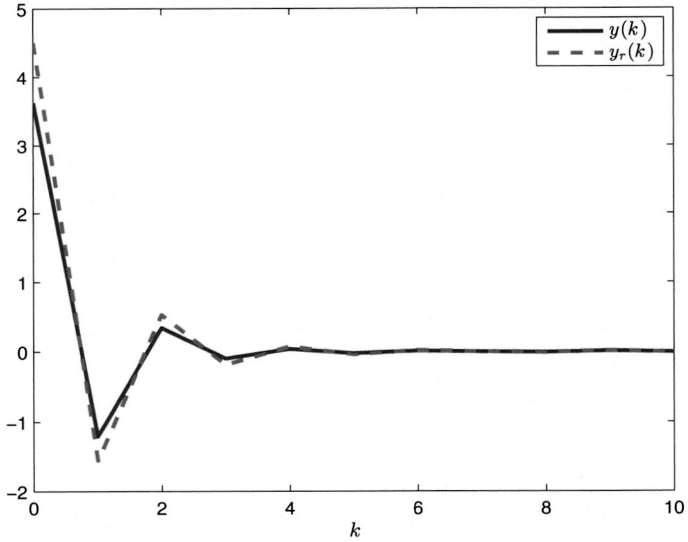

Fig. 13.6 Outputs $y(k)$ and $y_r(k)$ of the closed-loop system

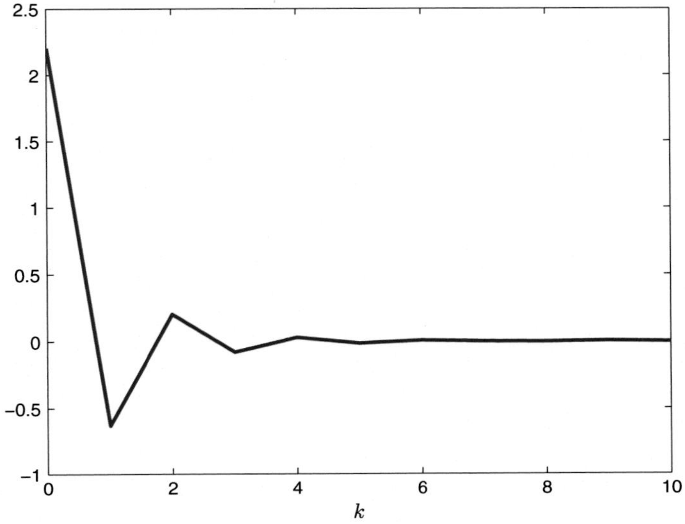

Fig. 13.7 The controlled input $u(k)$

13.5 Conclusion

In this chapter, the problem of guaranteed cost output tracking control has been studied for the discrete-time IT2 fuzzy system under imperfect premise matching. For the involved system, sufficient conditions have been given for the existence of fuzzy guaranteed cost control law and cost upper bound. The cost function minimization problem has been solved by the sufficient conditions in terms of LMIs. Furthermore, the tracking performance has been guaranteed with a small tracking error. The simulation results have been illustrated the effectiveness of the presented results.

Conclusion

Chapter 14
Conclusion and Further Work

This chapter draws conclusions on the book, and points out some possible research directions related to the work done in this book.

14.1 Conclusion

The focus of the book has been placed on modeling, analysis and synthesis problems for IT2 fuzzy-model-based systems (continuous-time systems and discrete-time systems). Specifically, several research problems have been investigated in detail.

1. The stability and stabilization problems have been investigated for both continuous-time and discrete-time IT2 fuzzy-model-based systems. To facilitate the stability analysis and control synthesis, an IT2 T–S fuzzy model has been employed to represent the dynamics of nonlinear systems of which the parameter uncertainties are captured by IT2 membership functions characterized by the LMFs and UMFs. Some novel IT2 fuzzy controllers have been proposed to perform the control process, where the membership functions and number of fuzzy rules can be freely chosen and are different from those of the IT2 T–S fuzzy model. To relax the stability analysis for this class of IT2 FMB control systems, the information of footprint of uncertainties, and the LMFs and UMFs have been taken into account for the stability analysis. Based on the Lyapunov stability theory, some stability conditions in terms of linear matrix inequalities are obtained to determine the system stability and achieve the controller design.
2. The dynamic output-feedback control and static output-feedback control problems have been studied for IT2 fuzzy-model-based systems. The IT2 fuzzy model and the controllers do not share the same membership functions. Some performance indexes, such as H_∞, L_2–L_∞, passive and dissipativity performances, have been considered in the controller design process. On the basis of Lyapunov stability theory, the IT2 fuzzy static and dynamic output-feedback controllers are designed respectively to guarantee that the closed-loop system is asymptotically

H. Li et al., *Analysis and Synthesis for Interval Type-2 Fuzzy-Model-Based Systems*, DOI 10.1007/978-981-10-0593-0_14

stable with the desired performances. The existence conditions of the two kinds of controllers are obtained in terms of convex optimization problem.

3. The switched dynamic output-feedback control problem has been considered for continuous-time IT2 fuzzy systems. A switched output-feedback controller, which depends on the values of membership functions, has been constructed. The membership functions of IT2 fuzzy systems contain parameter uncertainties. Based on the IT2 fuzzy set theory, the parameter uncertainties can be effectively obtained by LMFs and UMFs. A novel IT2 switched output-feedback controller has been designed to ensure that the closed-loop system is asymptotically stable with an H_∞ performance.
4. The sampled-data control problem for IT2 fuzzy systems with actuator fault has been addressed. The IT2 fuzzy system and the IT2 state-feedback controller share different membership functions. By considering the mismatched membership functions, the IT2 fuzzy model and the IT2 state-feedback sampled-data controller have been constructed. Based on Lyapunov stability theory, an IT2 state-feedback sampled-data controller has been designed such that the closed-loop system is asymptotically stable. The actuator failure has been considered in the control systems. The resulting closed-loop system is reliable since the designed controller can guarantee the asymptotic stability and H_∞ performance when the actuator experiences failure.
5. The problem of fault-tolerant control has been investigated for discrete-time IT2 fuzzy time delay system with actuator faults under imperfect premise matching. The time-varying delay and actuator fault have been taken into account for the IT2 fuzzy discrete-time systems. The novel IT2 state-feedback controller and the IT2 fuzzy model shared different LMFs and UMFs. The fault-tolerant controller has been designed to compensate for the effect of faults by stabilizing the closed-loop system under the actuator failures. Furthermore, the standard IT2 state-feedback controller has been designed such that the closed-loop system is asymptotically stable and has an H_∞ performance. The obtained conditions of the fault-tolerant controller and the standard IT2 controller can be expressed by the convex optimization problem.
6. The output tracking problem has been addressed for both continuous-time and discrete-time IT2 fuzzy-model-based systems. An IT2 fuzzy state-feedback controller is designed to perform the tracking control problem, where the membership functions can be freely chosen since the number of fuzzy rules is different from that of the IT2 T–S fuzzy model. Based on Lyapunov stability theory, some existence conditions of IT2 fuzzy systems and designed output tracking controller have been obtained to guarantee that the output of the closed-loop IT2 control system can track the output of a given reference model well.
7. The problem of filter design has been addressed for continuous-time IT2 fuzzy systems with $\mathcal{D}$ stability constraint based on a new performance index. The H_∞, L_2–L_∞, passive and dissipativity fuzzy filter design problems have been addressed for IT2 fuzzy systems with $\mathcal{D}$ stability constraint in a unified frame. Under the new performance index frame, using Lyapunov stability theory, a novel IT2 filter has been designed such that the filtering error system guarantees the

prescribed H_∞, L_2–L_∞, passive and dissipativity performance levels with $\mathcal{D}$ stability constraint. The existence conditions of the IT2 filter have been expressed as the convex optimization problem and the filter parameters in the conditions can be solved by the standard software.

8. The fault detection problem has been considered for continuous-time IT2 fuzzy systems subject to sensor nonlinearities. By using a general observer-based fault detection filter as a residual generator, the fault detection problem has been described as a filter design problem. The fault detection filter has been designed to guarantee the prescribed H_∞ performance level. A decomposition approach has been employed to handle the characteristic of sensor saturation. Using Lyapunov stability theory, a novel IT2 fault detection filter has been designed to guarantee that the fault detection system is asymptotically stable with an H_∞ performance.
9. The problem of model reduction has been investigated for continuous-time IT2 fuzzy systems subject to $\mathcal{D}$ stability constraints. The membership functions and the number of rules can be freely chosen and they are different between the original system and the reduced-order system. By introducing some slack matrices and utilizing Lyapunov stability theory, the existence conditions of model reduction have been obtained to guarantee that the reduced-order model can approximate the original system with an H_∞ performance. The parameters of the reduced-order system in the conditions can be obtained by standard software.

14.2 Further Work

Related topics for the future research work are listed below:

1. The future research direction is to investigate the stability and control synthesis problems for some kinds of IT2 fuzzy systems, such as stochastic IT2 fuzzy systems, Markov jumping IT2 fuzzy systems, switched IT2 fuzzy systems and two dimensional IT2 fuzzy systems. Take switched IT2 fuzzy systems as an example, some advanced techniques (such as switched quadratic Lyapunov functions, piecewise Lyapunov functions, and average dwell time) and the method proposed in this book can be used to address the problems of analysis and design for switched IT2 fuzzy systems. Analysis and synthesis of these systems are of theoretical importance and significance.
2. Networked control systems have been applied widely in practical industrial systems, ranging from the factory automation and grid-connected photovoltaic generation plants to autonomous mobile robots and cascaded H-bridge converters. However, the introduction of communication networks brings considerable challenges such as modeling, analysis and synthesis of NCSs, including network-induced delay, data packet dropouts, limited width, quantization, and so on. How to use the IT2 fuzzy-model-based systems to modeling the nonlinear NCSs in the uniform frame and then consider the stability, filtering and controller design problems are our future work.

3. For discrete-time IT2 fuzzy systems with time delay considered in this book, the results on stability may have some conservativeness. Some recently developed methods such as free-weight matrix method, delay-partitioning method, small gain based input-output method, and reciprocally convex method can be utilized to further reduce the conservativeness caused by time-delay. In addition, these methods can be used to reduce the conservativeness of sampled-data controller design. Furthermore, these advanced methods can facilitate the controller and filter design of continuous-time IT2 fuzzy systems with time delay.

References

1. Abiyev RH, Okyay K (2010) Type 2 fuzzy neural structure for identification and control of time-varying plants. IEEE Trans Ind Electron 57(12):4147–4159
2. Aliasghary M, Eksin I, Guzelkaya M, Kumbasar T (2013) A design methodology and analysis for interval type-2 fuzzy PI/PD controllers. Int J Innov Comput Inf Control 9(10):4215–4230
3. Ario C, Sala A (2008) Extensions to "stability analysis of fuzzy control systems subject to uncertain grades of membership". IEEE Trans Syst Man Cybern Part B: Cybern 38(2):558–563
4. Assawinchaichote W, Nguang SK (2004) H_∞ filtering for fuzzy singularly perturbed systems with pole placement constraints: an LMI approach. IEEE Trans Signal Process 52(6):1659–1667
5. Assawinchaichote W, Nguang SK, Shi P, Boukas EK (2008) H_∞ fuzzy state-feedback control design for nonlinear systems with-stability constraints: an LMI approach. Math Comput Simul 78(4):514–531
6. Bai Y, Zhuang HQ, Roth ZS (2005) Fuzzy logic control to suppress noises and coupling effects in a laser tracking system. IEEE Trans Control Syst Technol 13(1):113–121
7. Balasubramaniam P, Banu-Jarina L (2015) Robust stability criterion for discrete-time nonlinear switched systems with randomly occurring delays via T-S fuzzy approach. Complexity 20(6):49–61
8. Barkat S, Tlemçani A, Nouri H (2011) Noninteracting adaptive control of PMSM using interval type-2 fuzzy logic systems. IEEE Trans Fuzzy Syst 19(5):925–936
9. Baturone I, Moreno-Velo FJ, Sanchez-Solano S, Ollero A (2004) Automatic design of fuzzy controllers for car-like autonomous robots. IEEE Trans Fuzzy Syst 12(4):447–465
10. Bergsten P, Palm R, Driankov D (2002) Observers for Takagi-Sugeno fuzzy systems. IEEE Trans Syst Man Cybern Part B Cybern 32(1):114–121
11. Biglarbegian M, Melek WW, Mendel JM (2010) On the stability of interval type-2 TSK fuzzy logic control systems. IEEE Trans Syst Man Cybern Part B Cybern 40(3):798–818
12. Bingul Z, Cook GE, Strauss AM (2000) Application of fuzzy logic to spatial thermal control in fusion welding. IEEE Trans Ind Appl 36(6):1523–1530
13. Bonissone PP, Badami V, Chiang KH, Khedkar PS, Marcelle KW, Schutten MJ (1995) Industrial applications of fuzzy logic at general electric. Proc IEEE 38(3):450–465
14. Boroushaki M, Ghofrani MB, Lucas C, Yazdanpanah MJ (2003) Identification and control of a nuclear reactor core (VVER), using recurrent neural networks and fuzzy systems. IEEE Trans Nucl Sci 50(1):159–174
15. Boukezzoula R, Galichet S, Foulloy L (2004) Observer-based fuzzy adaptive control for a class of nonlinear systems: real-time implementation for a robot wrist. IEEE Trans Control Syst Technol 12(3):340–351

H. Li et al., *Analysis and Synthesis for Interval Type-2 Fuzzy-Model-Based Systems*, DOI 10.1007/978-981-10-0593-0

16. Bououden S, Chadli M, Karimi H (2015) Control of uncertain highly nonlinear biological process based on Takagi-Sugeno fuzzy models. Signal Process 108:195–205
17. Boyd S, El Ghaoui L, Feron E, Balakrishnan V (1994) Linear matrix inequalities in system and control theory. Society for Industrial and Applied Mathematics (SIAM), Philadelphia
18. Campos V, Souza FO, Torres L, Palhares RM (2013) New stability conditions based on piecewise fuzzy Lyapunov functions and tensor product transformations. IEEE Trans Fuzzy Syst 21(4):748–760
19. Cao J, Li P, Liu H (2010) An interval fuzzy controller for vehicle active suspension systems. IEEE Trans Intell Transp Syst 11(4):885–895
20. Cao J, Liu H, Li P, Brown D (2008) An interval type-2 fuzzy logic controller for quarter-vehicle active suspensions. Proc Inst Mech Eng Part D J Automob Eng 222(8):1361–1373
21. Cao YY, Frank MP (2000) Analysis and synthesis of nonlinear time-delay systems via fuzzy control approach. IEEE Trans Fuzzy Syst 8(2):200–211
22. Cao YY, Frank PM (2000) Robust H_∞ disturbance attenuation for a class of uncertain discrete-time fuzzy systems. IEEE Trans Fuzzy Syst 8(4):406–415
23. Cao YY, Frank PM (2001) Stability analysis and synthesis of nonlinear time-delay systems via linear Takagi-Sugeno fuzzy models. Fuzzy Sets Syst 124(2):213–229
24. Castillo O, Melin P (2008) Type-2 fuzzy logic: theory and applications, vol 223. Springer, Heidelberg
25. Castillo O, Melin P (2012) Type-2 fuzzy logic systems. In: Recent advances in interval type-2 fuzzy systems. Springer, Berlin, pp 7–12
26. Cecati C, Ciancetta F, Siano P (2010) A multilevel inverter for photovoltaic systems with fuzzy logic control. IEEE Trans Ind Electron 57(12):4115–4125
27. Chadli M, Abdo A, Ding SX (2013) H_-/H_∞ fault detection filter design for discrete-time Takagi-Sugeno fuzzy system. Automatica 49(7):1996–2005
28. Chadli M, Guerra TM (2012) LMI solution for robust static output feedback control of discrete Takagi-Sugeno fuzzy models. IEEE Trans Fuzzy Syst 20(6):1160–1165
29. Chadli M, Karimi H, Shi P (2014) On stability and stabilization of singular uncertain Takagi-Sugeno fuzzy systems. J Frankl Inst 351(3):1453–1463
30. Chadli M, Karimi HR (2013) Robust observer design for unknown inputs Takagi-Sugeno models. IEEE Trans Fuzzy Syst 21(1):158–164
31. Chang XH (2012) Robust nonfragile filtering of fuzzy systems with linear fractional parametric uncertainties. IEEE Trans Fuzzy Syst 20(6):1001–1011
32. Chang YH, Chan WS, Chang CW (2013) T-S fuzzy model-based adaptive dynamic surface control for ball and beam system. IEEE Trans Ind Electron 60(6):2251–2263
33. Chen B, Liu X (2004) Reliable control design of fuzzy dynamic systems with time-varying delay. Fuzzy Sets Syst 146(3):349–374
34. Chen B, Liu X (2005) Fuzzy guaranteed cost control for nonlinear systems with time-varying delay. IEEE Trans Fuzzy Syst 13(2):238–249
35. Chen B, Liu X, Lin C, Liu K (2009) Robust H_∞ control of Takagi-Sugeno fuzzy systems with state and input time delays. Fuzzy Sets Syst 160(4):403–422
36. Chen B, Liu X, Tong S, Lin C (2008) Observer-based stabilization of T-S fuzzy systems with input delay. IEEE Trans Fuzzy Syst 16(3):652–663
37. Chen BS, Lee HC, Wu CF (2015) Pareto optimal filter design for nonlinear stochastic fuzzy systems via multiobjective optimization. IEEE Trans Fuzzy Syst 23(2):387–399
38. Chen C, Liu H, Guan X (2009) H_∞ filtering of time-delay T-S fuzzy systems based on piecewise Lyapunov-Krasovskii functional. Signal Process 89(10):1998–2005
39. Chen Y, Xue A, Zhou S (2009) New delay-dependent L_2-L_∞ filter design for stochastic time-delay systems. Signal Process 89(6):974–980
40. Chiu CS (2010) T-S fuzzy maximum power point tracking control of solar power generation systems. IEEE Trans Energy Convers 25(4):1123–1132
41. Choi DJ, Park P (2004) Guaranteed cost controller design for discrete-time switching fuzzy systems. IEEE Trans Syst Man Cybern Part B Cybern 34(1):110–119

42. Derakhshan SF, Fatehi A (2014) Non-monotonic Lyapunov functions for stability analysis and stabilization of discrete time Takagi-Sugeno fuzzy systems. Int J Innov Comput Inf Control 10(4):1567–1586
43. Dong H, Wang Z, Gao H (2009) H_∞ fuzzy control for systems with repeated scalar nonlinearities and random packet losses. IEEE Trans Fuzzy Syst 17(2):440–450
44. Dong H, Wang Z, Lam J, Gao H (2012) Fuzzy-model-based robust fault detection with stochastic mixed time delays and successive packet dropouts. IEEE Trans Syst Man Cybern Part B Cybern 42(2):365–376
45. Dong J, Wang Y, Yang GH (2010) Output feedback fuzzy controller design with local nonlinear feedback laws for discrete-time nonlinear systems. IEEE Trans Syst Man Cybern Part B Cybern 40(6):1447–1459
46. Dong J, Yang GH (2008) Dynamic output feedback control synthesis for continuous-time T-S fuzzy systems via a switched fuzzy control scheme. IEEE Trans Syst Man Cybern Part B Cybern 38(4):1166–1175
47. Fang CH, Liu YS, Kau SW, Hong L, Lee CH (2006) A new LMI-based approach to relaxed quadratic stabilization of Takagi-Sugeno fuzzy control systems. IEEE Trans Fuzzy Syst 14(3):386–397
48. Feng G (2003) Controller synthesis of fuzzy dynamic systems based on piecewise Lyapunov functions. IEEE Trans Fuzzy Syst 11(5):605–612
49. Feng G (2006) A survey on analysis and design of model-based fuzzy control systems. IEEE Trans Fuzzy Syst 14(5):676–697
50. Feng G, Chen M, Sun D, Zhang T (2008) Approaches to robust filtering design of discrete time fuzzy dynamic systems. IEEE Trans Fuzzy Syst 16(3):331–340
51. Feng Z, Lam J (2012) Robust reliable dissipative filtering for discrete delay singular systems. Signal Process 92(12):3010–3025
52. Feng Z, Lam J, Gao H (2011) α-dissipativity analysis of singular time-delay systems. Automatica 47(11):2548–2552
53. Gao H, Chen T (2007) Stabilization of nonlinear systems under variable sampling: a fuzzy control approach. IEEE Trans Fuzzy Syst 15(5):972–983
54. Gao H, Chen T, Wang L (2008) Robust fault detection with missing measurements. Int J Control 81(5):804–819
55. Gao H, Lam J, Wang C (2006) Robust energy-to-peak filter design for stochastic time-delay systems. Syst Control Lett 55(2):101–111
56. Gao H, Liu X, Lam J (2009) Stability analysis and stabilization for discrete-time fuzzy systems with time-varying delay. IEEE Trans Syst Man Cybern Part B Cybern 39(2):306–317
57. Gao H, Zhao Y, Chen T (2009) H_∞ fuzzy control of nonlinear systems under unreliable communication links. IEEE Trans Fuzzy Syst 17(2):265–278
58. Gao Q, Feng G, Wang Y, Qiu J (2012) Universal fuzzy controllers based on generalized T-S fuzzy models. Fuzzy Sets Syst 201:55–70
59. Garcia G, Bernussou J (1995) Pole assignment for uncertain systems in a specified disk by state feedback. IEEE Trans Autom Control 40(1):184–190
60. Gassara H, El-Hajjaji A, Chaabane M (2010) Observer-based robust reliable control for uncertain T-S fuzzy systems with state time delay. IEEE Trans Fuzzy Syst 18(6):1027–1040
61. Hsiao FH, Chen C, Liang Y, Xu S, Chiang WL (2005) TS fuzzy controllers for nonlinear interconnected systems with multiple time delays. IEEE Trans Circuits Syst I Regul Pap 52(9):1883–1893
62. Grigoriadis KM, Watson-Jr JT (1997) Reduced-order H_∞ and L_2-L_∞ filtering via linear matrix inequalities. IEEE Trans Aerosp Electron Syst 33(4):1326–1338
63. Guerra T-M, Kruszewski A, Vermeiren L, Tirmant H (2006) Conditions of output stabilization for nonlinear models in the Takagi-Sugeno's form. Fuzzy Sets Syst 157(9):1248–1259
64. Guerra T-M, Vermeiren L (2004) LMI-based relaxed nonquadratic stabilization conditions for nonlinear systems in the Takagi-Sugeno's form. Automatica 40(5):823–829
65. Hagras H (2007) Type-2 FLCs: a new generation of fuzzy controllers. IEEE Comput Intell Mag 2(1):30–43

66. Hagras HA (2004) A hierarchical type-2 fuzzy logic control architecture for autonomous mobile robots. IEEE Trans Fuzzy Syst 12(4):524–539
67. He S, Liu CL (2012) Finite-time fuzzy control of nonlinear jump systems with time delays via dynamic observer-based state feedback. IEEE Trans Fuzzy Syst 20(4):605–614
68. Hu J, Wang Z, Shen B, Gao H (2013) Quantised recursive filtering for a class of nonlinear systems with multiplicative noises and missing measurements. Int J Control 86(4):650–663
69. Hu S, Yue D, Du Z, Liu J (2012) Reliable H_∞ non-uniform sampling tracking control for continuous-time non-linear systems with stochastic actuator faults. IET Control Theory Appl 6(1):120–129
70. Huang CJ, Hu K, Cheng HM, Chang TK, Luo Y, Lien YJ (2012) Application of type-2 fuzzy logic to rule-based intrusion alert correlation detection. Int J Innov Comput Inf Control 8(4):2865–2874
71. Huang HP, Yan JL, Cheng TH (2010) Development and fuzzy control of a pipe inspection robot. IEEE Trans Ind Electron 57(3):1088–1095
72. Huang SJ, He XQ, Zhang NN (2011) New results on filter design for nonlinear systems with time delay via T-S fuzzy models. IEEE Trans Fuzzy Syst 19(1):193–199
73. Jafarzadeh S, Fadali S, Sonbol A (2011) Stability analysis and control of discrete type-1 and type-2 TSK fuzzy systems: part I stability analysis. IEEE Trans Fuzzy Syst 6(6):989–1000
74. Jafarzadeh S, Fadali S, Sonbol A (2011) Stability analysis and control of discrete type-1 and type-2 TSK fuzzy systems: part II control design. IEEE Trans Fuzzy Syst 19(6):1001–1013
75. Jammeh EA, Fleury M, Wagner C, Hagras H, Ghanbari M (2009) Interval type-2 fuzzy logic congestion control for video streaming across IP networks. IEEE Trans Fuzzy Syst 17(5):1123–1142
76. Jiang B, Gao Z, Shi P, Xu Y (2010) Adaptive fault-tolerant tracking control of near-space vehicle using Takagi-Sugeno fuzzy models. IEEE Trans Fuzzy Syst 18(5):1000–1007
77. Johansson M, Rantzer A, Arzen K (1999) Piecewise quadratic stability of fuzzy systems. IEEE Trans Fuzzy Syst 7(6):713–722
78. Juang C, Chen C (2014) An interval type-2 neural fuzzy chip with on-chip incremental learning ability for time-varying data sequence prediction and system control. IEEE Trans Neural Netw Learn Syst 25(1):216–228
79. Juang C, Hsu C (2009) Reinforcement interval type-2 fuzzy controller design by online rule generation and Q-value-aided ant colony optimization. IEEE Trans Syst Man Cybern Part B Cybern 39(6):1528–1542
80. Jun Y (2009) H_∞ filtering for fuzzy systems with immeasurable premise variables: an uncertain system approach. Fuzzy Sets Syst 160(12):1738–1748
81. Karimi HR, Chadli M (2012) Robust observer design for Takagi-Sugeno fuzzy systems with mixed neutral and discrete delays and unknown inputs. Math Probl Eng doi:10.1155/2012/635709
82. Karnik NN, Mendel JM, Liang Q (1999) Type-2 fuzzy logic systems. IEEE Trans Fuzzy Syst 7(6):643–658
83. Kayacan E, Kaynak O (2012) Sliding mode control-based algorithm for online learning in type-2 fuzzy neural networks: application to velocity control of an electro hydraulic servo system. Int J Adapt Control Signal Process 26(7):645–659
84. Khanesar M, Kayacan E, Teshnehlab M, Kaynak O (2011) Analysis of the noise reduction property of type-2 fuzzy logic systems using a novel type-2 membership function. IEEE Trans Syst Man Cybern Part B Cybern 41(5):1395–1406
85. Khanesar MA, Kayacan E, Teshnehlab M, Kaynak O (2012) Extended Kalman filter based learning algorithm for type-2 fuzzy logic systems and its experimental evaluation. IEEE Trans Ind Electron 59(11):4443–4455
86. Khosla M, Sarin RK, Uddin M (2011) Design of an analog CMOS based interval type-2 fuzzy logic controller chip. Int J Artif Intell Expert Syst 2(4):167–183
87. Khosla M, Sarin RK, Uddin M (2012) Implementation of interval type-2 fuzzy systems with analog modules. In: IEEE Control and System Graduate Research Colloquium (ICSGRC). IEEE, pp 136–141

88. Kim E, Lee H (2000) New approaches to relaxed quadratic stability condition of fuzzy control systems. IEEE Trans Fuzzy Syst 8(5):523–534
89. Ko JW (2012) State-feedback H_∞ switching control for Takagi-Sugeno fuzzy systems based on partitioning the range of fuzzy weights. IET Control Theory Appl 6(15):2460–2466
90. Kruszewski A, Wang R, Guerra T-M (2008) Nonquadratic stabilization conditions for a class of uncertain nonlinear discrete time ts fuzzy models: a new approach. IEEE Trans Autom Control 53(2):606–611 (52(9):1883–1893 2005)
91. Lam HK (2009) Stability analysis of T-S fuzzy control systems using parameter-dependent Lyapunov function. IET Control Theory Appl 3(6):750–762
92. Lam HK, Leung FHF (2005) Stability analysis of fuzzy control systems subject to uncertain grades of membership. IEEE Trans Syst Man Cybern Part B Cybern 35(6):1322–1325
93. Lam HK, Leung FH (2008) Stability analysis of discrete-time fuzzy-model-based control systems with time delay: time delay-independent approach. Fuzzy Sets Syst 159(8):990–1000
94. Lam HK, Li H, Deters C, Secco E, Wurdemann HA, Althoefer K (2014) Control design for interval type-2 fuzzy systems under imperfect premise matching. IEEE Trans Ind Electron 61(2):956–968
95. Lam HK, Narimani M (2009) Stability analysis and performance design for fuzzy-model-based control system under imperfect premise matching. IEEE Trans Fuzzy Syst 17(4):949–961
96. Lam HK, Narimani M, Seneviratne LD (2011) LMI-based stability conditions for interval type-2 fuzzy-model-based control systems. In: IEEE International Conference on Fuzzy Systems, pp 298–303
97. Lam HK, Seneviratne LD (2008) Stability analysis of interval type-2 fuzzy-model-based control systems. IEEE Trans Syst Man Cybern Part B Cybern 38(3):617–628
98. Lam J, Zhou S (2007) Dynamic output feedback H_∞ control of discrete-time fuzzy systems: a fuzzy-basis-dependent Lyapunov function approach. Int J Syst Sci 38(1):25–37
99. Lee DH, Park JB, Joo YH (2011) A new fuzzy Lyapunov function for relaxed stability condition of continuous-time Takagi-Sugeno fuzzy systems. IEEE Trans Fuzzy Syst 19(4):785–791
100. Lee TH, Park JH, Ji D, Jung HY (2014) Leader-following consensus problem of heterogeneous multi-agent systems with nonlinear dynamics using fuzzy disturbance observer. Complexity 19(4):20–31
101. Lendek Z, Guerra T-M, Babuska R, Schutter BD (2010) Stability analysis and nonlinear observer design using Takagi-Sugeno fuzzy models, vol 262. Springer, Berlin
102. Lendek Z, Lauber J, Guerra T-M, Babuska R, Schutter BD (2010) Adaptive observers for TS fuzzy systems with unknown polynomial inputs. Fuzzy Sets Syst 161(15):2043–2065
103. Li H, Gao H, Shi P, Zhao X (2013) Fault-tolerant control of markovian jump stochastic system with augmented sliding mode observer approachs. Automatica 50(7):1825–1834
104. Li H, Gao Y, Wu L, Lam HK (2015) Fault detection for T-S fuzzy time-delay systems: delta operator and input-output methods. IEEE Trans Cybern 45(2):229–241
105. Li H, Jing X, Karimi HR (2014) Output-feedback based H_∞ control for active suspension systems with control delay. IEEE Trans Ind Electron 61(1):436–446
106. Li H, Liu H, Gao H, Shi P (2012) Reliable fuzzy control for active suspension systems with actuator delay and fault. IEEE Trans Fuzzy Syst 20(2):342–357
107. Li H, Pan Y, Zhou Q (2015) Filter design for interval type-2 fuzzy systems with D stability constraints under a unified frame. IEEE Trans Fuzzy Syst 23(3):719–725
108. Li H, Si Y, Wu L, Gao H (2011) Guaranteed cost control with poles assignment for a flexible air-breathing hypersonic vehicle. Int J Syst Sci 42(5):863–876
109. Li J, Wang HO, Niemann D, Tanaka K (2000) Dynamic parallel distributed compensation for Takagi-Sugeno fuzzy systems: an LMI approach. Inf Sci 123(3):201–221
110. Li XJ, Yang GH (2013) Switching-type H_∞ filter design for T-S fuzzy systems with unknown or partially unknown membership functions. IEEE Trans Fuzzy Syst 21(2):385–392
111. Li XJ, Yang GH (2014) Fault detection for T-S fuzzy systems with unknown membership functions. IEEE Trans Fuzzy Syst 22(1):139–152

112. Li Y, Tong S, Li T (2012) Adaptive fuzzy output feedback control of uncertain nonlinear systems with unknown backlash-like hysteresis. Inf Sci 198:130–146
113. Li Y, Tong S, Liu Y, Li T (2014) Adaptive fuzzy robust output feedback control of nonlinear systems with unknown dead zones based on small-gain approach. IEEE Trans Fuzzy Syst 22(1):164–176
114. Li Y, Tong S, Liu Y, Li T (2014) Adaptive fuzzy robust output feedback control of nonlinear systems with unknown dead zones based on small-gain approach. IEEE Trans Fuzzy Syst 22(1):164–176
115. Li Z, Gao H, Karimi HR (2014) Stability analysis and H_∞ controller synthesis of discrete-time switched systems with time delay. Syst Control Lett 66:85–93
116. Li Z, Wang J, Shao H (2002) Delay-dependent dissipative control for linear time-delay systems. J Frankl Inst 339(6):529–542
117. Liang Q, Mendel JM (2000) Equalization of nonlinear time-varying channels using type-2 fuzzy adaptive filters. IEEE Trans Fuzzy Syst 8(5):551–563
118. Liang Q, Mendel JM (2000) Interval type-2 fuzzy logic systems: theory and design. IEEE Trans Fuzzy Syst 8(5):535–550
119. Lien CH, Yu KW, Chang HC, Chung LY, Chen JD (2015) Robust reliable guaranteed cost control for uncertain T-S fuzzy neutral systems with interval time-varying delay and linear fractional perturbations. Optim Control Appl Methods 361(1):121–137
120. Lin C, Wang QG, Lee TH (2006) H_∞ output tracking control for nonlinear systems via T-S fuzzy model approach. IEEE Trans Syst Man Cybern Part B Cybern 36(2):450–457
121. Lin C, Wang QG, Lee TH, Chen B (2008) Filter design for nonlinear systems with time-delay through T-S fuzzy model approach. IEEE Trans Fuzzy Syst 16(3):739–746
122. Lin C, Wang QG, Lee TH, He Y, Chen B (2008) Observer-based H_∞ fuzzy control design for T-S fuzzy systems with state delays. Automatica 44(3):868–874
123. Lin CT, Lee CG (1994) Reinforcement structure/parameter learning for neural-network-based fuzzy logic control systems. IEEE Trans Fuzzy Syst 2(1):46–63
124. Lin CT, Lee CSG (1991) Neural-network-based fuzzy logic control and decision system. IEEE Trans Comput 40(12):1320–1336
125. Lin PZ, Lin CM, Hsu CF, Lee TT (2005) Type-2 fuzzy controller design using a sliding-mode approach for application to DC-DC converters. IEE Proc-Electr Power Appl 152(6):1482–1488
126. Liu H, Shi P, Karimi HR, Chadli M (2014) Finite-time stability and stabilisation for a class of nonlinear systems with time-varying delay. Int J Syst Sci 47(6):1433–1444
127. Liu M, Cao X, Shi P (2013) Fault estimation and tolerant control for fuzzy stochastic systems. IEEE Trans Fuzzy Syst 21(2):221–229
128. Liu M, Cao X, Shi P (2013) Fuzzy-model-based fault-tolerant design for nonlinear stochastic systems against simultaneous sensor and actuator faults. IEEE Trans Fuzzy Syst 21(5):789–799
129. Liu X, Sun Q, Hou X (2014) New approach on robust and reliable decentralized H_∞ tracking control for fuzzy interconnected systems with time-varying delay. ISRN Appl Math doi:10.1155/2014/705609
130. Liu YJ, Tong S, Chen CLP (2013) Adaptive fuzzy control via observer design for uncertain nonlinear systems with unmodeled dynamics. IEEE Trans Fuzzy Syst 21(2):314–327
131. Liu YJ, Tong SC, Li TS (2011) Observer-based adaptive fuzzy tracking control for a class of uncertain nonlinear MIMO systems. Fuzzy Sets Syst 164(1):25–44
132. Liu XD, Zhang QL (2003) New approaches to H_∞ controller designs based on fuzzy observers for TS fuzzy systems via LMI. Automatica 39(9):1571–1582
133. Lo JC, Lin ML (2003) Robust H_∞ nonlinear control via fuzzy static output feedback. IEEE Trans Syst Man Cybern Part B Cybern 50(11):1494–1502
134. Lozano R, Brogliato B, Egeland O, Maschke B (2000) Dissipative systems analysis and control: theory and applications. Springer, London
135. Lu R, Cheng H, Bai J (2015) uzzy-model-based quantized guaranteed cost control of nonlinear networked systems. IEEE Trans Fuzzy Syst 23(3):567–575

136. Mao Z, Jiang B, Shi P (2009) Protocol and fault detection design for nonlinear networked control systems. IEEE Trans Circuits Syst II Express Briefs 56(3):255–259
137. Martínez R, Castillo O, Aguilar LT (2009) Optimization of interval type-2 fuzzy logic controllers for a perturbed autonomous wheeled mobile robot using genetic algorithms. Inf Sci 179(13):2158–2174
138. Mathiyalagan K, Sakthivel R, Anthoni SM (2014) New stability criteria for stochastic Takagi-Sugeno fuzzy systems with time-varying delays. J Dyn Syst Meas Control 136(2):1–9
139. Melgarejo Rey MA, Bulla Blanco JO, Sierra Paez GK (2011) An embedded type-2 fuzzy processor for the inverted pendulum control problem. Lat Am Trans IEEE (Rev IEEE Am Lat) 9(3):240–246
140. Melin P, Castillo O (2004) A new method for adaptive control of non-linear plants using type-2 fuzzy logic and neural networks. Int J Gen Syst 33(2–3):289–304
141. Mendel JM, John RB (2002) Type-2 fuzzy sets made simple. IEEE Trans Fuzzy Syst 10(2):117–127
142. Mendel JM, John RI, Liu F (2006) Interval type-2 fuzzy logic systems made simple. IEEE Trans Fuzzy Syst 14(6):808–821
143. Mendel JM, Liu X (2013) Simplified interval type-2 fuzzy logic systems. IEEE Trans Fuzzy Syst 21(6):1056–1069
144. Mozelli LA, Palhares RM, Souza F, Mendes EM (2009) Reducing conservativeness in recent stability conditions of TS fuzzy systems. Automatica 45(6):1580–1583
145. Narimani M, Lam HK (2009) Relaxed LMI-based stability conditions for Takagi-Sugeno fuzzy control systems using regional-membership-function-shape-dependent analysis approach. IEEE Trans Fuzzy Syst 17(5):1221–1228
146. Nguang SK, Shi P (2006) Robust H_∞ output feedback control design for fuzzy dynamic systems with quadratic $\mathcal{D}$ stability constraints: an LMI approach. Inf Sci 176(15):2161–2191
147. Niu Y, Ho DW, Li CW (2011) H_∞ filtering for uncertain stochastic systems subject to sensor nonlinearities. Int J Syst Sci 42(5):737–749
148. Niu Y, Ho DW, Li CW (2010) Filtering for discrete fuzzy stochastic systems with sensor nonlinearities. IEEE Trans Fuzzy Syst 18(5):971–978
149. Ohtake H, Tanaka K, Wang HO (2006) Switching fuzzy controller design based on switching Lyapunov function for a class of nonlinear systems. IEEE Trans Syst Man Cybern Part B Cybern 36(1):13–23
150. Peng C, Yue D, Tian YC (2009) New approach on robust delay-dependent control for uncertain T-S fuzzy systems with interval time-varying delay. IEEE Trans Fuzzy Syst 17(4):890–900
151. Peng C, Han Q, Dong Y (2013) To transmit or not to transmit: a discrete event-triggered communication scheme for networked Takagi-Sugeno fuzzy systems. IEEE Trans Fuzzy Syst 21(1):164–170
152. Precup RE, Hellendoorn H (2011) A survey on industrial applications of fuzzy control. Comput Ind 62(3):213–226
153. Qi Y, Bao W, Chang J, Cui J (2014) Limit protection design: a guaranteed cost control method. In: 33rd Chinese Control Conference (CCC). IEEE, pp 3904–3908
154. Qiu J, Feng G, Gao H (2011) Asynchronous output-feedback control of networked nonlinear systems with multiple packet dropouts: T-S fuzzy affine model-based approach. IEEE Trans Fuzzy Syst 19(6):1014–1030
155. Qiu J, Feng G, Yang J (2009) A new design of delay-dependent robust H_∞ filtering for discrete-time T-S fuzzy systems with time-varying delay. IEEE Trans Fuzzy Syst 17(5):1044–1058
156. Rakkiyappan R, Sakthivel N (2014) Cluster synchronization for T-S fuzzy complex networks using pinning control with probabilistic time-varying delays. Complexity 21(1):59–77
157. Ranjbar-Sahraei B, Shabaninia F, Nemati A, Stan SD (2012) A novel robust decentralized adaptive fuzzy control for swarm formation of multiagent systems. IEEE Trans Ind Electron 59(8):3124–3134
158. Sakthivel R, Arunkumar A, Mathiyalagan K, Marshal Anthoni S (2011) Robust passivity analysis of fuzzy Cohen-Grossberg BAM neural networks with time-varying delays. Appl Math Comput 218(7):3799–3809

159. Sakthivel R, Vadivel P, Mathiyalagan K, Arunkumar A (2014) Fault-distribution dependent reliable H_∞ control for Takagi-Sugeno fuzzy systems. J Dyn Syst Meas Control 136(2):021021
160. Sala A, Ariño C (2007) Asymptotically necessary and sufficient conditions for stability and performance in fuzzy control: applications of Polya's theorem. Fuzzy Sets Syst 158(24):2671–2686
161. Sala A, Ariño C (2007) Relaxed stability and performance conditions for Takagi-Sugeno fuzzy systems with knowledge on membership function overlap. IEEE Trans Syst Man Cybern Part B Cybern 37(3):727–732
162. Sepúlveda R, Castillo O, Melin P, Rodríguez-Díaz A, Montiel O (2007) Experimental study of intelligent controllers under uncertainty using type-1 and type-2 fuzzy logic. Inf Sci 177(10):2023–2048
163. Shen H, Xu S, Zhou J, Lu J (2011) Fuzzy H_∞ filtering for nonlinear Markovian jump neutral systems. Int J Syst Sci 42(5):767–780
164. Shen Q, Jiang B, Shi P (2014) Adaptive fault diagnosis for T-S fuzzy systems with sensor faults and system performance analysis. IEEE Trans Fuzzy Syst 22(2):274–285
165. Sheng L, Ma X (2014) Stability analysis and controller design of interval type-2 fuzzy systems with time delay. Int J Syst Sci 45(5):977–993
166. Su X, Shi P, Wu L, Song YD (2012) A novel approach to filter design for T-S fuzzy discrete-time systems with time-varying delay. IEEE Trans Fuzzy Syst 20(6):1114–1129
167. Su X, Shi P, Wu L, Song YD (2013) A novel control design on discrete-time Takagi-Sugeno fuzzy systems with time-varying delays. IEEE Trans Fuzzy Syst 21(4):655–671
168. Su X, Wu L, Shi P (2013) Sensor networks with random link failures: distributed filtering for T-S fuzzy systems. IEEE Trans Ind Inform 9(3):1739–1750
169. Su X, Wu L, Shi P, Song YD (2012) Model reduction of Takagi-Sugeno fuzzy stochastic systems. IEEE Trans Syst Man Cybern Part B Cybern 42(6):1574–1585
170. Sugeno M, Kang G (1988) Structure identification of fuzzy model. Fuzzy Sets Syst 28(1):15–33
171. Tanaka K, Hori T, Wang HO (2003) A multiple Lyapunov function approach to stabilization of fuzzy control systems. IEEE Trans Fuzzy Syst 11(4):582–589
172. Tanaka K, Ikeda T, Wang HO (1998) Fuzzy regulators and fuzzy observers: relaxed stability conditions and LMI-based designs. IEEE Trans Fuzzy Syst 6(2):250–265
173. Tanaka K, Ohtake H, Wang HO (2007) A descriptor system approach to fuzzy control system design via fuzzy Lyapunov functions. IEEE Trans Fuzzy Syst 15(3):333–341
174. Takagi T, Sugeno M (1985) Fuzzy identification of systems and its applications to modeling and control. IEEE Trans Syst Man Cybern 15(1):116–132
175. Tanaka K, Ikeda T, Wang H (1996) Robust stabilization of a class of uncertain nonlinear systems via fuzzy control: quadratic stabilizability, H_∞ control theory, and linear matrix inequalities. IEEE Trans Fuzzy Syst 4(1):1–13
176. Tanaka K, Ohtake H, Wang HO (2009) Guaranteed cost control of polynomial fuzzy systems via a sum of squares approach. IEEE Trans Syst Man Cybern B Cybern 39(2):561–567
177. Tanaka K, Sano M (1995) Trajectory stabilization of a model car via fuzzy control. Fuzzy Sets Syst 70(2):155–170
178. Taniguchi T, Tanaka K, Wang HO (2000) Fuzzy descriptor systems and nonlinear model following control. IEEE Trans Fuzzy Syst 8(4):442–452
179. Takagi T, Sugeno M (1992) Stability analysis and design of fuzzy control systems. Fuzzy Set Syst 45(2):135–156
180. Tanaka K, Wang H (2001) Fuzzy control systems design and analysis: a linear matrix inequality approach. Wiley-Interscience, Hoboken
181. Teixeira MC, Zak SH (1999) Stabilizing controller design for uncertain nonlinear systems using fuzzy models. IEEE Trans Fuzzy Syst 7(2):133–144
182. Thumati BT, Feinstein M, Jagannathan S (2014) A model-based fault detection and prognostics scheme for Takagi-Sugeno fuzzy systems. IEEE Trans Fuzzy Syst 22(4):736–748
183. Tian E, Yue D (2013) Reliable H_∞ filter design for T-S fuzzy model-based networked control systems with random sensor failure. Int J Robust Nonlinear Control 23(1):15–32

184. Tian E, Yue D, Yang T, Gu Z, Lu G (2011) T-S fuzzy model-based robust stabilization for networked control systems with probabilistic sensor and actuator failure. IEEE Trans Fuzzy Syst 19(3):553–561
185. Tognetti ES, Oliveira RC, Peres PL (2011) Selective and stabilization of Takagi-Sugeno fuzzy systems. IEEE Trans Fuzzy Syst 19(5):890–900
186. Tong S, Huo B, Li Y (2014) Observer-based adaptive decentralized fuzzy fault-tolerant control of nonlinear large-scale systems with actuator failures. IEEE Trans Fuzzy Syst 22(1):1–15
187. Tseng CS, Chen BS, Uang HJ (2001) Fuzzy tracking control design for nonlinear dynamic systems via T-S fuzzy model. IEEE Trans Fuzzy Syst 9(3):381–392
188. Tsinias J (1997) Control Lyapunov functions, input-to-state stability applications to global feedback stabilization for composite systems. J Math Syst Estim Control 7:235–238
189. Vadivel P, Sakthivel R, Mathiyalagan K, Thangaraj P (2012) Robust stabilisation of nonlinear uncertain Takagi-Sugeno fuzzy systems by H_∞ control. IET Control Theory Appl 6(16):2556–2566
190. Veillette RJ, Medanic J, Perkins WR (1992) Design of reliable control systems. IEEE Trans Autom Control 37(3):290–304
191. Wang H, Shi P, Zhang J (2015) Event-triggered fuzzy filtering for a class of nonlinear networked control systems. Signal Process 113:159–168
192. Wang HO, Tanaka K, Griffin MF (1996) An approach to fuzzy control of nonlinear systems: stability design issues. IEEE Trans Fuzzy Syst 4(1):14–23
193. Wang WJ, Chen YJ, Sun CH (2007) Relaxed stabilization criteria for discrete-time T-S fuzzy control systems based on a switching fuzzy model piecewise Lyapunov function. IEEE Trans Syst Man Cybern Part B Cybern 37(3):551–559
194. Wu D (2012) On the fundamental differences between type-1 interval type-2 fuzzy logic controllers. IEEE Trans Fuzzy Syst 20(5):832–848
195. Wu D, Mendel JM (2007) Uncertainty measures for interval type-2 fuzzy sets. Inf Sci 177(23):5378–5393
196. Wu D, Tan WW (2004) A type-2 fuzzy logic controller for the liquid-level process. IEEE Int Conf Fuzzy Syst 2:953–958
197. Wu HN (2004) Reliable LQ fuzzy control for continuous-time nonlinear systems with actuator faults. IEEE Trans Syst Man Cybern Part B Cybern 34(4):1743–1752
198. Wu HN, Bai MZ (2009) Active fault-tolerant fuzzy control design of nonlinear model tracking with application to chaotic systems. IET Control Theory Appl 3(6):642–653
199. Wu HN, Li HX (2007) New approach to delay-dependent stability analysis stabilization for continuous-time fuzzy systems with time-varying delay. IEEE Trans Fuzzy Syst 15(3):482–493
200. Wu HN, Wang JW, Li HX (2012) Exponential stabilization for a class of nonlinear parabolic pde systems via fuzzy control approach. IEEE Trans Fuzzy Syst 20(2):318–329
201. Wu HN, Zhang HY (2005) Reliable mixed $\mathcal{L}_2/\mathcal{H}_\infty$ fuzzy static output feedback control for nonlinear systems with sensor faults. Automatica 41(11):1925–1932
202. Wu HW, Mendel JM (2002) Uncertainty bounds their use in the design of interval type-2 fuzzy logic systems. IEEE Trans Fuzzy Syst 10(5):622–639
203. Wu L, Ho D (2009) Fuzzy filter design for Itô stochastic systems with application to sensor fault detection. IEEE Trans Fuzzy Syst 17(1):233–242
204. Wu L, Su X, Shi P, Qiu J (2011) Model approximation for discrete-time state-delay systems in the T-S fuzzy framework. IEEE Trans Fuzzy Syst 19(2):366–378
205. Wu L, Yang X, Lam HK (2014) Dissipativity analysis synthesis for discrete-time TS fuzzy stochastic systems with time-varying delay. IEEE Trans Fuzzy Syst 22(2):380–394
206. Wu L, Zheng WX (2009) L_2-L_∞ control of nonlinear fuzzy Itô stochastic delay systems via dynamic output feedback. IEEE Trans Syst Man Cybern Part B Cybern 39(5):1308–1315
207. Wu X, Shen J, Li Y, Lee KY (2014) Fuzzy modeling stable model predictive tracking control of large-scale power plants. J Process Control 24(10):1609–1626
208. Wu Z, Shi P, Su H, Chu J (2012) Reliable control for discrete-time fuzzy systems with infinite-distributed delay. IEEE Trans Fuzzy Syst 20(1):22–31

209. Wu ZG, Shi P, Su H, Chu J (2014) Sampled-data fuzzy control of chaotic systems based on a T-S fuzzy model. IEEE Trans Fuzzy Syst 22(1):153–163
210. Xie L (1996) Output feedback H_∞ control of systems with parameter uncertainty. Int J Control 63(4):741–750
211. Xie L, de Souza CE (1992) Robust H_∞ control for linear systems with norm-bounded time-varying uncertainty. IEEE Trans Autom Control 37(8):1188–1191
212. Xie X, Yin S, Gao H, Kaynak O (2013) Asymptotic stability and stabilisation of uncertain delta operator systems with time-varying delays. IET Control Theory Appl 7(8):1071–1078
213. Xie XP, Yang DS, Zhu XL (2014) Relaxed observer design of discrete-time T-S fuzzy systems via a novel multi-instant fuzzy observer. Signal Process 102:296–303
214. Xu S, Lam J, Mao X (2007) Delay-dependent H_∞ control filtering for uncertain Markovian jump systems with time-varying delays. IEEE Trans Circuits Syst I Regul Pap 54(9):2070–2077
215. Xu S, Lam J, Zou Y, Lin Z, Paszke W (2005) Robust H_∞ filtering for uncertain 2-D continuous systems. IEEE Trans Signal Process 53(5):1731–1738
216. Xu S, Zheng WX, Zou Y (2009) Passivity analysis of neural networks with time-varying delays. IEEE Trans Circuits Syst II Express Briefs 56(4):325–329
217. Xu Y, Li Y, Tong S (2013) Fuzzy adaptive actuator failure compensation dynamic surface control of multi-input multi-output nonlinear systems. Int J Innov Comput Inf Control 9(12):4875–4888
218. Yang GH, Dong J (2008) Filtering for fuzzy singularly perturbed systems. IEEE Trans Syst Man Cybern Part B Cybern 38(5):1371–1389
219. Yang GH, Dong J (2010) Switching fuzzy dynamic output feedback control for nonlinear systems. IEEE Trans Syst Man Cybern Part B Cybern 40(2):505–516
220. Yang GH, Wang H (2014) Fault detection and isolation for a class of uncertain state-feedback fuzzy control systems. IEEE Trans Fuzzy Syst 23(1):139–151
221. Yang GH, Wang JL, Soh YC (2000) Reliable guaranteed cost control for uncertain nonlinear systems. IEEE Trans Autom Control 45(11):2188–2192
222. Yang H, Shi P, Li X, Li Z (2014) Fault-tolerant control for a class of T-S fuzzy systems via delta operator approach. Signal Process 98:166–173
223. Yang R, Liu GP, Shi P, Thomas C, Basin MV (2014) Predictive output feedback control for networked control systems. IEEE Trans Ind Electron 61(1):512–520
224. Yoneyama J (2007) Robust stability and stabilization for uncertain Takagi-Sugeno fuzzy time-delay systems. Fuzzy Sets Syst 158(2):115–134
225. Yin Y, Shi P, Liu F, Teo KL (2013) Fuzzy model-based robust H_∞ filtering for a class of nonlinear nonhomogeneous Markov jump systems. Signal Process 93(9):2381–2391
226. Zadeh L (1965) Fuzzy sets. Inf Control 8(3):338–353
227. Zadeh L (1973) Outline of a new approach to the analysis of complex systems decision processes. IEEE Trans Syst Man Cybern 3(1):28–44
228. Zhang B, Xu S, Zong G, Zou Y (2007) Delay-dependent stabilization for stochastic fuzzy systems with time delays. Fuzzy Sets Syst 158(20):2238–2250
229. Zhang B, Xu S, Zou Y (2008) Improved delay-dependent exponential stability criteria for discrete-time recurrent neural networks with time-varying delays. Neurocomputing 72(1):321–330
230. Zhang B, Zheng WX, Xu S (2013) Filtering of Markovian jump delay systems based on a new performance index. IEEE Trans Circuits Syst I 60(5):1250–1263
231. Zhang D, Cai W, Xie L, Wang Q, Non-fragile distributed filtering for TS fuzzy systems in sensor networks. IEEE Trans Fuzzy Syst 23(5):1883–1890
232. Zhang D, Wang QG, Yu L, Song H (2013) Protocol and fault detection design for nonlinear networked control systems. IEEE Trans Instrum Meas 62(12):3148–3159
233. Zhang H, Feng G (2008) Stability analysis controller design of discrete-time fuzzy large-scale systems based on piecewise Lyapunov functions. IEEE Trans Syst Man Cybern Part B Cybern 38(5):1390–1401

234. Zhang H, Shi Y, Mehr AS (2012) On filtering for discrete-time Takagi-Sugeno fuzzy systems. IEEE Trans Fuzzy Syst 20(2):396–401
235. Zhang H, Yan H, Yang F, Chen Q (2011) Quantized control design for impulsive fuzzy networked systems. IEEE Trans Fuzzy Syst 19(6):1153–1162
236. Zhang K, Jiang B, Shi P (2012) Fault estimation observer design for discrete-time Takagi-Sugeno fuzzy systems based on piecewise Lyapunov functions. IEEE Trans Fuzzy Syst 20(1):192–200
237. Zhao Y, Zhao L, Gao H (2010) Vibration control of seat suspension using H_∞ reliable control. J Vib Control 16(12):1859
238. Zheng F, Wang QG, Lee TH (2002) Output tracking control of MIMO fuzzy nonlinear systems using variable structure control approach. IEEE Trans Fuzzy Syst 10(6):686–697
239. Zhou Q, Shi P, Liu H, Xu S (2012) Neural-network-based decentralized adaptive output-feedback control for large-scale stochastic nonlinear systems. IEEE Trans Syst Man Cybern Part B Cybern 42(6):1608–1619
240. Zhou Q, Shi P, Lu J, Xu S (2011) Adaptive output-feedback fuzzy tracking control for a class of nonlinear systems. IEEE Trans Fuzzy Syst 19(5):972–982
241. Zhu X, Chen B, Yue D, Wang Y (2012) An improved input delay approach to stabilization of fuzzy systems under variable sampling. IEEE Trans Fuzzy Syst 20(2):330–341